mTOR Signaling in Metabolism and Cancer

mTOR Signaling in Metabolism and Cancer

Editor

Shile Huang

MDPI • Basel • Beijing • Wuhan • Barcelona • Belgrade • Manchester • Tokyo • Cluj • Tianjin

Editor
Shile Huang
Department of Biochemistry
and Molecular Biology,
Louisiana State University
Health Sciences Center
USA

Editorial Office
MDPI
St. Alban-Anlage 66
4052 Basel, Switzerland

This is a reprint of articles from the Special Issue published online in the open access journal *Cells* (ISSN 2073-4409) (available at: https://www.mdpi.com/journal/cells/special_issues/mTOR_Signaling).

For citation purposes, cite each article independently as indicated on the article page online and as indicated below:

LastName, A.A.; LastName, B.B.; LastName, C.C. Article Title. *Journal Name* **Year**, *Article Number*, Page Range.

ISBN 978-3-03943-553-1 (Hbk)
ISBN 978-3-03943-554-8 (PDF)

Contents

About the Editor

Shile Huang focuses on studying the role of mTOR signaling in human diseases, particularly in cancer. Since 1998, he has been studying how mTOR regulates cell survival, cell motility, and lymphangiogenesis; how some small molecules (e.g. curcumin, cryptotanshinone, dihydroartimisinin, and ciclopirox) act as anticancer agents; and how the heavy metal cadmium induces neuronal apoptosis.

Preface to "mTOR Signaling in Metabolism and Cancer"

The mechanistic/mammalian target of rapamycin (mTOR), a serine/threonine kinase, integrates environmental cues (hormones, growth factors, nutrients, oxygen, and energy), regulating cell growth, proliferation, survival and motility as well as metabolism. Dysregulation of mTOR signaling is implicated in a variety of disorders, such as cancer, obesity, diabetes, and neurodegenerative diseases. The articles published in this Special Issue reprint book summarize the current understanding of the mTOR pathway and its role in the regulation of tissue regeneration, regulatory T cell differentiation and function, and different types of cancer.

I am very thankful to all authors for their kind cooperation and wonderful contribution. I am also very grateful to the Managing Editor Beatty Teng for selection of this Special Issue as a reprint book.

Shile Huang
Editor

Editorial

mTOR Signaling in Metabolism and Cancer

Shile Huang [1,2]

[1] Department of Biochemistry and Molecular Biology, Louisiana State University Health Sciences Center, 1501 Kings Highway, Shreveport, LA 71130-3932, USA; shuan1@lsuhsc.edu; Tel.: +1-318-675-7759

[2] Feist-Weiller Cancer Center, Louisiana State University Health Sciences Center, 1501 Kings Highway, Shreveport, LA 71130-3932, USA

Received: 10 October 2020; Accepted: 13 October 2020; Published: 13 October 2020

Abstract: The mechanistic/mammalian target of rapamycin (mTOR), a serine/threonine kinase, is a central regulator for human physiological activity. Deregulated mTOR signaling is implicated in a variety of disorders, such as cancer, obesity, diabetes, and neurodegenerative diseases. The papers published in this special issue summarize the current understanding of the mTOR pathway and its role in the regulation of tissue regeneration, regulatory T cell differentiation and function, and different types of cancer including hematologic malignancies, skin, prostate, breast, and head and neck cancer. The findings highlight that targeting the mTOR pathway is a promising strategy to fight against certain human diseases.

Keywords: mTOR; PI3K; Akt; tissue regeneration; regulatory T cells; tumor; photodynamic therapy

---

The mechanistic/mammalian target of rapamycin (mTOR), a serine/threonine kinase, integrates environmental cues such as hormones, growth factors, nutrients, oxygen, and energy, regulating cell growth, proliferation, survival, motility and differentiation as well as metabolism (reviewed in [1,2]). Evidence has demonstrated that deregulated mTOR signaling is implicated in a variety of disorders, such as cancer, obesity, diabetes, and neurodegenerative diseases (reviewed in [1,2]). Current knowledge indicates that mTOR functions at least as two distinct complexes (mTORC1 and mTORC2) in mammalian cells. mTORC1 consists of mTOR, mLST8 (also termed G-protein β-subunit-like protein, GβL, a yeast homolog of LST8), raptor (regulatory-associated protein of mTOR), PRAS40 (proline-rich Akt substrate 40 kDa) and DEPTOR (DEP domain containing mTOR interacting protein), whereas mTORC2 is composed of mTOR, mLST8, rictor (rapamycin insensitive companion of mTOR), mSin1 (mammalian stress-activated protein kinase-interacting protein 1), protor (protein observed with rictor, also named PRR5, proline-rich protein 5), and DEPTOR [1,2]. mTORC1 regulates the phosphorylation or expression of p70 S6 kinase (S6K1), eukaryotic initiation factor 4E (eIF4E) binding protein 1 (4E-BP1), lipin1, ULK1 (Unc-51 like autophagy activating kinase 1), TFEB (transcription factor EB), ATF4 (activating transcription factor 4), HIF1α (hypoxia-inducible factor 1 alpha), etc., and mediates the protein synthesis/turnover, lipid synthesis and nucleotide synthesis, thus controlling cell growth, proliferation, autophagy, and metabolism (reviewed in [1,2]). mTORC2 regulates the phosphorylation/activity of Akt, serum/glucocorticoid regulated kinase (SGK), protein kinase C (PKC), etc., thereby controlling cell migration, apoptosis and metabolism (reviewed in [1,2]). These findings not only reveal the crucial role of mTOR in physiology and pathology, but also reflect the complexity of the mTOR signaling network.

The papers published in this special issue summarize the current understanding of the mTOR pathway and its role in the regulation of tissue regeneration, regulatory T cell differentiation and function, and diverse types of cancer including hematologic malignancies, skin, prostate, breast, and head and neck cancer.

Wei et al. (2019) discussed the role of mTOR signaling in the regeneration of tissues in the optic nerve, spinal cord, muscles, the liver and the intestine [3]. Activated mTOR enhances the regeneration of adult retinal ganglion cells after optic nerve injury. However, hyperactivation of mTOR in astrocytes promotes glial scar formation, leading to inhibition of spinal cord regeneration after spinal cord injury. These findings suggest that mTOR signaling exhibits opposite functions in the optic nerve and spinal cord regeneration. In mice, conditional knockout of *MTOR* or *RAPTOR* in muscle stem cells effectively inhibits activation, proliferation, and differentiation of satellite cells, impairing skeletal muscle regeneration, whereas *RICTOR* knockout in embryonic and adult satellite cells has no effect on skeletal muscle regeneration. Inhibition of mTORC1 with rapamycin inhibits the formation of nascent myofibers and the growth of regenerating myofibers during skeletal muscle regeneration. Furthermore, inhibition of mTORC1 not only suppresses the growth and proliferation of hepatocytes, but also blocks the proliferation of cholangiocytes and the formation of bipotential progenitor cells, which are essential for liver regeneration. These findings suggest that mTORC1, but not mTORC2, regulates skeletal muscle and liver regeneration. Similarly, mTORC1 is required for intestinal regeneration by controlling the proliferation and maintenance of intestinal stem cells. The development of novel drugs for tissue-specific activation or inhibition of mTOR may be beneficial to patients needing specific tissue regeneration.

Regulatory T cells (Tregs), a subset of T cells, suppress activation of the immune system and prevent autoimmune disease [4]. Chen et al. (2019) summarized the role of mTOR signaling in regulating the differentiation and function of Tregs [5]. Inhibition of mTOR with rapamycin decreases the production of effector T cells, but increases the generation and expansion of Tregs. Loss of mTORC1 signaling prevents naïve CD4+ T cells from differentiation to Th17 cells. However, disruption of either mTORC1 or mTORC2 has no effect on the differentiation of naïve CD4+ T cells into Foxp3+ Tregs. In addition, inhibition of mTORC1 attenuates the function of Tregs, while inhibition of mTORC2 increases Tregs function via promoting the activity of mTORC1, suggesting that mTORC1 and mTORC2 play opposite roles in mediating the function of Tregs. Furthermore, mTORC2 promotes the migration of Tregs to inflammatory sites. It is unclear if mTORC2 and mTORC1 are important for the expansion and migration of Tregs, respectively.

Acute lymphoblastic leukemia (ALL) is one of the aggressive hematologic malignancies that occurs in both children and adults [6]. Simioni et al. (2019) reviewed the advances in targeted therapy for ALL using mTOR inhibitors [7]. Constitutive activation of mTOR pathway is associated with deregulated production of malignant lymphoid cells and chemotherapeutic resistance in ALL. Overall, rapalogs (rapamycin, everolimus, temsirolimus) alone are primarily cytostatic, but they are synergistic with either conventional chemotherapeutic agents (doxorubicin, cyclophosphamide, dexamethasone) or other targeted therapies for ALL treatment. Treatment with dual PI3K/mTOR inhibitors (e.g., PKI-587 and BEZ235) or mTOR kinase inhibitors (e.g., AZD8055 and OSI-027) alone or in combination with chemotherapeutic agents not only inhibits cell proliferation but also induces apoptosis of ALL cells. The authors also briefly summarized clinical trials of some of these mTOR inhibitors for treatment of both T- and B-ALL.

The Warburg effect is associated with increased glycolysis, and has been implicated in chemoresistance in cancer therapy [8]. Mirabilii et al. (2020) discussed how hyperactivated mTOR, in concert with other metabolic modulators (AMPK and HIF1α) and microenvironmental stimuli, results in the acquisition of new glycolytic phenotype by directly and indirectly regulating the activity of certain key glycolytic enzymes in various hematologic malignancies [9]. For instance, in acute myeloid leukemia (AML) cells, mTOR upregulates the expression of PFKFB3 (6-phosphofructo-2-kinase/fructose-2,6-biphosphatase 3), increasing aerobic glycolysis. In chronic myeloid leukemia (CML) cells, mTOR, along with Bcr-Abl, upregulates the expression of pyruvate kinase isozymes M1/M2 (PKM1/2), enhancing aerobic glycolysis and reducing oxidative phosphorylation (OXPHOS). In acute lymphoblastic leukemia (ALL) cells, mTOR positively regulates the expression of hexokinase II, thus increasing lactate generation. The authors also discussed how these features could be targeted for therapeutic purposes.

Tan et al. (2019) reviewed genetic and epigenetic alterations of multiple genes related to the dysregulation of mTOR signaling, and discussed certain potential targets for therapeutic intervention in head and neck cancer, especially head and neck squamous cell carcinoma (HNSCC) [10]. Gain-of-function alterations (overexpression or mutations) of oncogenes (e.g., *EGFR*, *PIK3CA*, and *HRAS*) and loss-of-function mutations of tumor suppressor genes (e.g., *TP53* and *PTEN*) occur frequently in HNSCC, resulting in hyperactivation of mTOR signaling. The Cancer Genome Atlas (TCGA) database shows that mutations of *EIF4G1*, *RAC1*, *SZT2*, and *PLD1* in HNSCC also lead to aberrant mTOR signaling. Some of these mutated genes may be used as biomarkers to predict drug response. In addition, human papillomavirus (HPV) infection can activate mTOR pathway and inactivate p53 and Rb, promoting HNSCC development and progression. Accordingly, a number of clinical trials have been and are currently being conducted to evaluate the anticancer efficacy of mTOR inhibitors alone or in combination with chemotherapeutics or other kinase inhibitors.

Ayuk and Abrahamse (2019) discussed the mTOR signaling in cancer and the advances in photodynamic therapy (PDT) for cancer [11]. Mechanistically, PDT involves the application of a non-toxic photosensitizer to a specific tissue/organ, where the photosensitizer can be activated by a laser light at specific wavelengths in the presence of oxygen to generate reactive oxygen species (ROS), resulting in cell death. The effectiveness of PDT depends on the oxygen concentration, wavelength, types of photosensitizer and the genotype of tumor cells. Photosensitizers currently used include naturally occurring macrocycles (e.g., hemoglobin, vitamin B12, and chlorophyll) and tetrapyrroles (e.g., bacteriochlorins, chlorins, porphyrins, and phthalocyanines) as well as synthetic dyes. Induction of ROS by PDT inhibits the mTOR pathway. Also, PDT is synergistic with PI3K/mTOR inhibitors in cancer therapy.

Chamcheu et al. (2019) summarized the recent advances in the role of PI3K/Akt/mTOR signaling in the development and progression of skin cancers [12]. The skin comprises epidermis, dermis, and hypodermis, infiltrated by sweat glands, sensory cells, fibroblasts, macrophages, and lymphocytes. Genetic alterations or ultraviolet (UV) exposure results in the dysregulation of PI3K/Akt/mTOR pathway in melanocytes, basal cells, squamous cells, or Merkel cells, leading to melanoma, basal cell carcinoma, cutaneous squamous cell carcinoma, or Merkel cell carcinoma. The authors also discussed the current progress in preclinical and clinical studies for the development of PI3K/Akt/mTOR targeted therapies with natural phytochemicals (e.g., curcumin, epigallocatechin gallate, fisetin, resveratrol, and honokiol) and synthetic small molecule inhibitors including rapalogs and PI3K/mTOR kinase inhibitors. Some of these inhibitors are being tested in early-stage clinical trials, but their applications in the treatment of skin cancers need further testing.

Makarević et al. (2018) studied whether the inhibition of histone deacetylase (HDAC) counteracts the resistance to the mTOR inhibitor temsirolimus in a prostate cancer cell model [13]. For this, parental and temsirolimus-resistant PC3 prostate cancer cells were treated with the HDAC inhibitor valproic acid (VPA), followed by assays for tumor cell adhesion, migration, and invasion. The results indicate that treatment with temsirolimus (10 nM) inhibits the binding to human umbilical vein endothelial cells (HUVECs) or cell matrix (collagen, fibronectin, and laminin), cell migration and invasion in the parental cells, but not in the temsirolimus-resistant cells. However, treatment with VPA is able to suppress the cell adhesion, migration and invasion in both parental cells and temsirolimus-resistant cells. This is at least partly associated with a significant downregulation of integrin $\alpha5$ in the resistant tumor cells. The findings suggest that inhibition of HDAC is able to block the metastatic activity in temsirolimus-resistant prostate cancer cells.

Hyperactive PI3K/AKT/mTOR signaling has been implicated in triple negative breast cancer (TNBC), which contributes to resistance to chemotherapeutic agents, including microtubule-targeting agents [14]. Eribulin mesylate, a microtubule depolymerizing agent, has been approved by the US Food and Drug Administration (FDA) to treat taxane and anthracycline refractory breast cancer. Wen et al. (2019) investigated whether eribulin enhances the anticancer activity of the mTOR inhibitor everolimus in TNBC [15]. The results indicate that treatment with eribulin, like vinblastine (a microtubule depolymerizing agent), inhibits the phosphorylation of Akt, while treatment with paclitaxel (a microtubule stabilizing agent) or cisplatin (a DNA damaging agent) has the opposite effect. Inhibition of mTORC1 with everolimus induces phosphorylation of Akt, which is blocked by eribulin. Importantly, eribulin synergizes with everolimus in reducing cell viability in vitro and inhibiting tumor growth in two orthotopic xenograft mouse models of breast cancer (MDA-MB-468 and 4T1). The findings demonstrate that combination therapy with eribulin and everolimus has a great potential to combat refractory TNBC.

Funding: This research received no external funding.

Conflicts of Interest: The author declares no conflict of interest.

References

1. Liu, G.Y.; Sabatini, D.M. mTOR at the nexus of nutrition, growth, ageing and disease. *Nat. Rev. Mol. Cell Biol.* **2020**, *21*, 183–203. [PubMed]
2. Mossmann, D.; Park, S.; Hall, M.N. mTOR signalling and cellular metabolism are mutual determinants in cancer. *Nat. Rev. Cancer* **2018**, *18*, 744–757. [PubMed]
3. Wei, X.; Luo, L.; Chen, J. Roles of mTOR Signaling in Tissue Regeneration. *Cells* **2019**, *8*, 1075. [CrossRef] [PubMed]
4. Raffin, C.; Vo, L.T.; Bluestone, J.A. T_{reg} cell-based therapies: Challenges and perspectives. *Nat. Rev. Immunol.* **2020**, *20*, 158–172. [PubMed]
5. Chen, Y.; Colello, J.; Jarjour, W.; Zheng, S.G. Cellular Metabolic Regulation in the Differentiation and Function of Regulatory T Cells. *Cells* **2019**, *8*, 188. [CrossRef] [PubMed]
6. Roberts, K.G.; Mullighan, C.G. Genomics in acute lymphoblastic leukaemia: Insights and treatment implications. *Nat. Rev. Clin. Oncol.* **2015**, *12*, 344–357. [PubMed]
7. Simioni, C.; Martelli, A.M.; Zauli, G.; Melloni, E.; Neri, L.M. Targeting mTOR in Acute Lymphoblastic Leukemia. *Cells* **2019**, *8*, 190. [CrossRef] [PubMed]
8. Icard, P.; Shulman, S.; Farhat, D.; Steyaert, J.M.; Alifano, M.; Lincet, H. How the Warburg effect supports aggressiveness and drug resistance of cancer cells? *Drug Resist. Updat.* **2018**, *38*, 1–11. [CrossRef] [PubMed]
9. Mirabilii, S.; Ricciardi, M.R.; Tafuri, A. mTOR Regulation of Metabolism in Hematologic Malignancies. *Cells* **2020**, *9*, 404. [CrossRef] [PubMed]
10. Tan, F.H.; Bai, Y.; Saintigny, P.; Darido, C. mTOR Signalling in Head and Neck Cancer: Heads Up. *Cells* **2019**, *8*, 333. [CrossRef] [PubMed]
11. Ayuk, S.M.; Abrahamse, H. mTOR Signaling Pathway in Cancer Targets Photodynamic Therapy In Vitro. *Cells* **2019**, *8*, 431. [CrossRef] [PubMed]
12. Chamcheu, J.C.; Roy, T.; Uddin, M.B.; Banang-Mbeumi, S.; Chamcheu, R.N.; Walker, A.L.; Liu, Y.Y.; Huang, S. Role and Therapeutic Targeting of the PI3K/Akt/mTOR Signaling Pathway in Skin Cancer: A Review of Current Status and Future Trends on Natural and Synthetic Agents Therapy. *Cells* **2019**, *8*, 803. [CrossRef] [PubMed]
13. Makarević, J.; Rutz, J.; Juengel, E.; Maxeiner, S.; Mani, J.; Vallo, S.; Tsaur, I.; Roos, F.; Chun, F.K.; Blaheta, R.A. HDAC Inhibition Counteracts Metastatic Re-Activation of Prostate Cancer Cells Induced by Chronic mTOR Suppression. *Cells* **2018**, *7*, 129. [CrossRef] [PubMed]
14. Khan, M.A.; Jain, V.K.; Rizwanullah, M.; Ahmad, J.; Jain, K. PI3K/AKT/mTOR pathway inhibitors in triple-negative breast cancer: A review on drug discovery and future challenges. *Drug Discov. Today* **2019**, *24*, 2181–2191. [PubMed]
15. Wen, W.; Marcinkowski, E.; Luyimbazi, D.; Luu, T.; Xing, Q.; Yan, J.; Wang, Y.; Wu, J.; Guo, Y.; Tully, D.; et al. Eribulin Synergistically Increases Anti-Tumor Activity of an mTOR Inhibitor by Inhibiting pAKT/pS6K/pS6 in Triple Negative Breast Cancer. *Cells* **2019**, *8*, 1010. [CrossRef] [PubMed]

Review

Roles of mTOR Signaling in Tissue Regeneration

Xiangyong Wei, Lingfei Luo * and Jinzi Chen *

Laboratory of Molecular Developmental Biology, School of Life Sciences, Southwest University, Beibei, Chongqing 400715, China; xiangyongwei2016@outlook.com

* Correspondence: lluo@swu.edu.cn (L.L.); chjz2012@email.swu.edu.cn (J.C.); Tel.: +86-23-68367957 (L.L.); Fax: +86-23-68367958 (L.L.)

Received: 9 August 2019; Accepted: 7 September 2019; Published: 12 September 2019

Abstract: The mammalian target of rapamycin (mTOR), is a serine/threonine protein kinase and belongs to the phosphatidylinositol 3-kinase (PI3K)-related kinase (PIKK) family. mTOR interacts with other subunits to form two distinct complexes, mTORC1 and mTORC2. mTORC1 coordinates cell growth and metabolism in response to environmental input, including growth factors, amino acid, energy and stress. mTORC2 mainly controls cell survival and migration through phosphorylating glucocorticoid-regulated kinase (SGK), protein kinase B (Akt), and protein kinase C (PKC) kinase families. The dysregulation of mTOR is involved in human diseases including cancer, cardiovascular diseases, neurodegenerative diseases, and epilepsy. Tissue damage caused by trauma, diseases or aging disrupt the tissue functions. Tissue regeneration after injuries is of significance for recovering the tissue homeostasis and functions. Mammals have very limited regenerative capacity in multiple tissues and organs, such as the heart and central nervous system (CNS). Thereby, understanding the mechanisms underlying tissue regeneration is crucial for tissue repair and regenerative medicine. mTOR is activated in multiple tissue injuries. In this review, we summarize the roles of mTOR signaling in tissue regeneration such as neurons, muscles, the liver and the intestine.

Keywords: mTOR signaling; metabolism; tissue regeneration; neuron; muscle; liver; intestine

1. Introduction

In the 1970s, a new antifungal, was discovered in soil samples on the Polynesian island of Rapa Nui, which was isolated from *Streptomyces hygroscopicus* and called rapamycin [1,2]. Afterwards, FK506-binding protein 12 (FKBP12) was found to repress cell growth and proliferation [3]. During the 1990s, the target of rapamycin (TOR) and the mammalian target of rapamycin (mTOR) were discovered in yeast and mammals respectively [2]. Brown et al. reported that mTOR is a target of the FKBP12-rapamycin complex [4]. mTOR is a serine/threonine protein kinase, which recruits other proteins to form two different complexes, named mTOR complex 1 (mTORC1) and complex 2 (mTORC2). mTOR is conserved in the evolution from yeast to mammal [1]. mTORC1 and mTORC2 contain the same subunits: mTOR, mammalian lethal with Sec13 protein 8 (mLST8) and DEP domain-containing mTOR-interacting protein (DEPTOR). However, regulatory-associated protein of mTOR (Raptor) and 40 kDa proline-rich Akt substrate (PRAS40) are specific to mTORC1, while rapamycin-insensitive companion of mTOR (Rictor), Protor1/2 and mammalian stress-activated protein kinase(SAPK)-interacting protein 1 (mSin1) are specific to mTORC2 [1,5]. mTOR signaling plays crucial roles in the regulation of cell growth, metabolism, cell survival and migration. In response to growth factors, energy, amino acid, and oxygen, mTORC1 controls cell growth and metabolism through mRNA translation, synthesis of protein, lipid and nucleotide, and repression of catabolic processes such as autophagy [6]. The ribosomal S6 kinase (S6K) and eIF4E-binding protein 1 (4EBP1) are the main effectors of mTORC1. Unlike mTORC1, studies on mTORC2 are limited. mTORC2 mainly controls cell survival and migration through phosphorylation and activation of the downstream-effectors SGK1, Akt, and the PKC kinase families [5]. The mTORC2 is an effector of the insulin/PI3K pathway and is a key regulator of Akt [5,7]. mTOR signaling is the central pathway in response to the environment, and the disruption of mTOR signaling is associated with developmental defects, cancer, neurodegenerative diseases, type 2 diabetes, autoimmune diseases, and aging-related diseases [8–12]. Thus, mTOR is therefore a therapeutic target of these diseases [13].

Tissue damage caused by trauma, diseases, and aging, etc. can result in organ dysfunction. Afterward, tissue regeneration is critical for the restoration of organ functions and maintenance of homeostasis [14]. In adult humans, although the regenerative capacity of some organs, like the central nervous system (CNS) and heart, is weak, other organs, including the liver, intestines, muscles, and skin, do maintain the intrinsic ability to regenerate [15]. The key reasons why different organs obtain distinct regenerative capacities and different species obtain distinct regenerative capacities in the same organ remain to be elucidated. So, mechanistic insights into tissue regeneration are essential for tissue repair and regenerative medicine [14]. mTOR is one of the central regulatory signaling pathways between injuries and physiological reactions such as tissue regeneration. For example, in the CNS with very weak regenerative capacity, activated mTOR through the inactivation of PTEN (phosphatase and tensin homolog) or TSC1 (tuberous sclerosis complex 1) can robustly promote axonal regeneration [16]. mTOR is also vital in the regeneration of the intestines, liver and muscles.

In this review, we first briefly describe the structures, regulatory mechanisms, and physiological functions of mTORC1 and mTORC2. Then, we put our efforts toward summarizing the roles of mTOR signaling in the regeneration of neurons, muscles, the liver, and intestine. At the end, the development strategy of tissue-specific agonist or inhibitor of mTORC1 in regenerative medicine is discussed.

2. The Structure and Regulation of mTORC1

mTOR is a serine/threonine protein kinase and a member of the PI3K-related kinase (PIKK) family, which forms the mTORC1 and mTORC2 complexes with other proteins [17,18]. The mTORC1, a heterotrimeric protein kinase, is mainly composed of three core components including mTOR, Raptor, and mLST8 [19–22]. mTORC1 also contains two inhibitory subunits, PRAS40 and DEPTOR [23–25]. After acute rapamycin treatment, the FKBP12-rapamycin complex binds to the FKBP12-rapamycin-binding (FRB) domain of mTOR and blocks mTORC1 activation [26] (Figure 1A). The mTORC1 plays important roles in metabolism and cell growth in response to nutrients and is regulated by many factors including growth factors, amino acids, energy, oxygen, and DNA damage [1,27]. Insulin/insulin-like growth factors (IGFs) inhibit the TSC complex, an inhibitory heterotrimeric complex of mTOR containing TSC1, TSC2, and Tre2-Bub2-Cdc16 (TBC) 1 domain family, member 7 (TBC1D7) [28], thus activating mTORC1. This mTORC1 activation is dependent on the Akt-mediated phosphorylation of TSC, which dissociates TSC from the lysosomal membrane [29]. The Ras signaling activates mTORC1 through extracellular signal-regulated kinase (Erk) and $p90^{RSK}$, both of which phosphorylate and inhibit TSC2 [30] (Figure 1A). It is worth mentioning that Ras homolog enriched in brain (Rheb) is indispensable for mTORC1 activation. Some papers reported that Rheb activates mTORC1 through interruption of the FKBP38-mTOR interaction or directly binding to mTOR [31,32], however, the detailed mechanisms underlying activation of mTORC1 by Rheb remain to be fully elucidated.

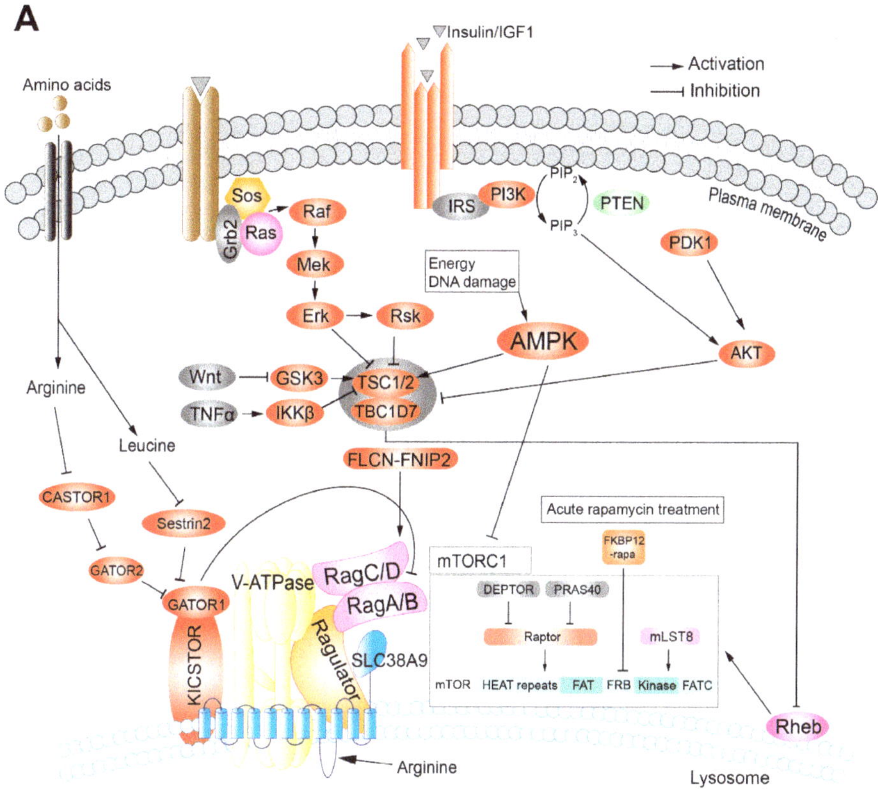

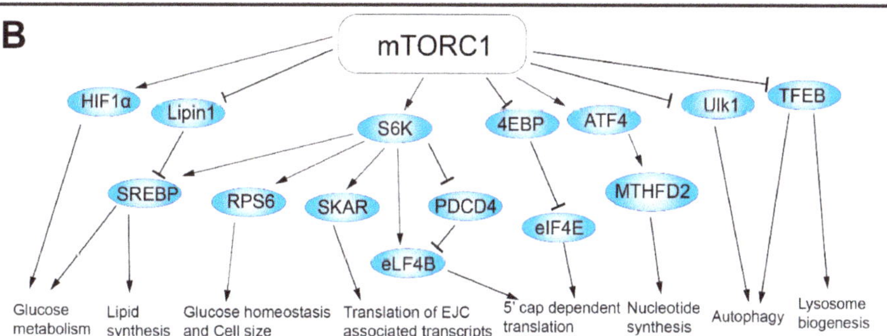

Figure 1. The regulatory mechanism and function of the mammalian target of rapamycin complex 1 (mTORC1). (**A**) The structures and regulatory mechanism of mTORC1. (**B**) The downstream functions of mTORC1.

The mTORC1 activation is closely related to the variation of amino acid concentrations. Different types of amino acid stimulate mTORC1 through different sensors. For example, cytosolic leucine, cytosolic arginine, and the lysosomal arginine are sensed by Sestrin2, CASTOR1 (Cellular Arginine Sensor for mTORC1) complex, and a candidate lysosomal amino acid sensor SLC38A9, respectively [1,33–36]. Amino acids activate mTORC1 through an amino acid sensing cascade involving the vacuolar H^+-ATPase (v-ATPase), RAG GTPases (small guanosine triphosphatases) and Ragulator. Unlike other stimulators, mTORC1 activation by amino acids is independent of the TSC-Rheb signaling axis [37] (Figure 1A). In contrast to leucine and arginine, glutamine also promotes mTORC1 activation, which is dependent on the related Arf family GTPases rather than Rag GTPase [38]. Folliculin-FNIP2 (folliculin interacting protein 2) complex, a Rag-interacting protein with GAP (GTPase-activating protein) activity for RagC/D, was recently reported to activate mTORC1 in the existence of amino acids [36,39]. Except cytosolic arginine, leucine and lysosomal arginine, whether other amino acids activate mTORC1, and the identity of their sensors remains unknown. Furthermore, energy, oxygen, and DNA damage negatively regulate mTORC1 through AMPK (5′ AMP-activated kinase), which indirectly inhibits mTORC1 activation via phosphorylation of TSC2 or direct phosphorylation of Raptor [40–42]. Moreover, both wingless-type MMTV integration site family (Wnt) signaling and tumor necrosis factor α (TNFα) activate mTORC1 through inhibition of TSC1 [43,44].

The activated mTORC1 enhances protein synthesis through direct phosphorylation of the ribosomal S6 kinase (S6K) and 4E-BP1 [45]. Then, the phosphorylated S6K (pS6K) promotes mRNA translation initiation through phosphorylation and activation of eIF4B, a positive regulator of the 5′ cap-binding eIF4F complex, and promotion of the degradation of PDCD4 (programmed cell death protein 4), an inhibitor of elF4B [46,47]. pS6K also regulates glucose homeostasis and cell size through phosphorylation of ribosomal protein s6 (rps6) [48]. Moreover, the interaction of pS6K and SKAR (S6K1 Aly/REF-like substrate) improves the translation efficiency of spliced mRNAs [49]. The phosphorylated 4E-BP1 dissociates its binding to eIF4E, which allows eIF4E to join in the eIF4F complex together with eIF4G, thus permitting the cap-dependent translation [50]. All the regulations above finally promote protein synthesis. The mTORC1-dependent anabolism is mediated by phosphorylation of S6K, inhibition of lipin1, an inhibitor of lipid synthesis, [51] and activation of ATF4 (activating transcription factor 4), a promoter of nucleotide synthesis [52]. mTORC1 also augments the glycolytic pathway through increasing the translation of hypoxia inducible factor 1α (HIFα), which drives the expression of phospho-fructo kinase (PFK) [53]. Furthermore, mTORC1 suppresses the catabolism such as autophagy and lysosome biogenesis through phosphorylation of ULK1 (unc-51 like autophagy activating kinase 1) and the transcription factor EB (TFEB) [54,55] (Figure 1B). In conclusion, mTORC1 regulates cell growth and metabolism in response to environmental inputs such as growth factors, nutrients, and DNA damage. It plays significant roles in development, physiological processes, and diseases.

3. The Structure and Regulation of mTORC2

Like mTORC1, mTORC2 also contains mTOR and mLST8 subunits. But Raptor in mTORC1 is replaced by Rictor in mTORC2 [56]. mTORC2 also includes DEPTOR, the regulatory subunits mSin1 and Protor1/2 [25,57,58]. mTORC2 can be impeded by prolonged rapamycin treatment [59]. Unlike mTORC1, the upstream and downstream activity of mTORC2 are not well-defined. mTORC2, as an effector of insulin/PI3K signaling, is inhibited by the pleckstrin homology domain of mSin1 when there is a lack of insulin. This autoinhibition by mSin1 is relieved upon its binding to phosphatidylinositol (3,4,5)-trisphosphate (PIP3) on the plasma membrane [7]. Akt activates mTORC2 through phosphorylation of mSin1 at T86, in turn the activated mTORC2 stimulates Akt through phosphorylation of Akt at S473, which forms a positive feedback regulatory loop [5,60]. In contrast to Akt, S6K suppresses mTORC2 via promoting the degradation of insulin receptor substrate 1 (IRS1) [61].

The mTORC2 mainly controls cell migration through phosphorylation of the AGC (protein kinase A/G/C) protein kinase family members such as PKCα [56], PKCδ [62], PKCξ [63], PKCγ and PKCε [64], all of which regulate cell migration through modulations of various aspects of cytoskeletal remodeling. Furthermore, another important function of mTORC2 is phosphorylation and activation of Akt, which in turn phosphorylates and inhibits forkhead box O1/3a (FoxO1/3a), TSC2 and the metabolic regulator glycogen synthase kinase 3β (GSK3β) [65,66], thus promoting cell survival and proliferation. In addition, mTORC2 can phosphorylate and activate SGK1, which regulates ion transport for cell survival [67] (Figure 2). mTORC2 is also involved in cancer, Alzheimer's disease (AD) [10,68].

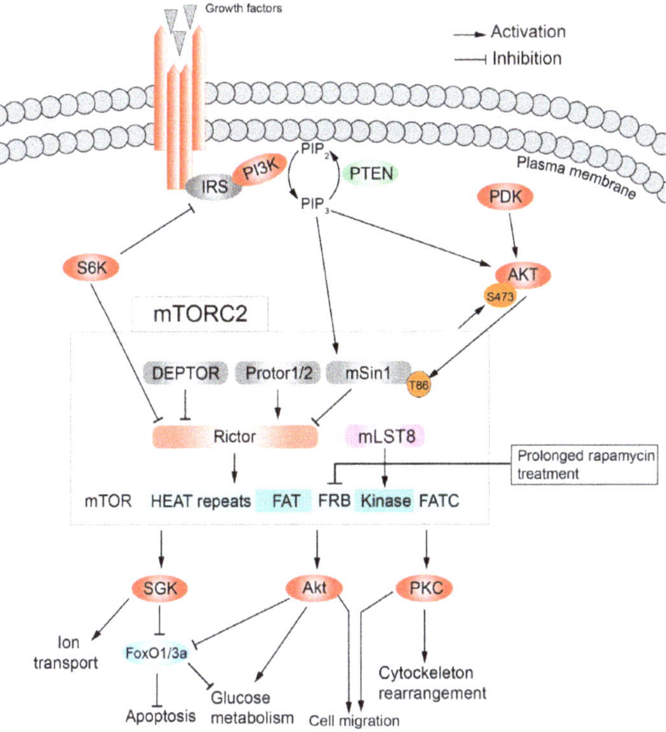

Figure 2. The structures, regulatory mechanism and functions of mTORC2.

4. Roles of mTOR in Neuronal Regeneration

The blood-brain barrier (BBB) is formed by endothelial cells, pericytes and astrocytes. These cells together form the neurovascular unit (NVU), which serves as an interface between the blood and the neural tissue. Impairment of BBB function is associated with neurodegenerative diseases [69]. The brain endothelial cells are vital for the function of BBB [70]. Brain vascular damage or occlusion can cause cerebrovascular diseases such as microbleeding, hemorrhagic stroke, and ischemic stroke. Macrophages and lymphatic vessels are important for the repair of brain blood vessels and the restoration of BBB functions [71,72]. The nervous system is comprised of the central nervous system (CNS) and peripheral nervous system (PNS) [73]. The PNS has a unique ability to regenerate [74–76]. However, the CNS of adult mammals including the brain and spinal cord obtains very limited regenerative capacity, which might partially result from abundant inhibitory growth factors in the CNS [77–80]. Effective therapeutic approaches are still missing for a wide variety of human neurodegenerative diseases including Parkinson's disease (PD), amyotrophic lateral sclerosis (ALS), Huntington's disease (HD), and Alzheimer's disease (AD) [81–86]. Therefore, the neuronal regeneration of the CNS of adult mammals constantly remains an important research topic, being of great significance for clinical treatment.

In general, axons after injury do not spontaneously regenerate in adult mammalian CNS because of a diminished intrinsic regenerative capacity and extrinsic growth-inhibitory factors [77,87,88]. Inhibitory factors from myelin including Nogo protein families, Oligodendrocyte myelin glycoprotein (OMgp), myelin-associated glycoprotein (Mag), ephrin B3, and transmembrane semaphorin 4D (Sema4D/CD100) block CNS axonal regrowth [79,89,90]. Chondroitin sulphate proteoglycans (CSPGs) produced by the reactive astrocytes in the glial scar become main inhibitory extracellular matrix (ECM) molecules at the lesion site of a mature CNS [90,91]. Therefore, both promotion of the intrinsic regenerative capacity and suppression of the inhibitory environment are required for efficient axon regeneration [77,92]. Either overexpression of osteopontin, IGF1 and ciliary neurotrophic factor (CNTF), or genetic inactivation of PTEN, a negative regulator of mTOR signaling, TSC1/2 and suppressor of cytokine signaling 3 (SOCS3), can promote axon regeneration. Signaling pathways like Janus kinase/signal transducers and activators of transcription (JAK/STAT3), canonical bone morphogenetic protein/drosophila mothers against decapentaplegic (BMP/Smad), non-canonical BMP pathway, Jun N-terminal kinase/p38 MAP kinase (JNK/MAPK), and mTOR signaling, have been reported to promote axonal regrowth during neuronal regeneration [16,93–95]. Here, we summarize the roles of mTOR in axonal regeneration of the CNS and PNS.

Injuries to the optic nerves and spinal cord have been widely used to study axonal regeneration of the CNS. mTOR signaling exhibits distinct functions in the optic nerve and spinal cord regeneration [76]. mTOR signaling is highly activated at the embryonic stage and diminished in the adult retinal ganglion cells (RGCs), suggesting the different functions at various developmental stages of RGCs. Activation of mTOR via genetic deletion of TSC1 or PTEN enhances axonal regeneration of RGCs after optic nerve injury [16]. Similarly, axon regeneration robustly occurs when PTEN is deficient in mouse cortical motor neurons [96], Drosophila sensory neurons [97], and Caenorhabditis elegans motor neurons [98]. The melanopsin/GPCR (cell-type-specific G protein-coupled receptor) signaling enhances axonal regeneration of RGCs through promotion of mTORC1 signaling. Interestingly, the regenerative activation is in a light-dependent manner [99]. Class I histone deacetylation enzyme HDACs that allow histones to wrap DNA more tightly, has been shown to repress the RGCs survival and regeneration after optic nerve injury. Dual deletions of HDAC1 and HDAC2 or HDAC3 deficiency robustly promotes RGCs regeneration [100,101]. Recently, dual functions of HDAC5 have been shown in the regeneration of dorsal root ganglions (DRGs) of the PNS [102,103]. Increases in the HDAC5 cytoplasmic localization by overexpressing the mutant $HDAC5^{AA}$ can stimulate RGCs regenerative ability after optic nerve injury. This enhanced RGC regeneration is dependent on mTORC1 [104]. Another study revealed that Wnt10b from fibroblast-derived exosomes (FD exosomes) enhances neurite regrowth through regulation of GSK3β and TSC2 to boost mTOR activation [105]. After optic nerve injury, the mTOR is unnecessary for the initial step of RGCs to enter into the regenerative status but required for long axon regeneration under inflammatory stimulation [106]. Furthermore, the elevation of mTOR either by the overexpression of Rheb, or by double deletion of Pten and Socs3, or by Pten deletion combined with the injection of cAMP (cyclic adenosine monophosphate) analogue 4-(chlorophenylthio) adenosine (CPT)-cAMP or inflammatory molecules (oncomodulin or zymosan) can enhance RGCs growth into the brain [107–110]. The combinatory treatments after optic nerve injury are able to achieve visual functional recovery.

Although studies above suggest that mTOR signaling promotes RGC regeneration, Li et al. showed that mTOR is a negative regulator of RGCs survival and activates astrocytes after optic nerve injury. In the rat retinal ischemia-reperfusion (I/R) injury model, neuroprotective effects of rapamycin can reduce the loss of RGCs after optic nerve injury. Thus, the neuroprotective effects of rapamycin supply a therapeutic method for optic neurodegeneration [111]. Overall, activated mTOR robustly enhances RGCs regeneration after optic nerve injury and also promotes astrocyte activation to form a physical and biochemical barrier glial scar for inhibiting axonal regeneration (Figure 3A). Specific activation of mTOR in RGCs or inhibition in astrocytes after injury promotes RGC regrowth and regeneration, suggesting mTOR as a potential therapeutic target for optic nerve injury and neurodegenerative disease. In summary, activated mTOR enhances RGCs regeneration after optic nerve injury and also promotes astrocytes activation to form physical and biochemical barrier glial scar that could inhibit axonal regeneration (Figure 3A).

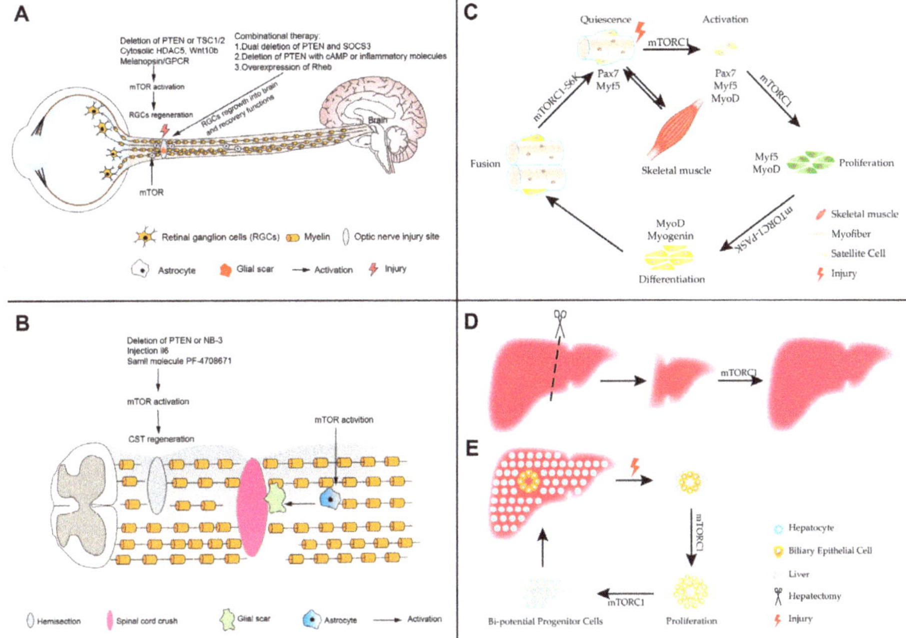

Figure 3. The roles of mTOR in the regeneration of neurons, muscles and the liver. (**A**) The optic nerve injury model. Activation of mTOR by deleting phosphatase and tensin homolog (PTEN) or tuberous sclerosis complex 1/2 (TSC1/2), the upstream cytosolic HDAC5 (histone deacetylase 5), Wnt10b (wingless-type MMTV integration site family, member 10b), and melanopsin/GPCR (cell-type-specific G protein-coupled receptor) robustly enhances the regeneration of retinal ganglion cells (RGCs). Combinational therapies augment the RGCs long-distance regeneration for visual function recovery through overexpression of Ras homolog enriched in brain (Rheb); the dual deletion of PTEN and suppressor of cytokine signaling 3 (SOCS3); deficiency of PTEN combined with injecting of cyclic adenosine monophosphate (cAMP) or inflammatory molecules (oncomodulin or zymosan). However, the excessive mTOR activation of astrocytes contributes to forming glial scar to inhibit axonal regeneration. (**B**) The spinal cord injury (SCI) model. In spinal cord crush model, the activated mTOR of astrocytes facilitates glial scar formation resulting in impeding the spinal cord regeneration after SCI. In the hemisection model, the stimulation of mTOR promotes the corticospinal tract (CST) regeneration post-SCI. (**C**) A schematic representation of skeletal muscle regeneration. mTORC1 stimulates satellite cells activation and proliferation, and their progenies differentiate into myoblasts under mTORC1 regulation. mTORC1 also promotes the fusion of myoblasts to form myofibers. (**D**) Partial hepatectomy (PH) model. Liver regeneration after PH is via the self-replication of existing hepatocytes, and mTORC1 regulates hepatocyte proliferation. (**E**) The severe liver injury model. Liver regeneration is via the trans-differentiation of cholangiocytes. In the process, mTORC1 regulates the proliferation of cholangiocytes and the formation of Bi-potential Progenitor Cells.

The spinal cord connects the brain with the PNS. Spinal cord injuries (SCI) are divided into traumatic and non-traumatic aetiologies [112]. Mechanisms of spinal cord regeneration are of great significance for post-SCI clinic therapy. In the animal model of SCI, the epidermal growth factor (EGF) receptor promotes transformation of inactivated astrocytes into a reactive status, which forms a glial scar to restrict neuronal recovery. The activation of astrocytes by EGF is dependent on the Rheb/mTOR pathway [113,114]. Chen et al. also showed that the PI3K/Akt/mTOR pathway contributes to the formation of glial scars by reactive astrocytes. Moreover, the overexpression of PTEN can attenuate gliosis at three days after SCI and enhances motor functional recovery [115]. In conclusion, hyperactive mTOR in astrocytes is a negative regulator for the functional recovery following SCI [114,116]. However, in different injury models like spinal cord hemisection, mTOR activation, stimulated by the deficiency of PTEN or contactin-6 (NB-3) as well as by injection with interleukin (IL-6) or small molecule PF-4708671 (an inhibitor of downstream substrate S6 kinase 1) efficiently promotes regrowth and regeneration of the corticospinal tract [117–121] (Figure 3B). mTOR also increases DRGs regrowth and functional recovery following PNS injury [122–126]. In summary, functions of mTOR in different neuronal injury models can be different or even opposing, which might be caused by mTOR activities in different cell types. Promotion of axonal regeneration can be achieved either by the inhibition of mTOR in astrocytes to attenuate glial scar formation, or by the activation of mTOR in neurons.

5. Roles of mTOR in Skeletal Muscle Regeneration

Muscle can quickly regenerate to recover its function after nearly complete myofiber destruction. It is important to maintain the physiological homeostasis of muscle throughout life [127]. Many studies have demonstrated that the quiescent satellite cells, which are believed to be resident muscle stem cells (MuSCs) and localized between the plasmalemma of myofibers and the basement membrane, mainly contribute to the skeletal muscle regeneration in response to muscle injury [128–132]. After muscle injury, satellite cells are activated to proliferate. Their progenies either maintain the satellite cells pool through self-renewal or differentiate into myoblasts expressing myogenic markers Myf5 (myogenic factor 5), MyoD (myogenic differentiation 1), myogenin and MRF4 (myogenic regulatory factor 4) [130,133] (Figure 3C). Ultimately, nascent myoblasts fuse together or fuse to existing myofibers to form new myofibers to accomplish skeletal muscle regeneration [130]. In different models of muscle injury, this process is regulated by multiple pathways such as AMPK, IGF-1/Akt, TGFβ/Smad and mTOR [127,130,133–137]. Here we focus on the roles of mTOR in the regulation of skeletal muscle regeneration.

In mice, conditional knockout (cKO) of *mtor* or *raptor* in MuSCs effectively inhibits activation, proliferation, and differentiation of satellite cells, which impairs skeletal muscle regeneration [133–137] (Figure 3C). Rapamycin treatment also confirms that mTOR inhibition indeed blocks the formation of nascent myofibers and the growth of regenerating myofibers during skeletal muscle regeneration [134,136]. Rapamycin-resistant (RR) and RR/kinase-inactive (RR/KI) experiments demonstrate that mTORC1 signaling regulates muscle regeneration through both kinase-independent and kinase-dependent mechanisms [134]. S6K1 is dispensable for the initial formation of nascent myofiber during regeneration, but its ablation impairs later muscle growth [134]. Ablation of 4EBP1 facilitates myofiber growth, but does not affect the activation of satellite cells [137]. After injury, activation of Per-Arnt-Sim domain kinase (PASK), a downstream phosphorylation target of mTORC1, phosphorylates Wdr5 to induce the expression of *myogenin* and stimulates MuSCs differentiation into myoblasts [138] (Figure 3C). The mTORC1-S6K pathway is also required for myoblasts fusion to accomplish myofiber formation during muscle regeneration [138]. micro-RNAs (miRNAs) have been reported as important modulators of myoblasts formation and fusion during muscle regeneration [139–142]. micro-RNA-1 (miR-1) stimulated by mTORC1 increases myoblast differentiation and enhances muscle regeneration through the HDAC4-follistatin axis [141]. Similar to miRNAs, long non-coding RNAs (lncRNAs) also regulate skeletal muscle regeneration. The lncRNA LINC00961 encodes a polypeptide small regulatory polypeptide of amino acid response (SPAR), which negatively regulates mTORC1 activation by interacting with lysosomal v-ATPase. The downregulation of SPAR after skeletal muscle injury activates mTORC1 to enhance muscle regeneration [143]. In contrast to mTORC1, *rictor* knockout in embryonic and adult satellite cells is ineffective in skeletal muscle regeneration [136,144]. After trauma, the presence of bone-derived and cardiac muscle-derived tissue ECM scaffolds in damaged muscle recruits more immune cells and form immune microenvironment, which facilitates muscle regeneration. Transplanted WT CD4$^+$ T cells rather than *rictor*$^{-/-}$ CD4$^+$ T cells promote muscle regeneration in the

Rag1$^{-/-}$ mice [145]. Taken together, these studies illustrate that mTORC1, but not mTORC2, acts as a key regulator of skeletal muscle regeneration.

In muscle injury-regeneration models induced by cardiotoxin (CTX) [146], $BaCl_2$ [142] and ischemia/reperfusion (I/R) [147], some pathways regulate muscle regeneration through mTOR signaling. In I/R-induced muscle injury, skeletal muscle protection by activated Sonic hedgehog (Shh) is blocked by the inhibition of AKT/mTOR/p70S6K [147]. This result implies that Shh stimulates skeletal muscle regeneration through the AKT/mTOR/p70S6K pathway. In CTX-induced muscle injury, Ca^{2+} influx flows into cells through T-type Ca^{2+} and Trpc1 channels, which enhances the activation of PI3K to activate the Akt/mTOR/p70S6K pathway and ultimately improves muscle regeneration [146]. The IGF pathway is essential for skeletal muscle regeneration [148,149]. mTORC1 inhibits miR-125b, which is a negative regulator of IGF2, thus promoting muscle regeneration after $BaCl_2$ treatment [142]. Furthermore, nutmeg extract stimulates soleus muscle regeneration through IGF1-AKT-mTOR and inhibits autophagy in aged rats [150].

Muscle-mass weakness and wasting are often caused by various pathological conditions, such as sarcopenia, diabetes and chronic obstructive pulmonary disease (COPD). On average, geriatric patients with sarcopenia lose 30% of their muscle mass and 35% of myofibers [151], and have decreased muscle protein synthesis [152]. Studies on aged rodents clarify that aging decreases the muscle regenerative capacity after injury [152–154]. Comparing to the aged rodents, young rodents with leucine supplements strongly improve skeletal muscle regeneration. During muscle injury, both aged and young rodents activate Akt/mTOR pathway, p70S6K and 4EBP1, but the young rodents have stronger activation than aged ones. Leucine supplementation increases the activation of the PI3K/Akt/mTOR pathway, then improves muscle regeneration in aged rodents [155–159]. However, homeostatic maintenance of MuSCs pool is also critical to ensure muscle regeneration upon re-injury in aged muscles [160,161]. Upon repeated injuries, differentiation of MuSCs at the expense of MuSCs is promoted by mTORC1 in aged mice, which can be inhibited by rapamycin [162]. In mice, skeletal muscle atrophy can be induced by diabetes. Acupuncture with low-frequency-electric-stimulation (Acu-LFES) promotes muscle regeneration through the IGF-1-Akt-mTOR pathway [163]. The muscle regeneration stimulated by the Acu-LFES induced miR-1 may also act through the mTORC1-miR-1-HDAC4-follistatin axis [141,163]. In mice with hypoxic treatment, mTOR is inhibited by the activated AMPK or REDD1 (regulated in development and DNA response 1), both of which impair muscle regeneration [164]. Muscle injury and the decreased mTOR activity are observed in COPD patients with hypoxemia [165]. At present, there are drugs/targets to be used in clinical trials to treat muscle wasting. However, drugs targeting mTOR signaling have not been trialed yet [166], which may be of future clinical interest to inhibit muscle atrophy and promote muscle regeneration.

6. Roles of mTOR in Liver Regeneration

The liver possesses extraordinary regenerative capacity comparing to other internal organs [167–169] Partial hepatectomy (PH) of rats and mice is one of the most widely used models to study liver regeneration [170,171]. In 1931, Higgins et al. first proposed a rat model with 2/3 hepatectomy, in which the remaining liver restores its original weight after 5–7 days of surgical resection [168] (Figure 3D). Other acute injury models include tetrachloride (CCl_4)- and thioacetamide (TAA)-induced chemical liver injury. Remaining hepatocytes after PH or chemical liver injury accomplish liver regeneration via self-replication without the participation of progenitor cells [168,172]. However, chronic liver diseases in humans cause severe liver injury and impair hepatocyte proliferation [173]. After extreme liver damage, regeneration is achieved via cholangiocytes transdifferentiation rather than self-replication, in zebrafish [174,175] and mice [176,177] (Figure 3E). The mTOR signaling pathway mediates liver regeneration after PH and initiates the transdifferentiation of biliary epithelial cells (BECs) after extreme liver injury [178,179].

A 2/3 hepatectomy in rodents promotes the release of cytokines and growth factors [171], which simulate hepatocyte proliferation via transducing activation of PI3K/Akt signaling pathway [180].

The "pif-pocket" mutant of PDK1 showed that the PI3K-dependent PDK1 kinase specially phosphorylates Akt during regeneration after PH [181]. In mice, deletion of vitamin D3 up-regulated protein 1 (VDUP1) enhances regeneration after PH via the HGF and TGF-α responsive activation of ERK1/2 and Akt [182,183]. In VDUP1 KO mice, 70% hepatectomy significantly increases the hepatic proliferative response, in accordance with CCl_4 treatment [183]. mTOR activated by Akt activation is well-conserved in the regulation of cell cycle and cell proliferation during liver regeneration [184] (Figure 3D). Hepatocyte proliferation in liver regeneration is regulated by cell cycle-related proteins, such as cyclin dependent kinase (CDK), cyclin D, cyclin E [185]. However, phosphorylation and activation of S6K1 by mTOR is vital to regulate the expression of cell cycle-related proteins, particularly cyclin D1, during liver regeneration [186]. After PH, rapamycin significantly reduces the rate of hepatic proliferation through inhibition of p70S6K activation, but not inhibition of p4EBP1 [187]. Meanwhile, it increases the rate of hepatic apoptosis [188] and significantly eliminates bleeding-induced hepatocyte hypertrophy [189]. After 2/3 hepatectomy, Cyclin D1 translation promoted by microRNA-21 (miR-21) enhances liver regeneration via eliminating the inhibition of Ras homolog gene family member B (Rhob) on Akt1/mTORC1 [190]. Cyclin D1 and CDK4 form an activation complex to promote liver regeneration via stimulation of G1 to S transition in hepatocytes [185]. Deletion of *sirtuin6 (sirt6)*, one of the sirtuin family of class III NAD^+-dependent histone deacetylase, delays G1 to S phase transition and impairs activation of the Akt/mTOR pathway during liver regeneration [191]. However, SIRT1 overexpression in mice impairs liver regeneration, which can be reversed by the leucine-activated mTORC1 [192]. Genetic manipulation or chemical drugs treatment, such as inactivation of apoptosis-stimulating protein two of p53 (ASPP2) [193], *let-7* deletion [194], rosmarinic acid [195], carbamazepine [196], and panax notoginseng saponins [197], stimulate liver regeneration by activation of mTOR after PH, which is blocked by the mTOR inhibition [194–196]. Severe injury in PH such as a 90% hepatectomy suppresses the capacity of liver regeneration, which is accompanied by the inactive mTOR signaling [198]. In addition to hepatocyte proliferation, increases in the hepatocyte size could account for liver regeneration [181,184,199]. In liver-specific STAT3-knockout (LS3-KO) mice, although hepatocyte proliferation after injury is impaired by the decreased cyclin D1 expression, increases in hepatocyte size during early liver regeneration may occur via the activation of the PI3K/Akt/mTOR pathway [181,199].

Contributions of non-parenchymal cells (NPCs) can also be important for liver regeneration [174,176,177,179,200]. Expansion of BECs during ductular reaction (DR) is promoted by mTORC1 signaling after extreme liver injury in zebrafish and mice [179,200] (Figure 3E). Inhibition of mTORC1 remarkably suppresses the dedifferentiation of BECs and the proliferation of bi-potential progenitor cells (BP-PCs), thus leading to a reduced number of BP-PCs for re-differentiation and impairing liver regeneration after extreme injury [179,200]. Estrogen promotes liver regeneration through the activation of mTORC1 signaling in zebrafish [201]. In contrast, the *rictor* mutant only loses 20% of regenerating liver mass compared to the wild-type [179]. These studies suggest that liver regeneration is mainly mediated by mTORC1 rather than mTORC2 signaling [179].

In addition to endogenous liver regeneration, liver transplantation is also an effective way to restore liver mass and functions after liver failure or cirrhosis [202,203]. In patients, associating liver partition with portal vein ligation during staged hepatectomy (ALPPS) activates mTORC1 signaling [204]. Small-for-size mouse liver transplantation (30% grafts) significantly downregulate mTORC1 signaling and suppress liver regeneration [205,206]. However, half-size transplantation (50% grafts) increases the mTORC1 activity. 30% grafts treated with amphiregulin restore mTORC1 activation and p70S6K phosphorylation to the level of 50% grafts [206]. These studies imply that activation of mTORC1 may be a promising therapeutic approach to stimulate liver regeneration during liver transplantation.

7. Roles of mTOR in Intestinal Regeneration

The small intestinal epithelium is a single cell layer with rapid self-renewing and strong regenerative capabilities. The mammalian intestinal epithelium mainly includes enterocytes, Goblet cells, enteroendocrine cells, and Paneth cells [207]. In flies, esg^+ (escargot) intestinal stem cells

(ISCs) and enteroblasts (EBs) are responsible for gut homeostasis and drive the regeneration following injury [208,209]. The TSC/TOR signaling pathway regulates proliferation and maintenance of ISCs in response to nutritional conditions [162,210]. Also, mTORC1 is the main component of mTOR pathway that is decisive for the intestinal development in zebrafish [211]. Similarly, *Lgr5* marks the ISCs at the base of intestinal crypt which maintain the intestinal homeostasis and intestinal regeneration in mouse [212]. Furthermore, *Dll1*$^{+}$ secretory progenitor cells, *Alpi*$^{+}$ enterocyte progenitor cells, and *Lyz*$^{+}$ Paneth cells can be induced to ISC after specific ablation of *Lgr5*$^{+}$ ISC [213–215]. mTORC1 takes an pivotal part in increasing the maintenance of ISCs activity and proliferation of intestinal epithelium [216]. In addition, the cooperation of mTORC1 and SIRT1 promotes the expansion of ISCs during the calorie restriction (CR) [217]. Taken together, mTORC1 regulates the intestinal development and maintains intestinal hemostasis and ISC, in which mTOR is conserved in drosophila, zebrafish and mouse.

The intestinal epithelium, which has a powerful regenerative capacity after injury, provides a model to study ISCs, cancer, and intestinal regeneration [218]. A few studies demonstrate that mTORC1 signaling is required for intestinal regeneration and conserved in *Drosophila* and mammals. In inflammatory bowel disease (IBD) mouse models induced by 2,4,6-trinitrobenzene sulfonic acid (TNBS) or dextran sodium sulfate (DSS), mTORC1 is required for intestinal regeneration against IBD. The deficiency of Regnase-1 promotes the intestinal regeneration in the acute IBD via controlling the mTOR and purine metabolism [219]. Akt/mTOR activated by the focal adhesion kinase (FAK) is required for the Wnt/myc-mediated intestinal regeneration and tumorigenesis [218,220]. mTORC1-S6k1/2 axis, but not eiF4ebp1/2, enhances crypt regeneration after DNA damage [221]. These studies imply that activation of mTORC1 may become a therapeutic approach for IBD and other intestinal injuries [222]. Besides, the mTOR-dependent autophagy is impaired in patients with ulcerative colitis (UC), implicating mTOR also as a therapeutic target for autoimmune diseases [223].

8. Perspectives

In this review, we summarize the roles of mTOR in the regeneration of neurons, muscles, the liver, and the intestine, which are mainly mediated by mTORC1 rather than mTORC2 signaling. The mTORC1 signaling network may be disrupted in various diseases, such as ALS, AD, PD, COPD, sarcopenia, liver failure, and IBD, which is harmful to tissue regeneration [151,165,203,222,224–226]. Although regeneration and recovery of injured tissues in human may be improved by drugs promoting mTOR such as carbamazepine and nutmeg [150,196,224,225], mechanisms underlying these drugs remain largely unknown. And the curative effects of these drugs still need improvements. Additionally, the agonist and inhibitor of mTOR lack tissue specificities, like leucine and rapamycin [2,225]. For example, in addition to the inhibition of mTORC1, rapamycin can also inhibit mTORC2 during a long-term treatment [225]. A large number of studies have reported that drug delivery systems can increase tissue specificity. For example, a bone-targeted nanoparticle (NP) delivery system can carry a β-catenin agonist or GSK3β inhibitor to fractured bone with concentrated accumulation, which enhances bone regeneration [227]. A tissue-specific agonist or antagonist of mTORC1 might also be achieved with this strategy. Tissue-specific mTORC1-antagonist may effectively treat neurodegenerative diseases resulting in nerve injury, and tissue-specific mTORC1-agonist may effectively promote muscle recovery in muscle injury. Recently, sirolimus in phase 1/2 trial has been shown to effectively treat patients with systemic lupus erythematosus (SLE) against tissue injury, improving the expansion of naïve T-cell populations and decreasing CD8^{+} memory T cells. But in this process, sirolimus causes reversible oral ulcers, headaches, and cytopenia [228]. The tissue-specific inhibitor of mTORC1 may prevent these side-effects in the clinic. Except organs depicted in this review, roles of mTOR signaling in the regeneration of other organs like the pancreas, heart, and kidney are rarely reported. Ultimately, a comprehensive understanding of the mTOR signaling network during different tissue regeneration processes, and the development of new drugs for tissue-specific mTOR activation or inhibition are of great scientific and clinical interest.

Funding: This work was supported by the National Key Basic Research Program of China (2015CB942800), National Natural Science Foundation of China (31730060, 91539201, and 91739304), and the 111 Program (B14037).

Acknowledgments: We would like to thank Mengzhu Lv for pictures and Xinmiao Tan, Rui Ye and Guozhen Wu for help and discussions.

Conflicts of Interest: The authors declare no conflict of interest.

References

1. Saxton, R.A.; Sabatini, D.M. mTOR Signaling in Growth, Metabolism, and Disease. *Cell* **2017**, *168*, 960–976. [CrossRef] [PubMed]
2. Johnson, S.C.; Rabinovitch, P.S.; Kaeberlein, M. mTOR is a key modulator of ageing and age-related disease. *Nature* **2013**, *493*, 338–345. [CrossRef] [PubMed]
3. Chung, J.; Kuo, C.J.; Crabtree, G.R.; Blenis, J. Rapamycin-FKBP specifically blocks growth-dependent activation of and signaling by the 70 kd S6 protein kinases. *Cell* **1992**, *69*, 1227–1236. [CrossRef]
4. Brown, E.J.; Albers, M.W.; Shin, T.B.; Keith, C.T.; Lane, W.S.; Schreiber, S.L. A mammalian protein targeted by G1-arresting rapamycin-receptor complex. *Nature* **1994**, *369*, 756–758. [CrossRef] [PubMed]
5. Oh, W.J.; Jacinto, E. mTOR complex 2 signaling and functions. *Cell Cycle* **2011**, *10*, 2305–2316. [CrossRef] [PubMed]
6. Kim, J.; Guan, K.L. mTOR as a central hub of nutrient signalling and cell growth. *Nat. Cell Biol.* **2019**, *21*, 63–71. [CrossRef]
7. Liu, P.; Gan, W.; Chin, Y.R.; Ogura, K.; Guo, J.; Zhang, J.; Wang, B.; Blenis, J.; Cantley, L.C.; Toker, A.; et al. PtdIns(3,4,5)P3-Dependent Activation of the mTORC2 Kinase Complex. *Cancer Discov.* **2015**, *5*, 1194–1209. [CrossRef]
8. Simioni, C.; Martelli, A.M.; Zauli, G.; Melloni, E.; Neri, L.M. Targeting mTOR in Acute Lymphoblastic Leukemia. *Cells* **2019**, *8*, 190. [CrossRef]
9. Kou, X.; Chen, D.; Chen, N. Physical Activity Alleviates Cognitive Dysfunction of Alzheimer's Disease through Regulating the mTOR Signaling Pathway. *Int. J. Mol. Sci.* **2019**, *20*, 1591. [CrossRef]
10. Kim, L.C.; Cook, R.S.; Chen, J. mTORC1 and mTORC2 in cancer and the tumor microenvironment. *Oncogene* **2017**, *36*, 2191–2201. [CrossRef]
11. Chakrabarti, P.; Kandror, K.V. The role of mTOR in lipid homeostasis and diabetes progression. *Curr. Opin. Endocrinol. Diabetes Obes.* **2015**, *22*, 340–346. [CrossRef] [PubMed]
12. Perl, A. mTOR activation is a biomarker and a central pathway to autoimmune disorders, cancer, obesity, and aging. *Ann. N. Y. Acad. Sci.* **2015**, *1346*, 33–44. [CrossRef] [PubMed]
13. Martelli, A.M.; Buontempo, F.; McCubrey, J.A. Drug discovery targeting the mTOR pathway. *Clin. Sci. (Lond.)* **2018**, *132*, 543–568. [CrossRef] [PubMed]
14. Moya, I.M.; Halder, G. Hippo-YAP/TAZ signalling in organ regeneration and regenerative medicine. *Nat. Rev. Mol. Cell Biol.* **2019**, *20*, 211–226. [CrossRef] [PubMed]
15. Baddour, J.A.; Sousounis, K.; Tsonis, P.A. Organ repair and regeneration: An overview. *Birth Defects Res. C Embryo Today* **2012**, *96*, 1–29. [CrossRef] [PubMed]
16. Park, K.K.; Liu, K.; Hu, Y.; Smith, P.D.; Wang, C.; Cai, B.; Xu, B.; Connolly, L.; Kramvis, I.; Sahin, M.; et al. Promoting axon regeneration in the adult CNS by modulation of the PTEN/mTOR pathway. *Science* **2008**, *322*, 963–966. [CrossRef]
17. Shimobayashi, M.; Hall, M.N. Making new contacts: The mTOR network in metabolism and signalling crosstalk. *Nat. Rev. Mol. Cell Biol.* **2014**, *15*, 155–162. [CrossRef] [PubMed]
18. Sciarretta, S.; Forte, M.; Frati, G.; Sadoshima, J. New Insights Into the Role of mTOR Signaling in the Cardiovascular System. *Circ. Res.* **2018**, *122*, 489–505. [CrossRef]
19. Yang, Q.; Guan, K.L. Expanding mTOR signaling. *Cell Res.* **2007**, *17*, 666–681. [CrossRef]
20. Kim, D.H.; Sarbassov, D.D.; Ali, S.M.; King, J.E.; Latek, R.R.; Erdjument-Bromage, H.; Tempst, P.; Sabatini, D.M. mTOR interacts with raptor to form a nutrient-sensitive complex that signals to the cell growth machinery. *Cell* **2002**, *110*, 163–175. [CrossRef]
21. Kim, D.H.; Sarbassov, D.D.; Ali, S.M.; Latek, R.R.; Guntur, K.V.; Erdjument-Bromage, H.; Tempst, P.; Sabatini, D.M. GbetaL, a positive regulator of the rapamycin-sensitive pathway required for the nutrient-sensitive interaction between raptor and mTOR. *Mol. Cell* **2003**, *11*, 895–904. [CrossRef]

22. Hara, K.; Maruki, Y.; Long, X.; Yoshino, K.; Oshiro, N.; Hidayat, S.; Tokunaga, C.; Avruch, J.; Yonezawa, K. Raptor, a binding partner of target of rapamycin (TOR), mediates TOR action. *Cell* **2002**, *110*, 177–189. [CrossRef]
23. Sancak, Y.; Thoreen, C.C.; Peterson, T.R.; Lindquist, R.A.; Kang, S.A.; Spooner, E.; Carr, S.A.; Sabatini, D.M. PRAS40 is an insulin-regulated inhibitor of the mTORC1 protein kinase. *Mol. Cell* **2007**, *25*, 903–915. [CrossRef] [PubMed]
24. Vander Haar, E.; Lee, S.I.; Bandhakavi, S.; Griffin, T.J.; Kim, D.H. Insulin signalling to mTOR mediated by the Akt/PKB substrate PRAS40. *Nat. Cell Biol.* **2007**, *9*, 316–323. [CrossRef] [PubMed]
25. Peterson, T.R.; Laplante, M.; Thoreen, C.C.; Sancak, Y.; Kang, S.A.; Kuehl, W.M.; Gray, N.S.; Sabatini, D.M. DEPTOR is an mTOR inhibitor frequently overexpressed in multiple myeloma cells and required for their survival. *Cell* **2009**, *137*, 873–886. [CrossRef] [PubMed]
26. Yang, H.; Rudge, D.G.; Koos, J.D.; Vaidialingam, B.; Yang, H.J.; Pavletich, N.P. mTOR kinase structure, mechanism and regulation. *Nature* **2013**, *497*, 217–223. [CrossRef] [PubMed]
27. Sabatini, D.M. Twenty-five years of mTOR: Uncovering the link from nutrients to growth. *Proc. Natl. Acad. Sci. USA* **2017**, *114*, 11818–11825. [CrossRef]
28. Dibble, C.C.; Elis, W.; Menon, S.; Qin, W.; Klekota, J.; Asara, J.M.; Finan, P.M.; Kwiatkowski, D.J.; Murphy, L.O.; Manning, B.D. TBC1D7 is a third subunit of the TSC1-TSC2 complex upstream of mTORC1. *Mol. Cell* **2012**, *47*, 535–546. [CrossRef]
29. Menon, S.; Dibble, C.C.; Talbott, G.; Hoxhaj, G.; Valvezan, A.J.; Takahashi, H.; Cantley, L.C.; Manning, B.D. Spatial control of the TSC complex integrates insulin and nutrient regulation of mTORC1 at the lysosome. *Cell* **2014**, *156*, 771–785. [CrossRef]
30. Ma, L.; Chen, Z.; Erdjument-Bromage, H.; Tempst, P.; Pandolfi, P.P. Phosphorylation and functional inactivation of TSC2 by Erk implications for tuberous sclerosis and cancer pathogenesis. *Cell* **2005**, *121*, 179–193. [CrossRef]
31. Bai, X.; Ma, D.; Liu, A.; Shen, X.; Wang, Q.J.; Liu, Y.; Jiang, Y. Rheb activates mTOR by antagonizing its endogenous inhibitor, FKBP38. *Science* **2007**, *318*, 977–980. [CrossRef] [PubMed]
32. Yang, H.; Jiang, X.; Li, B.; Yang, H.J.; Miller, M.; Yang, A.; Dhar, A.; Pavletich, N.P. Mechanisms of mTORC1 activation by RHEB and inhibition by PRAS40. *Nature* **2017**, *552*, 368–373. [CrossRef] [PubMed]
33. Ye, J.; Palm, W.; Peng, M.; King, B.; Lindsten, T.; Li, M.O.; Koumenis, C.; Thompson, C.B. GCN2 sustains mTORC1 suppression upon amino acid deprivation by inducing Sestrin2. *Genes Dev.* **2015**, *29*, 2331–2336. [CrossRef] [PubMed]
34. Chantranupong, L.; Scaria, S.M.; Saxton, R.A.; Gygi, M.P.; Shen, K.; Wyant, G.A.; Wang, T.; Harper, J.W.; Gygi, S.P.; Sabatini, D.M. The CASTOR Proteins Are Arginine Sensors for the mTORC1 Pathway. *Cell* **2016**, *165*, 153–164. [CrossRef] [PubMed]
35. Rebsamen, M.; Pochini, L.; Stasyk, T.; de Araujo, M.E.; Galluccio, M.; Kandasamy, R.K.; Snijder, B.; Fauster, A.; Rudashevskaya, E.L.; Bruckner, M.; et al. SLC38A9 is a component of the lysosomal amino acid sensing machinery that controls mTORC1. *Nature* **2015**, *519*, 477–481. [CrossRef] [PubMed]
36. Goberdhan, D.C.; Wilson, C.; Harris, A.L. Amino Acid Sensing by mTORC1: Intracellular Transporters Mark the Spot. *Cell Metab.* **2016**, *23*, 580–589. [CrossRef] [PubMed]
37. Jewell, J.L.; Russell, R.C.; Guan, K.L. Amino acid signalling upstream of mTOR. *Nat. Rev. Mol. Cell Biol.* **2013**, *14*, 133–139. [CrossRef] [PubMed]
38. Jewell, J.L.; Kim, Y.C.; Russell, R.C.; Yu, F.X.; Park, H.W.; Plouffe, S.W.; Tagliabracci, V.S.; Guan, K.L. Metabolism. Differential regulation of mTORC1 by leucine and glutamine. *Science* **2015**, *347*, 194–198. [CrossRef]
39. Tsun, Z.Y.; Bar-Peled, L.; Chantranupong, L.; Zoncu, R.; Wang, T.; Kim, C.; Spooner, E.; Sabatini, D.M. The folliculin tumor suppressor is a GAP for the RagC/D GTPases that signal amino acid levels to mTORC1. *Mol. Cell* **2013**, *52*, 495–505. [CrossRef]
40. Feng, Z.; Hu, W.; de Stanchina, E.; Teresky, A.K.; Jin, S.; Lowe, S.; Levine, A.J. The regulation of AMPK beta1, TSC2, and PTEN expression by p53: Stress, cell and tissue specificity, and the role of these gene products in modulating the IGF-1-AKT-mTOR pathways. *Cancer Res.* **2007**, *67*, 3043–3053. [CrossRef]
41. Gwinn, D.M.; Shackelford, D.B.; Egan, D.F.; Mihaylova, M.M.; Mery, A.; Vasquez, D.S.; Turk, B.E.; Shaw, R.J. AMPK phosphorylation of raptor mediates a metabolic checkpoint. *Mol. Cell* **2008**, *30*, 214–226. [CrossRef] [PubMed]

42. Inoki, K.; Zhu, T.; Guan, K.-L. TSC2 Mediates Cellular Energy Response to Control Cell Growth and Survival. *Cell* **2003**, *115*, 577–590. [CrossRef]
43. Inoki, K.; Ouyang, H.; Zhu, T.; Lindvall, C.; Wang, Y.; Zhang, X.; Yang, Q.; Bennett, C.; Harada, Y.; Stankunas, K.; et al. TSC2 integrates Wnt and energy signals via a coordinated phosphorylation by AMPK and GSK3 to regulate cell growth. *Cell* **2006**, *126*, 955–968. [CrossRef] [PubMed]
44. Lee, D.F.; Kuo, H.P.; Chen, C.T.; Hsu, J.M.; Chou, C.K.; Wei, Y.; Sun, H.L.; Li, L.Y.; Ping, B.; Huang, W.C.; et al. IKK beta suppression of TSC1 links inflammation and tumor angiogenesis via the mTOR pathway. *Cell* **2007**, *130*, 440–455. [CrossRef] [PubMed]
45. Nojima, H.; Tokunaga, C.; Eguchi, S.; Oshiro, N.; Hidayat, S.; Yoshino, K.; Hara, K.; Tanaka, N.; Avruch, J.; Yonezawa, K. The mammalian target of rapamycin (mTOR) partner, raptor, binds the mTOR substrates p70 S6 kinase and 4E-BP1 through their TOR signaling (TOS) motif. *J. Biol. Chem.* **2003**, *278*, 15461–15464. [CrossRef] [PubMed]
46. Holz, M.K.; Ballif, B.A.; Gygi, S.P.; Blenis, J. mTOR and S6K1 mediate assembly of the translation preinitiation complex through dynamic protein interchange and ordered phosphorylation events. *Cell* **2005**, *123*, 569–580. [CrossRef]
47. Dorrello, N.V.; Peschiaroli, A.; Guardavaccaro, D.; Colburn, N.H.; Sherman, N.E.; Pagano, M. S6K1- and betaTRCP-mediated degradation of PDCD4 promotes protein translation and cell growth. *Science* **2006**, *314*, 467–471. [CrossRef] [PubMed]
48. Ruvinsky, I.; Meyuhas, O. Ribosomal protein S6 phosphorylation: From protein synthesis to cell size. *Trends Biochem. Sci.* **2006**, *31*, 342–348. [CrossRef]
49. Ma, X.M.; Yoon, S.O.; Richardson, C.J.; Julich, K.; Blenis, J. SKAR links pre-mRNA splicing to mTOR/S6K1-mediated enhanced translation efficiency of spliced mRNAs. *Cell* **2008**, *133*, 303–313. [CrossRef]
50. Gingras, A.C.; Gygi, S.P.; Raught, B.; Polakiewicz, R.D.; Abraham, R.T.; Hoekstra, M.F.; Aebersold, R.; Sonenberg, N. Regulation of 4E-BP1 phosphorylation: A novel two-step mechanism. *Genes Dev.* **1999**, *13*, 1422–1437. [CrossRef]
51. Peterson, T.R.; Sengupta, S.S.; Harris, T.E.; Carmack, A.E.; Kang, S.A.; Balderas, E.; Guertin, D.A.; Madden, K.L.; Carpenter, A.E.; Finck, B.N.; et al. mTOR complex 1 regulates lipin 1 localization to control the SREBP pathway. *Cell* **2011**, *146*, 408–420. [CrossRef] [PubMed]
52. Ben-Sahra, I.; Hoxhaj, G.; Ricoult, S.J.H.; Asara, J.M.; Manning, B.D. mTORC1 induces purine synthesis through control of the mitochondrial tetrahydrofolate cycle. *Science* **2016**, *351*, 728–733. [CrossRef] [PubMed]
53. Duvel, K.; Yecies, J.L.; Menon, S.; Raman, P.; Lipovsky, A.I.; Souza, A.L.; Triantafellow, E.; Ma, Q.; Gorski, R.; Cleaver, S.; et al. Activation of a metabolic gene regulatory network downstream of mTOR complex 1. *Mol. Cell* **2010**, *39*, 171–183. [CrossRef] [PubMed]
54. Kim, J.; Kundu, M.; Viollet, B.; Guan, K.L. AMPK and mTOR regulate autophagy through direct phosphorylation of Ulk1. *Nat. Cell Biol.* **2011**, *13*, 132–141. [CrossRef] [PubMed]
55. Martina, J.A.; Chen, Y.; Gucek, M.; Puertollano, R. MTORC1 functions as a transcriptional regulator of autophagy by preventing nuclear transport of TFEB. *Autophagy* **2012**, *8*, 903–914. [CrossRef] [PubMed]
56. Sarbassov, D.D.; Ali, S.M.; Kim, D.H.; Guertin, D.A.; Latek, R.R.; Erdjument-Bromage, H.; Tempst, P.; Sabatini, D.M. Rictor, a novel binding partner of mTOR, defines a rapamycin-insensitive and raptor-independent pathway that regulates the cytoskeleton. *Curr. Biol.* **2004**, *14*, 1296–1302. [CrossRef] [PubMed]
57. Frias, M.A.; Thoreen, C.C.; Jaffe, J.D.; Schroder, W.; Sculley, T.; Carr, S.A.; Sabatini, D.M. mSin1 is necessary for Akt/PKB phosphorylation, and its isoforms define three distinct mTORC2s. *Curr. Biol.* **2006**, *16*, 1865–1870. [CrossRef]
58. Woo, S.Y.; Kim, D.H.; Jun, C.B.; Kim, Y.M.; Haar, E.V.; Lee, S.I.; Hegg, J.W.; Bandhakavi, S.; Griffin, T.J.; Kim, D.H. PRR5, a novel component of mTOR complex 2, regulates platelet-derived growth factor receptor beta expression and signaling. *J. Biol. Chem.* **2007**, *282*, 25604–25612. [CrossRef]
59. Lamming, D.W.; Ye, L.; Katajisto, P.; Goncalves, M.D.; Saitoh, M.; Stevens, D.M.; Davis, J.G.; Salmon, A.B.; Richardson, A.; Ahima, R.S.; et al. Rapamycin-induced insulin resistance is mediated by mTORC2 loss and uncoupled from longevity. *Science* **2012**, *335*, 1638–1643. [CrossRef]
60. Yang, G.; Murashige, D.S.; Humphrey, S.J.; James, D.E. A Positive Feedback Loop between Akt and mTORC2 via SIN1 Phosphorylation. *Cell Rep.* **2015**, *12*, 937–943. [CrossRef]

61. Shah, O.J.; Wang, Z.; Hunter, T. Inappropriate activation of the TSC/Rheb/mTOR/S6K cassette induces IRS1/2 depletion, insulin resistance, and cell survival deficiencies. *Curr. Biol.* **2004**, *14*, 1650–1656. [CrossRef] [PubMed]
62. Gan, X.; Wang, J.; Wang, C.; Sommer, E.; Kozasa, T.; Srinivasula, S.; Alessi, D.; Offermanns, S.; Simon, M.I.; Wu, D. PRR5L degradation promotes mTORC2-mediated PKC-delta phosphorylation and cell migration downstream of Galpha12. *Nat. Cell Biol.* **2012**, *14*, 686–696. [CrossRef] [PubMed]
63. Li, X.; Gao, T. mTORC2 phosphorylates protein kinase Cζ to regulate its stability and activity. *EMBO Rep.* **2013**, *15*, 191–198. [CrossRef] [PubMed]
64. Thomanetz, V.; Angliker, N.; Cloetta, D.; Lustenberger, R.M.; Schweighauser, M.; Oliveri, F.; Suzuki, N.; Ruegg, M.A. Ablation of the mTORC2 component rictor in brain or Purkinje cells affects size and neuron morphology. *J. Cell Biol.* **2013**, *201*, 293–308. [CrossRef] [PubMed]
65. Guertin, D.A.; Stevens, D.M.; Thoreen, C.C.; Burds, A.A.; Kalaany, N.Y.; Moffat, J.; Brown, M.; Fitzgerald, K.J.; Sabatini, D.M. Ablation in mice of the mTORC components raptor, rictor, or mLST8 reveals that mTORC2 is required for signaling to Akt-FOXO and PKCalpha, but not S6K1. *Dev. Cell* **2006**, *11*, 859–871. [CrossRef] [PubMed]
66. Jacinto, E.; Facchinetti, V.; Liu, D.; Soto, N.; Wei, S.; Jung, S.Y.; Huang, Q.; Qin, J.; Su, B. SIN1/MIP1 maintains rictor-mTOR complex integrity and regulates Akt phosphorylation and substrate specificity. *Cell* **2006**, *127*, 125–137. [CrossRef] [PubMed]
67. Garcia-Martinez, J.M.; Alessi, D.R. mTOR complex 2 (mTORC2) controls hydrophobic motif phosphorylation and activation of serum- and glucocorticoid-induced protein kinase 1 (SGK1). *Biochem. J.* **2008**, *416*, 375–385. [CrossRef] [PubMed]
68. Lee, H.K.; Kwon, B.; Lemere, C.A.; de la Monte, S.; Itamura, K.; Ha, A.Y.; Querfurth, H.W. mTORC2 (Rictor) in Alzheimer's Disease and Reversal of Amyloid-beta Expression-Induced Insulin Resistance and Toxicity in Rat Primary Cortical Neurons. *J. Alzheimers Dis.* **2017**, *56*, 1015–1036. [CrossRef] [PubMed]
69. Langen, U.H.; Ayloo, S.; Gu, C. Development and Cell Biology of the Blood-Brain Barrier. *Annu. Rev. Cell Dev. Biol.* **2019**, *35*. [CrossRef] [PubMed]
70. Abbott, N.J.; Ronnback, L.; Hansson, E. Astrocyte-endothelial interactions at the blood-brain barrier. *Nat. Rev. Neurosci.* **2006**, *7*, 41–53. [CrossRef] [PubMed]
71. Liu, C.; Wu, C.; Yang, Q.; Gao, J.; Li, L.; Yang, D.; Luo, L. Macrophages Mediate the Repair of Brain Vascular Rupture through Direct Physical Adhesion and Mechanical Traction. *Immunity* **2016**, *44*, 1162–1176. [CrossRef] [PubMed]
72. Chen, J.; He, J.; Ni, R.; Yang, Q.; Zhang, Y.; Luo, L. Cerebrovascular Injuries Induce Lymphatic Invasion into Brain Parenchyma to Guide Vascular Regeneration in Zebrafish. *Dev. Cell* **2019**, *49*, 697–710. [CrossRef] [PubMed]
73. Sousa, A.M.M.; Meyer, K.A.; Santpere, G.; Gulden, F.O.; Sestan, N. Evolution of the Human Nervous System Function, Structure, and Development. *Cell* **2017**, *170*, 226–247. [CrossRef] [PubMed]
74. Richner, M.; Ulrichsen, M.; Elmegaard, S.L.; Dieu, R.; Pallesen, L.T.; Vaegter, C.B. Peripheral nerve injury modulates neurotrophin signaling in the peripheral and central nervous system. *Mol. Neurobiol.* **2014**, *50*, 945–970. [CrossRef] [PubMed]
75. Hoffman, P.N. A conditioning lesion induces changes in gene expression and axonal transport that enhance regeneration by increasing the intrinsic growth state of axons. *Exp. Neurol.* **2010**, *223*, 11–18. [CrossRef] [PubMed]
76. He, Z.; Jin, Y. Intrinsic Control of Axon Regeneration. *Neuron* **2016**, *90*, 437–451. [CrossRef] [PubMed]
77. Omura, T.; Omura, K.; Tedeschi, A.; Riva, P.; Painter, M.W.; Rojas, L.; Martin, J.; Lisi, V.; Huebner, E.A.; Latremoliere, A.; et al. Robust Axonal Regeneration Occurs in the Injured CAST/Ei Mouse CNS. *Neuron* **2015**, *86*, 1215–1227. [CrossRef] [PubMed]
78. Benowitz, L.I.; Yin, Y. Combinatorial treatments for promoting axon regeneration in the CNS: Strategies for overcoming inhibitory signals and activating neurons' intrinsic growth state. *Dev. Neurobiol.* **2007**, *67*, 1148–1165. [CrossRef] [PubMed]
79. Filbin, M.T. Myelin-associated inhibitors of axonal regeneration in the adult mammalian CNS. *Nat. Rev. Neurosci.* **2003**, *4*, 703–713. [CrossRef] [PubMed]
80. Fitch, M.T.; Silver, J. CNS injury, glial scars, and inflammation: Inhibitory extracellular matrices and regeneration failure. *Exp. Neurol.* **2008**, *209*, 294–301. [CrossRef] [PubMed]

81. Brown, R.H.; Al-Chalabi, A. Amyotrophic Lateral Sclerosis. *N. Engl. J. Med.* **2017**, *377*, 162–172. [CrossRef] [PubMed]
82. Kalia, L.V.; Lang, A.E. Parkinson's disease. *Lancet* **2015**, *386*, 896–912. [CrossRef]
83. Ballard, C.; Gauthier, S.; Corbett, A.; Brayne, C.; Aarsland, D.; Jones, E. Alzheimer's disease. *Lancet* **2011**, *377*, 1019–1031. [CrossRef]
84. Walker, F.O. Huntington's disease. *Lancet* **2007**, *369*, 218–228. [CrossRef]
85. Li, C.; Gotz, J. Tau-based therapies in neurodegeneration: Opportunities and challenges. *Nat. Rev. Drug Discov.* **2017**, *16*, 863–883. [CrossRef] [PubMed]
86. Goldman, S.A. Stem and Progenitor Cell-Based Therapy of the Central Nervous System: Hopes, Hype, and Wishful Thinking. *Cell Stem Cell* **2016**, *18*, 174–188. [CrossRef] [PubMed]
87. Miao, L.; Yang, L.; Huang, H.; Liang, F.; Ling, C.; Hu, Y. mTORC1 is necessary but mTORC2 and GSK3beta are inhibitory for AKT3-induced axon regeneration in the central nervous system. *Elife* **2016**, *5*, e14908. [CrossRef]
88. Kadoya, K.; Tsukada, S.; Lu, P.; Coppola, G.; Geschwind, D.; Filbin, M.T.; Blesch, A.; Tuszynski, M.H. Combined intrinsic and extrinsic neuronal mechanisms facilitate bridging axonal regeneration one year after spinal cord injury. *Neuron* **2009**, *64*, 165–172. [CrossRef]
89. Silver, J.; Schwab, M.E.; Popovich, P.G. Central nervous system regenerative failure: Role of oligodendrocytes, astrocytes, and microglia. *Cold Spring Harb. Perspect. Biol.* **2014**, *7*, a020602. [CrossRef] [PubMed]
90. Yiu, G.; He, Z. Glial inhibition of CNS axon regeneration. *Nat. Rev. Neurosci.* **2006**, *7*, 617–627. [CrossRef]
91. Silver, J.; Miller, J.H. Regeneration beyond the glial scar. *Nat. Rev. Neurosci.* **2004**, *5*, 146–156. [CrossRef] [PubMed]
92. Sun, F.; He, Z. Neuronal intrinsic barriers for axon regeneration in the adult CNS. *Curr. Opin. Neurobiol.* **2010**, *20*, 510–518. [CrossRef] [PubMed]
93. Park, K.K.; Hu, Y.; Muhling, J.; Pollett, M.A.; Dallimore, E.J.; Turnley, A.M.; Cui, Q.; Harvey, A.R. Cytokine-induced SOCS expression is inhibited by cAMP analogue: Impact on regeneration in injured retina. *Mol. Cell Neurosci.* **2009**, *41*, 313–324. [CrossRef] [PubMed]
94. Zhong, J.; Zou, H. BMP signaling in axon regeneration. *Curr. Opin. Neurobiol.* **2014**, *27*, 127–134. [CrossRef] [PubMed]
95. Li, C.; Hisamoto, N.; Nix, P.; Kanao, S.; Mizuno, T.; Bastiani, M.; Matsumoto, K. The growth factor SVH-1 regulates axon regeneration in C. elegans via the JNK MAPK cascade. *Nat. Neurosci.* **2012**, *15*, 551–557. [CrossRef] [PubMed]
96. Jin, D.; Liu, Y.; Sun, F.; Wang, X.; Liu, X.; He, Z. Restoration of skilled locomotion by sprouting corticospinal axons induced by co-deletion of PTEN and SOCS3. *Nat. Commun.* **2015**, *6*, 8074. [CrossRef] [PubMed]
97. Song, Y.; Ori-McKenney, K.M.; Zheng, Y.; Han, C.; Jan, L.Y.; Jan, Y.N. Regeneration of Drosophila sensory neuron axons and dendrites is regulated by the Akt pathway involving Pten and microRNA bantam. *Genes Dev.* **2012**, *26*, 1612–1625. [CrossRef] [PubMed]
98. Byrne, A.B.; Walradt, T.; Gardner, K.E.; Hubbert, A.; Reinke, V.; Hammarlund, M. Insulin/IGF1 signaling inhibits age-dependent axon regeneration. *Neuron* **2014**, *81*, 561–573. [CrossRef]
99. Li, S.; Yang, C.; Zhang, L.; Gao, X.; Wang, X.; Liu, W.; Wang, Y.; Jiang, S.; Wong, Y.H.; Zhang, Y.; et al. Promoting axon regeneration in the adult CNS by modulation of the melanopsin/GPCR signaling. *Proc. Natl. Acad. Sci. USA* **2016**, *113*, 1937–1942. [CrossRef]
100. Lebrun-Julien, F.; Suter, U. Combined HDAC1 and HDAC2 Depletion Promotes Retinal Ganglion Cell Survival After Injury Through Reduction of p53 Target Gene Expression. *ASN Neuro* **2015**, *7*, 1759091415593066. [CrossRef]
101. Schmitt, H.M.; Schlamp, C.L.; Nickells, R.W. Targeting HDAC3 Activity with RGFP966 Protects Against Retinal Ganglion Cell Nuclear Atrophy and Apoptosis After Optic Nerve Injury. *J. Ocul. Pharmacol. Ther.* **2018**, *34*, 260–273. [CrossRef]
102. Cho, Y.; Cavalli, V. HDAC5 is a novel injury-regulated tubulin deacetylase controlling axon regeneration. *EMBO J.* **2012**, *31*, 3063–3078. [CrossRef]
103. Cho, Y.; Sloutsky, R.; Naegle, K.M.; Cavalli, V. Injury-induced HDAC5 nuclear export is essential for axon regeneration. *Cell* **2013**, *155*, 894–908. [CrossRef]
104. Pita-Thomas, W.; Mahar, M.; Joshi, A.; Gan, D.; Cavalli, V. HDAC5 promotes optic nerve regeneration by activating the mTOR pathway. *Exp. Neurol.* **2019**, *317*, 271–283. [CrossRef]

105. Tassew, N.G.; Charish, J.; Shabanzadeh, A.P.; Luga, V.; Harada, H.; Farhani, N.; D'Onofrio, P.; Choi, B.; Ellabban, A.; Nickerson, P.E.B.; et al. Exosomes Mediate Mobilization of Autocrine Wnt10b to Promote Axonal Regeneration in the Injured CNS. *Cell Rep.* **2017**, *20*, 99–111. [CrossRef]
106. Leibinger, M.; Andreadaki, A.; Fischer, D. Role of mTOR in neuroprotection and axon regeneration after inflammatory stimulation. *Neurobiol. Dis.* **2012**, *46*, 314–324. [CrossRef]
107. Lim, J.-H.A.; Stafford, B.K.; Nguyen, P.L.; Lien, B.V.; Wang, C.; Zukor, K.; He, Z.; Huberman, A.D. Neural activity promotes long-distance, target-specific regeneration of adult retinal axons. *Nat. Neurosci.* **2016**, *19*, 1073–1084. [CrossRef]
108. Sun, F.; Park, K.K.; Belin, S.; Wang, D.; Lu, T.; Chen, G.; Zhang, K.; Yeung, C.; Feng, G.; Yankner, B.A.; et al. Sustained axon regeneration induced by co-deletion of PTEN and SOCS3. *Nature* **2011**, *480*, 372–375. [CrossRef]
109. de Lima, S.; Koriyama, Y.; Kurimoto, T.; Oliveira, J.T.; Yin, Y.; Li, Y.; Gilbert, H.Y.; Fagiolini, M.; Martinez, A.M.; Benowitz, L. Full-length axon regeneration in the adult mouse optic nerve and partial recovery of simple visual behaviors. *Proc. Natl. Acad. Sci. USA* **2012**, *109*, 9149–9154. [CrossRef]
110. Pernet, V.; Schwab, M.E. Lost in the jungle: New hurdles for optic nerve axon regeneration. *Trends Neurosci.* **2014**, *37*, 381–387. [CrossRef]
111. Li, N.; Wang, F.; Zhang, Q.; Jin, M.; Lu, Y.; Chen, S.; Guo, C.; Zhang, X. Rapamycin mediates mTOR signaling in reactive astrocytes and reduces retinal ganglion cell loss. *Exp. Eye Res.* **2018**, *176*, 10–19. [CrossRef]
112. Ahuja, C.S.; Wilson, J.R.; Nori, S.; Kotter, M.R.N.; Druschel, C.; Curt, A.; Fehlings, M.G. Traumatic spinal cord injury. *Nat. Rev. Dis. Prim.* **2017**, *3*, 17018. [CrossRef]
113. Codeluppi, S.; Svensson, C.I.; Hefferan, M.P.; Valencia, F.; Silldorff, M.D.; Oshiro, M.; Marsala, M.; Pasquale, E.B. The Rheb-mTOR pathway is upregulated in reactive astrocytes of the injured spinal cord. *J. Neurosci.* **2009**, *29*, 1093–1104. [CrossRef]
114. Kanno, H.; Ozawa, H.; Sekiguchi, A.; Yamaya, S.; Tateda, S.; Yahata, K.; Itoi, E. The role of mTOR signaling pathway in spinal cord injury. *Cell Cycle* **2012**, *11*, 3175–3179. [CrossRef]
115. Chen, C.-H.; Sung, C.-S.; Huang, S.-Y.; Feng, C.-W.; Hung, H.-C.; Yang, S.-N.; Chen, N.-F.; Tai, M.-H.; Wen, Z.-H.; Chen, W.-F. The role of the PI3K/Akt/mTOR pathway in glial scar formation following spinal cord injury. *Exp. Neurol.* **2016**, *278*, 27–41. [CrossRef]
116. Luan, Y.; Chen, M.; Zhou, L. MiR-17 targets PTEN and facilitates glial scar formation after spinal cord injuries via the PI3K/Akt/mTOR pathway. *Brain Res. Bull.* **2017**, *128*, 68–75. [CrossRef]
117. Liu, K.; Lu, Y.; Lee, J.K.; Samara, R.; Willenberg, R.; Sears-Kraxberger, I.; Tedeschi, A.; Park, K.K.; Jin, D.; Cai, B.; et al. PTEN deletion enhances the regenerative ability of adult corticospinal neurons. *Nat. Neurosci.* **2010**, *13*, 1075–1081. [CrossRef]
118. Al-Ali, H.; Ding, Y.; Slepak, T.; Wu, W.; Sun, Y.; Martinez, Y.; Xu, X.M.; Lemmon, V.P.; Bixby, J.L. The mTOR Substrate S6 Kinase 1 (S6K1) Is a Negative Regulator of Axon Regeneration and a Potential Drug Target for Central Nervous System Injury. *J. Neurosci.* **2017**, *37*, 7079–7095. [CrossRef]
119. Du, K.; Zheng, S.; Zhang, Q.; Li, S.; Gao, X.; Wang, J.; Jiang, L.; Liu, K. Pten Deletion Promotes Regrowth of Corticospinal Tract Axons 1 Year after Spinal Cord Injury. *J. Neurosci.* **2015**, *35*, 9754–9763. [CrossRef]
120. Yang, P.; Wen, H.; Ou, S.; Cui, J.; Fan, D. IL-6 promotes regeneration and functional recovery after cortical spinal tract injury by reactivating intrinsic growth program of neurons and enhancing synapse formation. *Exp. Neurol.* **2012**, *236*, 19–27. [CrossRef]
121. Huang, Z.; Gao, Y.; Sun, Y.; Zhang, C.; Yin, Y.; Shimoda, Y.; Watanabe, K.; Liu, Y. NB-3 signaling mediates the cross-talk between post-traumatic spinal axons and scar-forming cells. *EMBO J.* **2016**, *35*, 1745–1765. [CrossRef]
122. Park, K.K.; Liu, K.; Hu, Y.; Kanter, J.L.; He, Z. PTEN/mTOR and axon regeneration. *Exp. Neurol.* **2010**, *223*, 45–50. [CrossRef]
123. Abe, N.; Borson, S.H.; Gambello, M.J.; Wang, F.; Cavalli, V. Mammalian target of rapamycin (mTOR) activation increases axonal growth capacity of injured peripheral nerves. *J. Biol. Chem.* **2010**, *285*, 28034–28043. [CrossRef]
124. Chen, W.; Lu, N.; Ding, Y.; Wang, Y.; Chan, L.T.; Wang, X.; Gao, X.; Jiang, S.; Liu, K. Rapamycin-Resistant mTOR Activity Is Required for Sensory Axon Regeneration Induced by a Conditioning Lesion. *eNeuro* **2016**, *3*, ENEURO.0358-16.2016. [CrossRef]

125. Thomson, S.E.; Charalambous, C.; Smith, C.A.; Tsimbouri, P.M.; Dejardin, T.; Kingham, P.J.; Hart, A.M.; Riehle, M.O. Microtopographical cues promote peripheral nerve regeneration via transient mTORC2 activation. *Acta Biomater.* **2017**, *60*, 220–231. [CrossRef]
126. Liu, Y.; Kelamangalath, L.; Kim, H.; Han, S.B.; Tang, X.; Zhai, J.; Hong, J.W.; Lin, S.; Son, Y.J.; Smith, G.M. NT-3 promotes proprioceptive axon regeneration when combined with activation of the mTor intrinsic growth pathway but not with reduction of myelin extrinsic inhibitors. *Exp. Neurol.* **2016**, *283*, 73–84. [CrossRef]
127. Baghdadi, M.B.; Tajbakhsh, S. Regulation and phylogeny of skeletal muscle regeneration. *Dev. Biol.* **2018**, *433*, 200–209. [CrossRef]
128. Hawke, T.J.; Garry, D.J. Myogenic satellite cells: Physiology to molecular biology. *J. Appl. Physiol.* **2001**, *91*, 534–551. [CrossRef]
129. Mauro, A. Satellite cell of skeletal muscle fibers. *J. Biophys. Biochem. Cytol.* **1961**, *9*, 493–495. [CrossRef]
130. Thomson, D.M. The Role of AMPK in the Regulation of Skeletal Muscle Size, Hypertrophy, and Regeneration. *Int. J. Mol. Sci.* **2018**, *19*, 3125. [CrossRef]
131. Kuang, S.; Rudnicki, M.A. The emerging biology of satellite cells and their therapeutic potential. *Trends Mol. Med.* **2008**, *14*, 82–91. [CrossRef]
132. Charge, S.B.; Rudnicki, M.A. Cellular and molecular regulation of muscle regeneration. *Physiol. Rev.* **2004**, *84*, 209–238. [CrossRef]
133. Zhang, P.; Liang, X.; Shan, T.; Jiang, Q.; Deng, C.; Zheng, R.; Kuang, S. mTOR is necessary for proper satellite cell activity and skeletal muscle regeneration. *Biochem. Biophys. Res. Commun.* **2015**, *463*, 102–108. [CrossRef]
134. Ge, Y.; Wu, A.L.; Warnes, C.; Liu, J.; Zhang, C.; Kawasome, H.; Terada, N.; Boppart, M.D.; Schoenherr, C.J.; Chen, J. mTOR regulates skeletal muscle regeneration in vivo through kinase-dependent and kinase-independent mechanisms. *Am. J. Physiol. Cell Physiol.* **2009**, *297*, C1434–C1444. [CrossRef]
135. Lepper, C.; Conway, S.J.; Fan, C.M. Adult satellite cells and embryonic muscle progenitors have distinct genetic requirements. *Nature* **2009**, *460*, 627–631. [CrossRef]
136. Rion, N.; Castets, P.; Lin, S.; Enderle, L.; Reinhard, J.R.; Eickhorst, C.; Ruegg, M.A. mTOR controls embryonic and adult myogenesis via mTORC1. *Development* **2019**, *146*, dev172460. [CrossRef]
137. Jash, S.; Dhar, G.; Ghosh, U.; Adhya, S. Role of the mTORC1 complex in satellite cell activation by RNA-induced mitochondrial restoration: Dual control of cyclin D1 through microRNAs. *Mol. Cell Biol.* **2014**, *34*, 3594–3606. [CrossRef]
138. Kikani, C.K.; Wu, X.; Fogarty, S.; Kang, S.A.W.; Dephoure, N.; Gygi, S.P.; Sabatini, D.M.; Rutter, J. Activation of PASK by mTORC1 is required for the onset of the terminal differentiation program. *Proc. Natl. Acad. Sci. USA* **2019**, *116*, 10382–10391. [CrossRef]
139. Greco, S.; De Simone, M.; Colussi, C.; Zaccagnini, G.; Fasanaro, P.; Pescatori, M.; Cardani, R.; Perbellini, R.; Isaia, E.; Sale, P.; et al. Common micro-RNA signature in skeletal muscle damage and regeneration induced by Duchenne muscular dystrophy and acute ischemia. *FASEB J.* **2009**, *23*, 3335–3346. [CrossRef]
140. Yuasa, K.; Hagiwara, Y.; Ando, M.; Nakamura, A.; Takeda, S.I.; Hijikata, T. MicroRNA-206 is highly expressed in newly formed muscle fibers implications regarding potential for muscle regeneration and maturation in muscular dystrophy.pdf. *Cell Struct. Funct.* **2008**, *33*, 163–169. [CrossRef]
141. Sun, Y.; Ge, Y.; Drnevich, J.; Zhao, Y.; Band, M.; Chen, J. Mammalian target of rapamycin regulates miRNA-1 and follistatin in skeletal myogenesis. *J. Cell Biol.* **2010**, *189*, 1157–1169. [CrossRef]
142. Ge, Y.; Sun, Y.; Chen, J. IGF-II is regulated by microRNA-125b in skeletal myogenesis. *J. Cell Biol.* **2011**, *192*, 69–81. [CrossRef]
143. Matsumoto, A.; Pasut, A.; Matsumoto, M.; Yamashita, R.; Fung, J.; Monteleone, E.; Saghatelian, A.; Nakayama, K.I.; Clohessy, J.G.; Pandolfi, P.P. mTORC1 and muscle regeneration are regulated by the LINC00961-encoded SPAR polypeptide. *Nature* **2017**, *541*, 228–232. [CrossRef]
144. Hung, C.M.; Calejman, C.M.; Sanchez-Gurmaches, J.; Li, H.; Clish, C.B.; Hettmer, S.; Wagers, A.J.; Guertin, D.A. Rictor/mTORC2 loss in the Myf5 lineage reprograms brown fat metabolism and protects mice against obesity and metabolic disease. *Cell Rep.* **2014**, *8*, 256–271. [CrossRef]
145. Sadtler, K.; Estrellas, K.; Allen, B.W.; Wolf, M.T.; Fan, H.; Tam, A.J.; Patel, C.H.; Luber, B.S.; Wang, H.; Wagner, K.R.; et al. Developing a pro-regenerative biomaterial scaffold microenvironment requires T helper 2 cells. *Science* **2016**, *352*, 366–370. [CrossRef]

146. Zanou, N.; Schakman, O.; Louis, P.; Ruegg, U.T.; Dietrich, A.; Birnbaumer, L.; Gailly, P. Trpc1 ion channel modulates phosphatidylinositol 3-kinase/Akt pathway during myoblast differentiation and muscle regeneration. *J. Biol. Chem.* **2012**, *287*, 14524–14534. [CrossRef]
147. Zeng, Q.; Fu, Q.; Wang, X.; Zhao, Y.; Liu, H.; Li, Z.; Li, F. Protective Effects of Sonic Hedgehog Against Ischemia/Reperfusion Injury in Mouse Skeletal Muscle via AKT/mTOR/p70S6K Signaling. *Cell. Physiol. Biochem. Int. J. Exp. Cell. Physiol. Biochem. Pharmacol.* **2017**, *43*, 1813–1828. [CrossRef]
148. Sharples, A.P.; Hughes, D.C.; Deane, C.S.; Saini, A.; Selman, C.; Stewart, C.E. Longevity and skeletal muscle mass: The role of IGF signalling, the sirtuins, dietary restriction and protein intake. *Aging Cell* **2015**, *14*, 511–523. [CrossRef]
149. Erbay, E.; Park, I.H.; Nuzzi, P.D.; Schoenherr, C.J.; Chen, J. IGF-II transcription in skeletal myogenesis is controlled by mTOR and nutrients. *J. Cell Biol.* **2003**, *163*, 931–936. [CrossRef]
150. Pratiwi, Y.S.; Lesmana, R.; Goenawan, H.; Sylviana, N.; Setiawan, I.; Tarawan, V.M.; Lestari, K.; Abdulah, R.; Dwipa, L.; Purba, A.; et al. Nutmeg Extract Increases Skeletal Muscle Mass in Aging Rats Partly via IGF1-AKT-mTOR Pathway and Inhibition of Autophagy. *Evid.-Based Complement. Altern. Med. eCAM* **2018**, *2018*, 2810840. [CrossRef]
151. Frontera, W.R.; Hughes, V.A.; Fielding, R.A.; Fiatarone, M.A.; Evans, W.J.; Roubenoff, R. Aging of skeletal muscle: A 12-yr longitudinal study. *J. Appl. Physiol.* **2000**, *88*, 1321–1326. [CrossRef]
152. Zacharewicz, E.; Della Gatta, P.; Reynolds, J.; Garnham, A.; Crowley, T.; Russell, A.P.; Lamon, S. Identification of microRNAs linked to regulators of muscle protein synthesis and regeneration in young and old skeletal muscle. *PLoS ONE* **2014**, *9*, e114009. [CrossRef]
153. Conboy, I.M.; Conboy, M.J.; Smythe, G.M.; Rando, T.A. Notch-mediated restoration of regenerative potential to aged muscle. *Science* **2003**, *302*, 1575–1577. [CrossRef]
154. Shavlakadze, T.; McGeachie, J.; Grounds, M.D. Delayed but excellent myogenic stem cell response of regenerating geriatric skeletal muscles in mice. *Biogerontology* **2010**, *11*, 363–376. [CrossRef]
155. Pereira, M.G.; Silva, M.T.; da Cunha, F.M.; Moriscot, A.S.; Aoki, M.S.; Miyabara, E.H. Leucine supplementation improves regeneration of skeletal muscles from old rats. *Exp. Gerontol.* **2015**, *72*, 269–277. [CrossRef]
156. Perry, R.A., Jr.; Brown, L.A.; Lee, D.E.; Brown, J.L.; Baum, J.I.; Greene, N.P.; Washington, T.A. Differential effects of leucine supplementation in young and aged mice at the onset of skeletal muscle regeneration. *Mech. Ageing Dev.* **2016**, *157*, 7–16. [CrossRef]
157. Pereira, M.G.; Baptista, I.L.; Carlassara, E.O.; Moriscot, A.S.; Aoki, M.S.; Miyabara, E.H. Leucine supplementation improves skeletal muscle regeneration after cryolesion in rats. *PLoS ONE* **2014**, *9*, e85283. [CrossRef]
158. Perry, R.A., Jr.; Brown, L.A.; Lee, D.E.; Brown, J.L.; Baum, J.I.; Greene, N.P.; Washington, T.A. The Akt/mTOR pathway: Data comparing young and aged mice with leucine supplementation at the onset of skeletal muscle regeneration. *Data Brief* **2016**, *8*, 1426–1432. [CrossRef]
159. Mouisel, E.; Vignaud, A.; Hourde, C.; Butler-Browne, G.; Ferry, A. Muscle weakness and atrophy are associated with decreased regenerative capacity and changes in mTOR signaling in skeletal muscles of venerable (18-24-month-old) dystrophic mdx mice. *Muscle Nerve* **2010**, *41*, 809–818. [CrossRef]
160. Jones, D.L.; Rando, T.A. Emerging models and paradigms for stem cell ageing. *Nat. Cell Biol.* **2011**, *13*, 506–512. [CrossRef]
161. Chandel, N.S.; Jasper, H.; Ho, T.T.; Passegue, E. Metabolic regulation of stem cell function in tissue homeostasis and organismal ageing. *Nat. Cell Biol.* **2016**, *18*, 823–832. [CrossRef]
162. Haller, S.; Kapuria, S.; Riley, R.R.; O'Leary, M.N.; Schreiber, K.H.; Andersen, J.K.; Melov, S.; Que, J.; Rando, T.A.; Rock, J.; et al. mTORC1 Activation during Repeated Regeneration Impairs Somatic Stem Cell Maintenance. *Cell Stem Cell* **2017**, *21*, 806–818. [CrossRef]
163. Su, Z.; Robinson, A.; Hu, L.; Klein, J.D.; Hassounah, F.; Li, M.; Wang, H.; Cai, H.; Wang, X.H. Acupuncture plus Low-Frequency Electrical Stimulation (Acu-LFES) Attenuates Diabetic Myopathy by Enhancing Muscle Regeneration. *PLoS ONE* **2015**, *10*, e0134511. [CrossRef]
164. Chaillou, T.; Koulmann, N.; Meunier, A.; Pugniere, P.; McCarthy, J.J.; Beaudry, M.; Bigard, X. Ambient hypoxia enhances the loss of muscle mass after extensive injury. *Pflug. Archiv Eur. J. Physiol.* **2014**, *466*, 587–598. [CrossRef]

165. Favier, F.B.; Costes, F.; Defour, A.; Bonnefoy, R.; Lefai, E.; Bauge, S.; Peinnequin, A.; Benoit, H.; Freyssenet, D. Downregulation of Akt/mammalian target of rapamycin pathway in skeletal muscle is associated with increased REDD1 expression in response to chronic hypoxia. *Am. J. Physiol. Regul. Integr. Comp. Physiol.* **2010**, *298*, R1659–R1666. [CrossRef]
166. Cohen, S.; Nathan, J.A.; Goldberg, A.L. Muscle wasting in disease: Molecular mechanisms and promising therapies. *Nat. Rev. Drug Discov.* **2015**, *14*, 58–74. [CrossRef]
167. Miyajima, A.; Tanaka, M.; Itoh, T. Stem/progenitor cells in liver development, homeostasis, regeneration, and reprogramming. *Cell Stem Cell* **2014**, *14*, 561–574. [CrossRef]
168. Mao, S.A.; Glorioso, J.M.; Nyberg, S.L. Liver regeneration. *Transl. Res.* **2014**, *163*, 352–362. [CrossRef]
169. Michalopoulos, G.K.; DeFrances, M.C. Liver regeneration. *Science* **1997**, *276*, 60–66. [CrossRef]
170. Greene, A.K.; Puder, M. Partial hepatectomy in the mouse: Technique and perioperative management. *J. Investig. Surg.* **2003**, *16*, 99–102. [CrossRef]
171. Bohm, F.; Kohler, U.A.; Speicher, T.; Werner, S. Regulation of liver regeneration by growth factors and cytokines. *EMBO Mol. Med.* **2010**, *2*, 294–305. [CrossRef]
172. Yanger, K.; Knigin, D.; Zong, Y.; Maggs, L.; Gu, G.; Akiyama, H.; Pikarsky, E.; Stanger, B.Z. Adult hepatocytes are generated by self-duplication rather than stem cell differentiation. *Cell Stem Cell* **2014**, *15*, 340–349. [CrossRef]
173. Itoh, T.; Miyajima, A. Liver regeneration by stem/progenitor cells. *Hepatology* **2014**, *59*, 1617–1626. [CrossRef]
174. He, J.; Lu, H.; Zou, Q.; Luo, L. Regeneration of liver after extreme hepatocyte loss occurs mainly via biliary transdifferentiation in zebrafish. *Gastroenterology* **2014**, *146*, 789–800. [CrossRef]
175. Choi, T.Y.; Ninov, N.; Stainier, D.Y.; Shin, D. Extensive conversion of hepatic biliary epithelial cells to hepatocytes after near total loss of hepatocytes in zebrafish. *Gastroenterology* **2014**, *146*, 776–788. [CrossRef]
176. Raven, A.; Lu, W.Y.; Man, T.Y.; Ferreira-Gonzalez, S.; O'Duibhir, E.; Dwyer, B.J.; Thomson, J.P.; Meehan, R.R.; Bogorad, R.; Koteliansky, V.; et al. Cholangiocytes act as facultative liver stem cells during impaired hepatocyte regeneration. *Nature* **2017**, *547*, 350–354. [CrossRef]
177. Deng, X.; Zhang, X.; Li, W.; Feng, R.X.; Li, L.; Yi, G.R.; Zhang, X.N.; Yin, C.; Yu, H.Y.; Zhang, J.P.; et al. Chronic Liver Injury Induces Conversion of Biliary Epithelial Cells into Hepatocytes. *Cell Stem Cell* **2018**, *23*, 114–122. [CrossRef]
178. Panasyuk, G.; Patitucci, C.; Espeillac, C.; Pende, M. The role of the mTOR pathway during liver regeneration and tumorigenesis. *Ann. D'endocrinologie* **2013**, *74*, 121–122. [CrossRef]
179. He, J.; Chen, J.; Wei, X.; Leng, H.; Mu, H.; Cai, P.; Luo, L. mTORC1 Signaling is Required for the Dedifferentiation from Biliary Cell to Bi-potential Progenitor Cell in Zebrafish Liver Regeneration. *Hepatology* **2019**. [CrossRef]
180. Fujiyoshi, M.; Ozaki, M. Molecular mechanisms of liver regeneration and protection for treatment of liver dysfunction and diseases. *J. Hepato-Biliary-Pancreat. Sci.* **2011**, *18*, 13–22. [CrossRef]
181. Haga, S.; Ozaki, M.; Inoue, H.; Okamoto, Y.; Ogawa, W.; Takeda, K.; Akira, S.; Todo, S. The survival pathways phosphatidylinositol-3 kinase (PI3-K)/phosphoinositide-dependent protein kinase 1 (PDK1)/Akt modulate liver regeneration through hepatocyte size rather than proliferation. *Hepatology* **2009**, *49*, 204–214. [CrossRef]
182. Fouraschen, S.M.; de Ruiter, P.E.; Kwekkeboom, J.; de Bruin, R.W.; Kazemier, G.; Metselaar, H.J.; Tilanus, H.W.; van der Laan, L.J.; de Jonge, J. mTOR signaling in liver regeneration: Rapamycin combined with growth factor treatment. *World J. Transplant.* **2013**, *3*, 36–47. [CrossRef]
183. Kwon, H.J.; Won, Y.S.; Yoon, Y.D.; Yoon, W.K.; Nam, K.H.; Choi, I.P.; Kim, D.Y.; Kim, H.C. Vitamin D3 up-regulated protein 1 deficiency accelerates liver regeneration after partial hepatectomy in mice. *J. Hepatol.* **2011**, *54*, 1168–1176. [CrossRef]
184. Chen, P.; Yan, H.; Chen, Y.; He, Z. The variation of AkT/TSC1–TSC1/mTOR signal pathway in hepatocytes after partial hepatectomy in rats. *Exp. Mol. Pathol.* **2009**, *86*, 101–107. [CrossRef]
185. Rickheim, D.G.; Nelsen, C.J.; Fassett, J.T.; Timchenko, N.A.; Hansen, L.K.; Albrecht, J.H. Differential regulation of cyclins D1 and D3 in hepatocyte proliferation. *Hepatology* **2002**, *36*, 30–38. [CrossRef]
186. Espeillac, C.; Mitchell, C.; Celton-Morizur, S.; Chauvin, C.; Koka, V.; Gillet, C.; Albrecht, J.H.; Desdouets, C.; Pende, M. S6 kinase 1 is required for rapamycin-sensitive liver proliferation after mouse hepatectomy. *J. Clin. Investig.* **2011**, *121*, 2821–2832. [CrossRef]
187. Jiang, Y.-P.; Ballou, L.M.; Lin, R.Z. Rapamycin-insensitive Regulation of 4E-BP1 in Regenerating Rat Liver. *J. Biol. Chem.* **2001**, *276*, 10943–10951. [CrossRef]

188. Palmes, D.; Zibert, A.; Budny, T.; Bahde, R.; Minin, E.; Kebschull, L.; Holzen, J.; Schmidt, H.; Spiegel, H.U. Impact of rapamycin on liver regeneration. *Virchows Arch.* **2008**, *452*, 545–557. [CrossRef]
189. Matot, I.; Nachmansson, N.; Duev, O.; Schulz, S.; Schroeder-Stein, K.; Frede, S.; Abramovitch, R. Impaired liver regeneration after hepatectomy and bleeding is associated with a shift from hepatocyte proliferation to hypertrophy. *FASEB J.* **2017**, *31*, 5283–5295. [CrossRef]
190. Ng, R.; Song, G.; Roll, G.R.; Frandsen, N.M.; Willenbring, H. A microRNA-21 surge facilitates rapid cyclin D1 translation and cell cycle progression in mouse liver regeneration. *J. Clin. Investig.* **2012**, *122*, 1097–1108. [CrossRef]
191. Liu, Q.; Pu, S.; Chen, L.; Shen, J.; Cheng, S.; Kuang, J.; Li, H.; Wu, T.; Li, R.; Jiang, W.; et al. Liver-specific Sirtuin6 ablation impairs liver regeneration after 2/3 partial hepatectomy. *Wound Repair Regen.* **2019**, *27*, 366–374. [CrossRef]
192. Garcia-Rodriguez, J.L.; Barbier-Torres, L.; Fernandez-Alvarez, S.; Gutierrez-de Juan, V.; Monte, M.J.; Halilbasic, E.; Herranz, D.; Alvarez, L.; Aspichueta, P.; Marin, J.J.; et al. SIRT1 controls liver regeneration by regulating bile acid metabolism through farnesoid X receptor and mammalian target of rapamycin signaling. *Hepatology* **2014**, *59*, 1972–1983. [CrossRef]
193. Shi, H.; Zhang, Y.; Ji, J.; Xu, P.; Shi, H.; Yue, X.; Ren, F.; Chen, Y.; Duan, Z.; Chen, D. Deficiency of apoptosis-stimulating protein two of p53 promotes liver regeneration in mice by activating mammalian target of rapamycin. *Sci. Rep.* **2018**, *8*, 17927. [CrossRef]
194. Wu, L.; Nguyen, L.H.; Zhou, K.; de Soysa, T.Y.; Li, L.; Miller, J.B.; Tian, J.; Locker, J.; Zhang, S.; Shinoda, G.; et al. Precise let-7 expression levels balance organ regeneration against tumor suppression. *Elife* **2015**, *4*, e09431. [CrossRef]
195. Lou, K.; Yang, M.; Duan, E.; Zhao, J.; Yu, C.; Zhang, R.; Zhang, L.; Zhang, M.; Xiao, Z.; Hu, W.; et al. Rosmarinic acid stimulates liver regeneration through the mTOR pathway. *Phytomedicine* **2016**, *23*, 1574–1582. [CrossRef]
196. Kawaguchi, T.; Kodama, T.; Hikita, H.; Tanaka, S.; Shigekawa, M.; Nawa, T.; Shimizu, S.; Li, W.; Miyagi, T.; Hiramatsu, N.; et al. Carbamazepine promotes liver regeneration and survival in mice. *J. Hepatol.* **2013**, *59*, 1239–1245. [CrossRef]
197. Zhong, H.; Wu, H.; Bai, H.; Wang, M.; Wen, J.; Gong, J.; Miao, M.; Yuan, F. Panax notoginseng saponins promote liver regeneration through activation of the PI3K/AKT/mTOR cell proliferation pathway and upregulation of the AKT/Bad cell survival pathway in mice. *BMC Complement. Altern Med.* **2019**, *19*, 122. [CrossRef]
198. Zhang, D.X.; Li, C.H.; Zhang, A.Q.; Jiang, S.; Lai, Y.H.; Ge, X.L.; Pan, K.; Dong, J.H. mTOR-Dependent Suppression of Remnant Liver Regeneration in Liver Failure After Massive Liver Resection in Rats. *Dig. Dis. Sci.* **2015**, *60*, 2718–2729. [CrossRef]
199. Haga, S.; Ogawa, W.; Inoue, H.; Terui, K.; Ogino, T.; Igarashi, R.; Takeda, K.; Akira, S.; Enosawa, S.; Furukawa, H.; et al. Compensatory recovery of liver mass by Akt-mediated hepatocellular hypertrophy in liver-specific STAT3-deficient mice. *J. Hepatol.* **2005**, *43*, 799–807. [CrossRef]
200. Planas-Paz, L.; Sun, T.; Pikiolek, M.; Cochran, N.R.; Bergling, S.; Orsini, V.; Yang, Z.; Sigoillot, F.; Jetzer, J.; Syed, M.; et al. YAP, but Not RSPO-LGR4/5, Signaling in Biliary Epithelial Cells Promotes a Ductular Reaction in Response to Liver Injury. *Cell Stem Cell* **2019**, *25*, 39–53. [CrossRef]
201. Chaturantabut, S.; Shwartz, A.; Evason, K.J.; Cox, A.G.; Labella, K.; Schepers, A.G.; Yang, S.; Acuna, M.; Houvras, Y.; Mancio-Silva, L.; et al. Estrogen Activation of G-Protein-Coupled Estrogen Receptor 1 Regulates Phosphoinositide 3-Kinase and mTOR Signaling to Promote Liver Growth in Zebrafish and Proliferation of Human Hepatocytes. *Gastroenterology* **2019**, *156*, 1788–1804. [CrossRef]
202. Fausto, N. Liver regeneration: From laboratory to clinic. *Liver Transpl.* **2001**, *7*, 835–844. [CrossRef]
203. Bataller, R.; Brenner, D.A. Liver fibrosis. *J. Clin. Investig.* **2005**, *115*, 209–218. [CrossRef]
204. Uribe, M.; Uribe-Echevarria, S.; Mandiola, C.; Zapata, M.I.; Riquelme, F.; Romanque, P. Insight on ALPPS—Associating Liver Partition and Portal Vein Ligation for Staged Hepatectomy—Mechanisms: Activation of mTOR pathway. *HPB (Oxf.)* **2018**, *20*, 729–738. [CrossRef]
205. Pan, N.; Lv, X.; Liang, R.; Wang, L.; Liu, Q. Suppression of graft regeneration, not ischemia/reperfusion injury, is the primary cause of small-for-size syndrome after partial liver transplantation in mice. *PLoS ONE* **2014**, *9*, e93636. [CrossRef]

206. Liu, Q.; Rehman, H.; Krishnasamy, Y.; Haque, K.; Schnellmann, R.G.; Lemasters, J.J.; Zhong, Z. Amphiregulin stimulates liver regeneration after small-for-size mouse liver transplantation. *Am. J. Transplant.* **2012**, *12*, 2052–2061. [CrossRef]
207. Gehart, H.; Clevers, H. Tales from the crypt: New insights into intestinal stem cells. *Nat. Rev. Gastroenterol. Hepatol.* **2019**, *16*, 19–34. [CrossRef]
208. Biteau, B.; Hochmuth, C.E.; Jasper, H. JNK activity in somatic stem cells causes loss of tissue homeostasis in the aging Drosophila gut. *Cell Stem Cell* **2008**, *3*, 442–455. [CrossRef]
209. Jiang, H.; Patel, P.H.; Kohlmaier, A.; Grenley, M.O.; McEwen, D.G.; Edgar, B.A. Cytokine/Jak/Stat signaling mediates regeneration and homeostasis in the Drosophila midgut. *Cell* **2009**, *137*, 1343–1355. [CrossRef]
210. Amcheslavsky, A.; Ito, N.; Jiang, J.; Ip, Y.T. Tuberous sclerosis complex and Myc coordinate the growth and division of Drosophila intestinal stem cells. *J. Cell Biol.* **2011**, *193*, 695–710. [CrossRef]
211. Makky, K.; Tekiela, J.; Mayer, A.N. Target of rapamycin (TOR) signaling controls epithelial morphogenesis in the vertebrate intestine. *Dev. Biol.* **2007**, *303*, 501–513. [CrossRef]
212. Metcalfe, C.; Kljavin, N.M.; Ybarra, R.; de Sauvage, F.J. Lgr5+ stem cells are indispensable for radiation-induced intestinal regeneration. *Cell Stem Cell* **2014**, *14*, 149–159. [CrossRef]
213. van Es, J.H.; Sato, T.; van de Wetering, M.; Lyubimova, A.; Yee Nee, A.N.; Gregorieff, A.; Sasaki, N.; Zeinstra, L.; van den Born, M.; Korving, J.; et al. Dll1+ secretory progenitor cells revert to stem cells upon crypt damage. *Nat. Cell Biol.* **2012**, *14*, 1099–1104. [CrossRef]
214. Tetteh, P.W.; Basak, O.; Farin, H.F.; Wiebrands, K.; Kretzschmar, K.; Begthel, H.; van den Born, M.; Korving, J.; de Sauvage, F.; van Es, J.H.; et al. Replacement of Lost Lgr5-Positive Stem Cells through Plasticity of Their Enterocyte-Lineage Daughters. *Cell Stem Cell* **2016**, *18*, 203–213. [CrossRef]
215. Yu, S.; Tong, K.; Zhao, Y.; Balasubramanian, I.; Yap, G.S.; Ferraris, R.P.; Bonder, E.M.; Verzi, M.P.; Gao, N. Paneth Cell Multipotency Induced by Notch Activation following Injury. *Cell Stem Cell* **2018**, *23*, 46–59. [CrossRef]
216. Zhou, J.Y.; Huang, D.G.; Qin, Y.C.; Li, X.G.; Gao, C.Q.; Yan, H.C.; Wang, X.Q. mTORC1 signaling activation increases intestinal stem cell activity and promotes epithelial cell proliferation. *J. Cell. Physiol.* **2019**, *234*, 19028–19038. [CrossRef]
217. Igarashi, M.; Guarente, L. mTORC1 and SIRT1 Cooperate to Foster Expansion of Gut Adult Stem Cells during Calorie Restriction. *Cell* **2016**, *166*, 436–450. [CrossRef]
218. Ashton, G.H.; Morton, J.P.; Myant, K.; Phesse, T.J.; Ridgway, R.A.; Marsh, V.; Wilkins, J.A.; Athineos, D.; Muncan, V.; Kemp, R.; et al. Focal adhesion kinase is required for intestinal regeneration and tumorigenesis downstream of Wnt/c-Myc signaling. *Dev. Cell* **2010**, *19*, 259–269. [CrossRef]
219. Nagahama, Y.; Shimoda, M.; Mao, G.; Singh, S.K.; Kozakai, Y.; Sun, X.; Motooka, D.; Nakamura, S.; Tanaka, H.; Satoh, T.; et al. Regnase-1 controls colon epithelial regeneration via regulation of mTOR and purine metabolism. *Proc. Natl. Acad. Sci. USA* **2018**, *115*, 11036–11041. [CrossRef]
220. Morton, J.P.; Myant, K.B.; Sansom, O.J. A FAK-PI-3K-mTOR axis is required for Wnt-Myc driven intestinal regeneration and tumorigenesis. *Cell Cycle* **2014**, *10*, 173–175. [CrossRef]
221. Faller, W.J.; Jackson, T.J.; Knight, J.R.; Ridgway, R.A.; Jamieson, T.; Karim, S.A.; Jones, C.; Radulescu, S.; Huels, D.J.; Myant, K.B.; et al. mTORC1-mediated translational elongation limits intestinal tumour initiation and growth. *Nature* **2015**, *517*, 497–500. [CrossRef]
222. Guan, Y.; Zhang, L.; Li, X.; Zhang, X.; Liu, S.; Gao, N.; Li, L.; Gao, G.; Wei, G.; Chen, Z.; et al. Repression of Mammalian Target of Rapamycin Complex 1 Inhibits Intestinal Regeneration in Acute Inflammatory Bowel Disease Models. *J. Immunol.* **2015**, *195*, 339–346. [CrossRef]
223. Perl, A. mTOR-dependent autophagy contributes to end-organ resistance and serves as target for treatment in autoimmune disease. *EBioMedicine* **2018**, *36*, 12–13. [CrossRef]
224. Maiese, K.; Chong, Z.Z.; Shang, Y.C.; Wang, S. mTOR: On target for novel therapeutic strategies in the nervous system. *Trends Mol. Med.* **2013**, *19*, 51–60. [CrossRef]
225. Laplante, M.; Sabatini, D.M. mTOR signaling in growth control and disease. *Cell* **2012**, *149*, 274–293. [CrossRef]
226. Arroyo, V.; Jalan, R. Acute-on-Chronic Liver Failure: Definition, Diagnosis, and Clinical Characteristics. *Semin. Liver Dis.* **2016**, *36*, 109–116. [CrossRef]

227. Wang, Y.; Newman, M.R.; Ackun-Farmmer, M.; Baranello, M.P.; Sheu, T.-J.; Puzas, J.E.; Benoit, D.S.W. Fracture-Targeted Delivery of β-Catenin Agonists via Peptide-Functionalized Nanoparticles Augments Fracture Healing. *ACS Nano* **2017**, *11*, 9445–9458. [CrossRef]
228. Lai, Z.-W.; Kelly, R.; Winans, T.; Marchena, I.; Shadakshari, A.; Yu, J.; Dawood, M.; Garcia, R.; Tily, H.; Francis, L.; et al. Sirolimus in patients with clinically active systemic lupus erythematosus resistant to, or intolerant of, conventional medications: A single-arm, open-label, phase 1/2 trial. *Lancet* **2018**, *391*, 1186–1196. [CrossRef]

 cells

Review

Cellular Metabolic Regulation in the Differentiation and Function of Regulatory T Cells

Ye Chen [1], Jacob Colello [2], Wael Jarjour [1] and Song Guo Zheng [1,*]

1 Division of Rheumatology and Immunology, Department of Internal Medicine at Ohio State University of Medicine and Wexner Medical Center, Columbus, OH 43201, USA; Ye.Chen@osumc.edu (Y.C.); Wael.Jarjour@osumc.edu (W.J.)

2 Division of Rheumatology, Department of Medicine at Penn State College of Medicine, Hershey, PA 17033, USA; jcolello@pennstatehealth.psu.edu

* Correspondence: SongGuo.Zheng@osumc.edu; Tel.: +1-614-293-7452

Received: 3 January 2019; Accepted: 20 February 2019; Published: 21 February 2019

Abstract: Regulatory T cells (Tregs) are essential for maintaining immune tolerance and preventing autoimmune and inflammatory diseases. The activity and function of Tregs are in large part determined by various intracellular metabolic processes. Recent findings have focused on how intracellular metabolism can shape the development, trafficking, and function of Tregs. In this review, we summarize and discuss current research that reveals how distinct metabolic pathways modulate Tregs differentiation, phenotype stabilization, and function. These advances highlight numerous opportunities to alter Tregs frequency and function in physiopathologic conditions via metabolic manipulation and have important translational implications.

Keywords: cell metabolism; T cells; Foxp3; mTOR

1. Overview of the Effects of Cellular Metabolism on Tregs

Regulatory T cells (Tregs) are crucial for immune homeostasis and the control of inflammatory disorders [1–4]. We mainly focus the discussion on CD4+CD25+Foxp3+ Tregs in this review. Tregs subsets include: thymus derived Tregs (tTregs), peripheral derived Tregs (pTregs) which are generated extrathymically at peripheral sites, and iTregs that are induced ex vivo following TCR stimulation in the presence of transforming growth factor β [5]. Tregs-mediated protection has been applied in numerous preclinical models of autoimmune diseases and transplantation, which informs on their therapeutic potential for human diseases [6–9].

Like conventional CD4+ T cells, Tregs also have a high degree of plasticity related to different transcriptional programs [10–13], which are in turn impacted by cellular metabolism [14]. Recent findings show that Tregs use glycolysis and fatty acid oxidation differently than effector conventional T cells and naïve T cells [15]. Compared to effector conventional T cells, mouse Tregs oxidize lipids at higher rates and exhibit low glycolytic flux in vitro. On the other hand, the modest energy and biosynthesis demands of naïve T cells are typically met by the tricarboxylic acid cycle, lipid oxidation, and glycolysis (Figure 1) [16]. Moreover, glycolysis inhibits Tregs differentiation and promotes Tregs expansion, whereas fatty-acid oxidation (FAO) promotes Tregs differentiation [15,17]. In addition, the expression of Foxp3 in Tregs inhibits Myc expression and reduces glycolysis, which can be suitable for Tregs in low glucose condition [17]. Foxp3, in turn, induces oxidate phosphorylation and increases the ratio of oxidized nicotinamide adenine dinucleotide (NAD) over the reduced form (NADH), allowing Tregs to survive in elevated lactate environments [17]. Furthermore, autophagy is one of the first responses when cells experience nutrient limitation and is critical for Tregs fitness. Deficient autophagy leads to the upregulation of the mechanistic target of rapamycin (mTOR) and c-Myc

(*Myc* proto-oncogene) expression as well as an increase in glycolysis, resulting in impaired Tregs function [18,19].

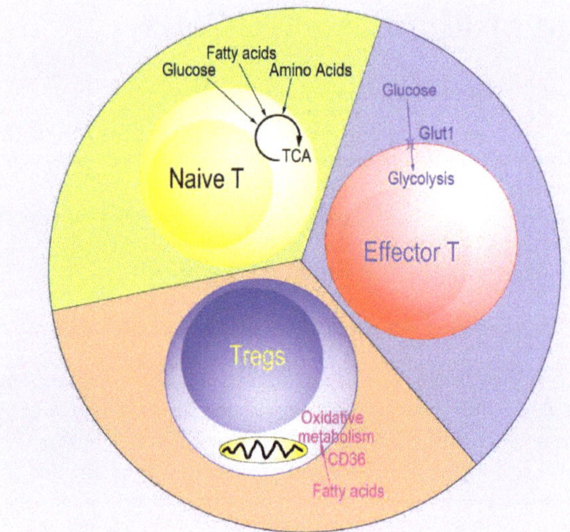

Figure 1. Model of energy usage by naïve T, effector T, and Regulatory T cells (Tregs). Naïve T cells use glucose, fatty acids, and amino acids as their energy source. Effector T cells have higher energy efficiency and use glucose as their primary energy source. In contrast, the glucose transporter 1 is absent in Tregs and Tregs use fatty-acid oxidation (FAO) as their main energy source.

Different chain lengths of fatty acids have dissimilar effects on Tregs differentiation. Adding a short chain fatty acid to mouse or human naïve CD4+ T cells enhances Tregs differentiation, while a long chain fatty acid (LC-FA) decreases Tregs differentiation [20]. Our previous work demonstrated that sodium butyrate, which belongs to the short-chain fatty acid family, promotes Tregs induction and displays therapeutic potential in several inflammatory disorders [21]. However, Raud et al. recently reported that Carnitine palmitoyltransferase 1a (Cpt1a), a critical regulator of LC-FA oxidation, is largely dispensable for Tregs generation [22].

mTOR is a 289 kDa serine/threonine protein kinase that is highly evolutionarily conserved and has two complexes mTORC1 and mTORC2 [23]. It can directly affect T cell proliferation and differentiation through the integration of environmental cues such as energy stores, nutrients, and growth factors; and can be selectively inhibited by rapamycin [24]. Generally, mTORC1 is more sensitive to rapamycin than mTORC2 [25], however, in naïve CD4+ T cells, mTORC1 and mTORC2 have essentially the same sensitivity to rapamycin [26]. This review describes the effects of mTOR signaling dependent cellular metabolic regulation on Tregs phenotype and differentiation/suppressive function. Moreover, we discuss the role of mTOR in its modulation of T cell metabolism, which could provide targets for metabolic manipulation.

2. mTOR

As a member of phosphatidylinositol-3 kinases (PI3K) family, mTOR contains two N-terminal HEAT domains (binding domain), which are important for protein–protein interactions. It also includes an FRB region (rapamycin binding domain of mTOR), a FAT domain (a domain in PI3K-related kinases), a structurally supportive C-terminal FATC domain (a domain in PI3K-related kinases), and a kinase domain [27]. During T cell activation, T cell receptor (TCR) stimulates the mTORC1 and mTORC2 via triggering the recruitment of PI3K to the TCR receptor [28]. The activation of PI3K leads to activation of the serine–threonine kinase AKT (also known as protein kinase B) by pyruvate dehydrogenase kinase 1 (PDK1), following the activation of mTOR signaling [29]. Additionally, PI3K can directly induce the activation of mTORC2 [30]. Diverse environmental inputs can be integrated into the mTOR pathway. For example, through mTOR, metabolic cues and immune signals have an ability to direct T cell fate decisions [31]. Moreover, co-stimulatory signals, TCR and cytokine can activate mTOR via PI3K-AKT signaling to meet energy demands and activate T cells.

2.1. mTOR and Tregs Differentiation

The most profound function of mTOR in Tregs generation was first revealed using the selective inhibitor of mTOR, rapamycin, which decreased the production of effector T cells and increased the generation of Tregs [32]. Furthermore, a lack of mTORC1 signaling may lead to a failure of differentiation from naïve CD4+ T cells to Th17 lineage. When mTORC2 and mTORC1 were both mutually absent, however, naïve CD4+ T cells were differentiated into Foxp3+ Tregs [33]. This research underscores the significant role of mTOR as a fundamental regulatory factor in the differentiation of Tregs and Th17 cells (Figure 2).

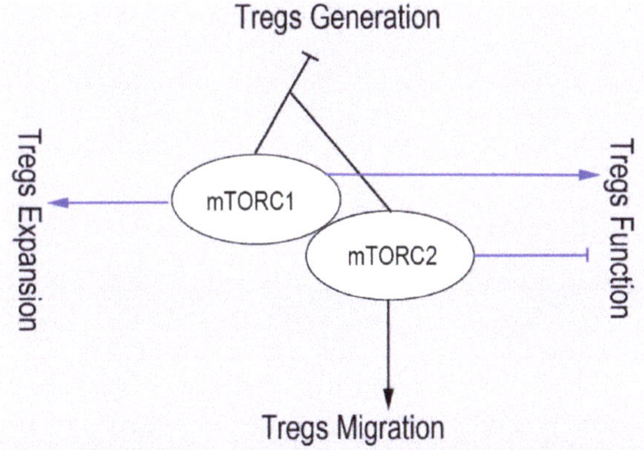

Figure 2. The roles of mTORC1 and mTORC2 on Tregs generation, expansion, function, and migration. The absence of mTOR signaling dramatically increase Tregs generation, while deleting either mTORC1 or mTORC2 signaling does not lead to the upregulation of Foxp3+ Tregs. mTORC1 and mTORC2 play opposite roles in Tregs function, the absence of main component Raptor of mTORC1 limits Tregs function, and lack of mTORC2 increases Tregs function via promoting the activity of mTORC1. mTORC2 promotes the migration of Tregs to inflammatory sites. However, the effects of mTORC1 on the Tregs migration remain unclear. mTOR signaling is essential for Tregs expansion. Consequently, Slc3a2-deficient Tregs have an impaired mTORC1 pathway and show lower proliferation ability. However, the role of mTORC2 on Tregs expansion remains unclear.

2.2. mTOR and Tregs Function

Tregs generation is enhanced during an immune response by inhibition of mTOR. Such activity is considered a required step in maintaining Tregs suppressive capabilities. Recent evidence has revealed a critical role of mTORC1 complex in the development of Tregs suppressive activity [34] (Figure 2). If Raptor is specifically deleted from Tregs, mTORC1 is disrupted. This leads not only to a profound loss of Tregs suppressive activity, but also causes the development of a fatal early-onset inflammatory disorder. Mechanistically, cholesterol/lipid metabolism is enhanced through Raptor/mTORC1 signaling in Tregs [34]. In order to establish Tregs functional competency, the mevalonate pathway can up-regulate the Tregs suppressive molecules CTLA-4 and ICOS, as well as coordinate Tregs proliferation. Inhibition of the mTORC2 pathway is partly involved in maintaining

Tregs function by mTORC1 [34]. Nevertheless, mTOR signaling is critical for properly programming activated Tregs function in order to protect tissue homeostasis and preserve immune tolerance. Tregs-specific deletion of mTOR impairs Tregs function and homeostasis, resulting in the spontaneous effector T cell activation and in the development of inflammation in barrier tissues, which is correlated with the reduction in the local tissues of both peripheral Tregs (pTregs) and thymic-derived Tregs (tTregs) [35]. In contrast, Toll-like receptor (TLR) signals enhance Tregs proliferation through mTORC1 signaling pathway, glucose transporter 1 (Glut1) upregulation, and glycolysis. However, these signals decrease the suppressive ability of Tregs [36]. It is likely that the TLR signal results in high levels of pro-inflammatory cytokines such as IL-6, IL-1, TNFα, and these pro-inflammatory cytokines decrease Tregs functionality even as Tregs maintain mTORC1 expression.

2.3. mTOR and Tregs Expansion

Although rapamycin is commonly used to block tumor cell growth, it is interesting that proliferation of CD4+ T cell and activation-induced cell death cannot be blocked by rapamycin in vitro and in kidney transplant rejection [37]. On the contrary, expanded Tregs may suppress the proliferation of effector T cells in vitro as well as prevent allograft rejection in vivo [37,38]. A new study has revealed that branched-chain amino acids (BCAAs) could be essential for the maintenance of Tregs profiling state, through metabolic reprogramming of the amino acid transporter solute carrier family 3 member 1 (Slc3a2) dependent pathway. Slc3a2-deficient Tregs impair the mTORC1 pathway and show lower proliferation ability [39] (Figure 2). The expansion of Tregs by rapamycin usually requires the addition of IL-2. Thus, even when mTORC1 is inhibited by rapamycin [40], IL-2 maintains the ability to expand Treg cells.

2.4. mTOR and Tregs Migration

As one of the complexes of mTOR, mTORC2 has been proven to have control of spatial aspects of cell growth via actin reorganization [41,42]. The immune-modulatory function is critically interrelated with the migration of activated Tregs to inflammatory tissue [43]. A recent study demonstrated that glycolysis was beneficial for their relocation. Migration of Tregs to inflamed tissue was initiated by pro-migratory stimuli through a PI3K-mTORC2-mediated signaling pathway, which culminated in stimulation of the enzyme glucokinase (GCK). Subsequently, GCK increases cytoskeletal rearrangements by interacting with actin. If Tregs lack this pathway, they will still be functionally suppressive but will fail to migrate to skin allografts as well as inhibit rejection [44] (Figure 2).

3. Promising Metabolic Targets to Manipulate Tregs Frequency and Function

mTOR is a very important regulator of cell survival and plays bidirectional roles in Tregs induction and function. The absence of mTOR signaling dramatically increases Tregs generation and inhibits the function of Tregs. To advance therapeutics and promote homeostasis of the immune system, it is necessary to identify more specific targets that modify Tregs function or induction. Next, we discuss the effects of some mTOR interrelated metabolic regulators on Tregs phenotype.

3.1. Hypoxia-Inducible Factor 1α (HIF1α)

The transcription factor hypoxia-inducible factor (HIF1α) is a necessary protein for sensing oxygen saturation and subsequently initiating the cellular response to hypoxia [45,46]. It also is closely associated with Tregs and the mTOR pathway. Furthermore, impeding glycolysis up-regulates production of Tregs that occurs through inhibition of mTOR-mediated induction of HIF1α [45]. Interestingly, HIF1α inhibits Tregs differentiation in a transcription-independent manner [47], while not affecting *Foxp3* mRNA levels. HIF1α exerts this inhibition through promoting the degradation of Foxp3 protein by ubiquitination (Figure 3). Moreover, lack of HIF1α promotes Tregs induction and protects mice from autoimmune neuro-inflammation [45].

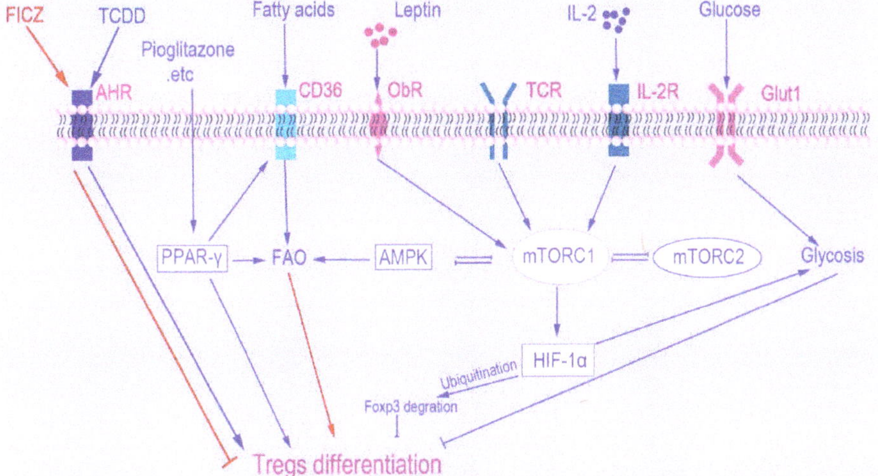

Figure 3. Metabolic and mTOR signaling control of Regulatory T cells' differentiation. Different ligand binding with aryl hydrocarbon receptor (AHR) leads to dissimilar outcome in Tregs induction. For example, 6-formylindolo [3,2-b] carbazole (FICZ) binding with AHR inhibits Tregs generation while 2,3,7,8-tetrachlorodibenzo-*p*-dioxin (TCDD) promotes Tregs generation. Peroxisome proliferator-activated receptor-γ (PPAR-γ) increase Tregs differentiation via regulating the balance between fatty-acid oxidation (FAO) and glycometabolism. AMP-activated protein kinase (AMPK) is an important metabolic checkpoint in Tregs differentiation, and there is controversy regarding the role of the AMPK in Tregs differentiation. mTOR signaling is the most important metabolic regulator of Tregs. If mTOR is absent in naïve CD4+ T cells, it can dramatically increase Tregs differentiation even under normal activating conditions. HIF1α does not directly regulate the expression of Foxp3. However, it promotes the degradation of Foxp3 protein by ubiquitination.

Inflammation and hypoxia are two independent factors regulating the balance between Th17 and Tregs. Our group reported that tTregs are unstable in the inflammatory environment and fail to suppress collagen-induced arthritis [48]. Mechanistically, the presence of the inflammatory cytokine IL-6 converts tTregs to Th17-like cells in vitro [2,11,12,49], and IL-6 also increases HIF1α expression in a stat3-dependent manner [47]. Up to this date, it is unclear whether IL-6 regulates Tregs stability via metabolic alteration through HIF1α. A number of published articles reported that HIF1α is a key regulator in inflammation and autoimmune diseases, such as systemic lupus erythematosus (SLE) [50], rheumatoid arthritis (RA) [51,52], type 1 diabetes (T1DM) [53], multiple sclerosis (MS) [54], psoriasis [55], and inflammatory bowel disease (IBD) [56]. These findings indicate that HIF1α is a potential target to manipulate Tregs phenotype in autoimmune diseases.

3.2. AMP-Activated Protein Kinase (AMPK)

AMPK senses the cellular AMP/ATP ratio and is activated by low energy balance (high AMP/ATP ratio) [57–59]. Activated AMPK promotes FAO via upregulating a series of lipid oxidation related genes, such as Acetyl-CoA carboxylase 1 (ACC1), Acetyl-CoA carboxylase 2 (ACC2), Cpt1a, and sterol regulatory element binding transcription factor 1 (SREVP-1c) [60]. In addition, AMPK also regulates glycolysis via adjusting the expression of Glut1 [61]. AMPK is responsible for Tregs differentiation via regulating the balance of FAO and glycolysis. Activation of AMPK by metformin increases Tregs induction; and in murine models inhibits the progression of experimental autoimmune encephalomyelitis (EAE), and inflammatory bowel disease [15,62,63]. TCR activates both AMPK and mTOR, the latter active kinase is a negative regulator for the former one under limited nutrient condition [64]. Interestingly, the absence of AMPK has no effect on Tregs differentiation even though it enhances mTORC1 activity. Liver kinase B1 (LKB1) is a best-studied upstream kinase of AMPK and is an important metabolic sensor of Tregs. The absence of LKB1 in Foxp3+ Tregs limits the number and function of Tregs, and the effects of LKB1 on Tregs generation and function are independent of AMPK and mTOR [65,66]. Moreover, a recent paper demonstrated that Tregs differentiation was independent of the AMPK-Driven LC-FAO [22].

3.3. *Leptin*

Leptin is a cytokine-like hormone and is structurally similar to IL-6. Leptin mediates metabolism and T cell function [67,68]. Chronic leptin- and leptin-receptor deficiency is correlated with resistance to autoimmunity and high susceptibility to infection [69,70]. For example, leptin levels in SLE patients are correlated with regulatory T cell frequency [71]. De Rosa et al. reported that both leptin and its receptor are constitutively expressed in freshly isolated human Tregs. Increased leptin signaling acts as an antagonist during Tregs proliferation (Figure 3) [72].

3.4. *Peroxisome Proliferator-Activated Receptors (PPARs)*

Peroxisome proliferator-activated receptors (PPARs) are nuclear hormone receptors that function to regulate cell growth, homeostasis, and differentiation. PPARα, β/δ, and γ are three primary isoforms, each with distinct functions and tissue distribution [73]. PPARs play an important role in peroxisomal mediated β-oxidation of FAO. When PPARs are activated with specific ligands, conformational changes occur, resulting in heterodimerization with retinoid X receptor (RXRα), which then binds to promoter regions of target genes involved in FAO [74,75]. PPARs also regulate glucose metabolism. Recent studies have demonstrated that agonists of PPARs inhibit inflammatory and immune responses in non-alcoholic fatty liver disease, at least in part, through increased expression of Foxp3 and induction of Tregs (Figure 3) [76,77]. Although studies have revealed a clear picture of anti-inflammatory function by PPAR-γ, the action of other PPARs on the Tregs population remains uncertain. Recent studies have focused on the relationship between PPARs and visceral adipose tissue (VAT) Tregs. VAT Tregs are a unique subset of Tregs that uniquely express PPARγ and are specifically recruited to adipose tissue to suppress the inflammatory process [78]. In mice that specifically lack PPARγ in Tregs, VAT Treg cell population is reduced. These mice display enhanced insulin resistance and increased susceptibility to diabetic pathology [79,80]. These findings suggest that PPARγ may be a promising target for obesity-associated insulin resistance (IR). However, the absence of PPARγ in VAT Tregs does not perturb their frequency in aged mice, and VAT Tregs show a gene expression profile more similar to fat effector conventional T cells than splenic Tregs [81]. One possible reason is that long term inflammation in aged mice may change the characteristics of Tregs and promote the transdifferentiation. These results highlight the importance of PPARγ in Treg differentiation, migration, and function although aging and chronic inflammation may affect the role of PPARγ in Treg biology.

3.5. *The Aryl Hydrocarbon Receptor (AHR)*

AHR exists as a receptor and transcription factor, which is essential for xenobiotic metabolism and shows a cital function in immunity [82]. AHR has a high-affinity ligand, TCDD (2,3,7,8-tetrachlorodibenzo-*p*-dioxin) and when activated in vivo, AHR–TCDD complex leads to the induction of CD4+CD25+Foxp3+ Tregs. Alternatively, 6-formylindolo [3,2-b] carbazole (FICZ) may activate AHR to interfere with Tregs differentiation, boosting Th17 cell differentiation and worsening experimental autoimmune encephalomyelitis (EAE). Therefore, AHR regulates the balance of Tregs and Th17 cell differentiation in a ligand-specific manner (Figure 3) [83] and can be a unique target for immunosuppression therapy.

3.6. *Interleukin 2 (IL-2)*

IL-2 was first found as a T cell growth factor and plays an important role in T cell proliferation and differentiation [84]. We confirmed, along with several groups, that IL-2 is still necessary to induce and expand Tregs [10,85]. Zeng et al. demonstrated that IL-2 enhancement of Tregs function was dependent on the activation of mTORC1 [34]. Interestingly, IL-2 may partner with rapamycin (mTOR inhibitor) in Tregs expansion in vitro (Box 1, Figure 3). Low dose IL-2 is a promising method for the treatment of autoimmune diseases like lupus [86,87], T1DM [88], and graft-versus-host (GVHD) [89,90].

4. Concluding Remarks and Perspectives

Recently, research has highlighted the complex roles of intrinsic metabolic pathways in the development and function of Tregs, which may have significant implications on immune diseases and responses. In this review, we briefly summarize the roles of metabolic sensors in the biological features of Tregs as well as their potential to be targets of clinical immune-modifying therapies (Figure 3). Immunometabolism is a promising new field, and profound questions remain to be answered (Box 1).

Box 1. Some unsolved questions in the metabolic regulation of Tregs.

- The different roles of mTORC1 and mTORC2 in Treg induction, migration, expansion, and function;
- The different metabolic profiles of Tregs during steady states and inflammatory conditions;
- Identification of metabolic factors that correlate Tregs development, function, and expansion with environment cues;
- The basis of the requirement of high doses of IL-2 (mTOR activator) along with rapamycin (inhibit mTOR) to expand Tregs in vitro.

Funding: This work was supported by NIH R01 AR059103, NIH Star award and NIH R61 AR073409.

Conflicts of Interest: The authors declare no financial or commercial conflicts of interest.

References

1. Qian, X.; Wang, K.; Wang, X.; Zheng, S.G.; Lu, L. Generation of human regulatory t cells de novo with suppressive function prevent xenogeneic graft versus host disease. *Int. Immunopharmacol.* **2011**, *11*, 630–637. [CrossRef] [PubMed]
2. Kong, N.; Lan, Q.; Chen, M.; Wang, J.; Shi, W.; Horwitz, D.A.; Quesniaux, V.; Ryffel, B.; Liu, Z.; Brand, D.; et al. Antigen-specific transforming growth factor beta-induced treg cells, but not natural treg cells, ameliorate autoimmune arthritis in mice by shifting the th17/treg cell balance from th17 predominance to treg cell predominance. *Arthritis Rheum.* **2012**, *64*, 2548–2558. [CrossRef] [PubMed]
3. Lan, Q.; Zhou, X.; Fan, H.; Chen, M.; Wang, J.; Ryffel, B.; Brand, D.; Ramalingam, R.; Kiela, P.R.; Horwitz, D.A.; et al. Polyclonal cd4+foxp3+ treg cells induce tgfbeta-dependent tolerogenic dendritic cells that suppress the murine lupus-like syndrome. *J. Mol. Cell Biol.* **2012**, *4*, 409–419. [CrossRef] [PubMed]
4. Li, B.; Zheng, S.G. How regulatory t cells sense and adapt to inflammation. *Cell. Mol. Immunol.* **2015**, *12*, 519–520. [CrossRef] [PubMed]
5. Shevach, E.M.; Thornton, A.M. Ttregs, ptregs, and itregs: Similarities and differences. *Immunol. Rev.* **2014**, *259*, 88–102. [CrossRef] [PubMed]
6. Bluestone, J.A.; Bour-Jordan, H. Current and future immunomodulation strategies to restore tolerance in autoimmune diseases. *Cold Spring Harb. Perspect. Biol.* **2012**, *4*. [CrossRef] [PubMed]
7. Juvet, S.C.; Whatcott, A.G.; Bushell, A.R.; Wood, K.J. Harnessing regulatory t cells for clinical use in transplantation: The end of the beginning. *Am. J. Transplant.* **2014**, *14*, 750–763. [CrossRef]
8. Horwitz, D.A.; Zheng, S.G.; Gray, J.D.; Wang, J.H.; Ohtsuka, K.; Yamagiwa, S. Regulatory t cells generated ex vivo as an approach for the therapy of autoimmune disease. *Semin. Immunol.* **2004**, *16*, 135–143. [CrossRef]
9. Horwitz, D.A.; Gray, J.D.; Zheng, S.G. The potential of human regulatory t cells generated ex vivo as a treatment for lupus and other chronic inflammatory diseases. *Arthritis Res.* **2002**, *4*, 241–246. [CrossRef]
10. Zheng, S.G.; Wang, J.; Wang, P.; Gray, J.D.; Horwitz, D.A. Il-2 is essential for tgf-beta to convert naive cd4+cd25- cells to cd25+foxp3+ regulatory t cells and for expansion of these cells. *J. Immunol.* **2007**, *178*, 2018–2027. [CrossRef]

11. Zheng, S.G.; Wang, J.; Horwitz, D.A. Cutting edge: Foxp3+cd4+cd25+ regulatory t cells induced by il-2 and tgf-beta are resistant to th17 conversion by il-6. *J. Immunol.* **2008**, *180*, 7112–7116. [CrossRef] [PubMed]
12. Zhou, X.; Kong, N.; Wang, J.; Fan, H.; Zou, H.; Horwitz, D.; Brand, D.; Liu, Z.; Zheng, S.G. Cutting edge: All-trans retinoic acid sustains the stability and function of natural regulatory t cells in an inflammatory milieu. *J. Immunol.* **2010**, *185*, 2675–2679. [CrossRef] [PubMed]
13. Gu, J.; Lu, L.; Chen, M.; Xu, L.; Lan, Q.; Li, Q.; Liu, Z.; Chen, G.; Wang, P.; Wang, X.; et al. Tgf-beta-induced cd4+foxp3+ t cells attenuate acute graft-versus-host disease by suppressing expansion and killing of effector cd8+ cells. *J. Immunol.* **2014**, *193*, 3388–3397. [CrossRef] [PubMed]
14. Procaccini, C.; Carbone, F.; Di Silvestre, D.; Brambilla, F.; De Rosa, V.; Galgani, M.; Faicchia, D.; Marone, G.; Tramontano, D.; Corona, M.; et al. The proteomic landscape of human ex vivo regulatory and conventional t cells reveals specific metabolic requirements. *Immunity* **2016**, *44*, 406–421. [CrossRef] [PubMed]
15. Michalek, R.D.; Gerriets, V.A.; Jacobs, S.R.; Macintyre, A.N.; MacIver, N.J.; Mason, E.F.; Sullivan, S.A.; Nichols, A.G.; Rathmell, J.C. Cutting edge: Distinct glycolytic and lipid oxidative metabolic programs are essential for effector and regulatory cd4+ t cell subsets. *J. Immunol.* **2011**, *186*, 3299–3303. [CrossRef] [PubMed]
16. Pearce, E.L. Metabolism in t cell activation and differentiation. *Curr. Opin. Immunol.* **2010**, *22*, 314–320. [CrossRef]
17. Angelin, A.; Gil-de-Gomez, L.; Dahiya, S.; Jiao, J.; Guo, L.; Levine, M.H.; Wang, Z.; Quinn, W.J., 3rd; Kopinski, P.K.; Wang, L.; et al. Foxp3 reprograms t cell metabolism to function in low-glucose, high-lactate environments. *Cell Metab.* **2017**, *25*, 1282–1293.e7. [CrossRef]
18. Kabat, A.M.; Harrison, O.J.; Riffelmacher, T.; Moghaddam, A.E.; Pearson, C.F.; Laing, A.; Abeler-Dorner, L.; Forman, S.P.; Grencis, R.K.; Sattentau, Q.; et al. The autophagy gene atg16l1 differentially regulates treg and th2 cells to control intestinal inflammation. *Elife* **2016**, *5*, e12444. [CrossRef]
19. Wei, J.; Long, L.; Yang, K.; Guy, C.; Shrestha, S.; Chen, Z.; Wu, C.; Vogel, P.; Neale, G.; Green, D.R.; et al. Autophagy enforces functional integrity of regulatory t cells by coupling environmental cues and metabolic homeostasis. *Nat. Immunol.* **2016**, *17*, 277–285. [CrossRef]
20. Haghikia, A.; Jorg, S.; Duscha, A.; Berg, J.; Manzel, A.; Waschbisch, A.; Hammer, A.; Lee, D.H.; May, C.; Wilck, N.; et al. Dietary fatty acids directly impact central nervous system autoimmunity via the small intestine. *Immunity* **2015**, *43*, 817–829. [CrossRef]
21. Chen, X.; Su, W.; Wan, T.; Yu, J.; Zhu, W.; Tang, F.; Liu, G.; Olsen, N.; Liang, D.; Zheng, S.G. Sodium butyrate regulates th17/treg cell balance to ameliorate uveitis via the nrf2/ho-1 pathway. *Biochem. Pharmacol.* **2017**, *142*, 111–119. [CrossRef] [PubMed]
22. Raud, B.; Roy, D.G.; Divakaruni, A.S.; Tarasenko, T.N.; Franke, R.; Ma, E.H.; Samborska, B.; Hsieh, W.Y.; Wong, A.H.; Stuve, P.; et al. Etomoxir actions on regulatory and memory t cells are independent of cpt1a-mediated fatty acid oxidation. *Cell Metab.* **2018**, *28*, 504–515.e7. [CrossRef] [PubMed]
23. Yang, Q.; Guan, K.L. Expanding mtor signaling. *Cell Res.* **2007**, *17*, 666–681. [CrossRef] [PubMed]
24. Blouet, C.; Ono, H.; Schwartz, G.J. Mediobasal hypothalamic p70 s6 kinase 1 modulates the control of energy homeostasis. *Cell Metab.* **2008**, *8*, 459–467. [CrossRef]
25. Verbist, K.C.; Guy, C.S.; Milasta, S.; Liedmann, S.; Kaminski, M.M.; Wang, R.; Green, D.R. Metabolic maintenance of cell asymmetry following division in activated t lymphocytes. *Nature* **2016**, *532*, 389–393. [CrossRef] [PubMed]
26. Delgoffe, G.M.; Pollizzi, K.N.; Waickman, A.T.; Heikamp, E.; Meyers, D.J.; Horton, M.R.; Xiao, B.; Worley, P.F.; Powell, J.D. The kinase mtor regulates the differentiation of helper t cells through the selective activation of signaling by mtorc1 and mtorc2. *Nat. Immunol.* **2011**, *12*, 295–303. [CrossRef]
27. Weichhart, T.; Hengstschlager, M.; Linke, M. Regulation of innate immune cell function by mtor. *Nat. Rev. Immunol.* **2015**, *15*, 599–614. [CrossRef] [PubMed]
28. Laplante, M.; Sabatini, D.M. Mtor signaling in growth control and disease. *Cell* **2012**, *149*, 274–293. [CrossRef]
29. Saxton, R.A.; Sabatini, D.M. Mtor signaling in growth, metabolism, and disease. *Cell* **2017**, *168*, 960–976. [CrossRef]
30. Shimobayashi, M.; Hall, M.N. Making new contacts: The mtor network in metabolism and signalling crosstalk. *Nat. Rev. Mol. Cell Biol.* **2014**, *15*, 155–162. [CrossRef]
31. Chi, H. Regulation and function of mtor signalling in t cell fate decisions. *Nat. Rev. Immunol.* **2012**, *12*, 325–338. [CrossRef]

32. Zheng, Y.; Collins, S.L.; Lutz, M.A.; Allen, A.N.; Kole, T.P.; Zarek, P.E.; Powell, J.D. A role for mammalian target of rapamycin in regulating t cell activation versus anergy. *J. Immunol.* **2007**, *178*, 2163–2170. [CrossRef]
33. Powell, J.D.; Pollizzi, K.N.; Heikamp, E.B.; Horton, M.R. Regulation of immune responses by mtor. *Annu. Rev. Immunol.* **2012**, *30*, 39–68. [CrossRef] [PubMed]
34. Zeng, H.; Yang, K.; Cloer, C.; Neale, G.; Vogel, P.; Chi, H. Mtorc1 couples immune signals and metabolic programming to establish t(reg)-cell function. *Nature* **2013**, *499*, 485–490. [CrossRef]
35. Chapman, N.M.; Zeng, H.; Nguyen, T.M.; Wang, Y.; Vogel, P.; Dhungana, Y.; Liu, X.; Neale, G.; Locasale, J.W.; Chi, H. Mtor coordinates transcriptional programs and mitochondrial metabolism of activated treg subsets to protect tissue homeostasis. *Nat. Commun.* **2018**, *9*, 2095. [CrossRef]
36. Gerriets, V.A.; Kishton, R.J.; Johnson, M.O.; Cohen, S.; Siska, P.J.; Nichols, A.G.; Warmoes, M.O.; de Cubas, A.A.; MacIver, N.J.; Locasale, J.W.; et al. Foxp3 and toll-like receptor signaling balance treg cell anabolic metabolism for suppression. *Nat. Immunol.* **2016**, *17*, 1459–1466. [CrossRef] [PubMed]
37. Battaglia, M.; Stabilini, A.; Roncarolo, M.G. Rapamycin selectively expands cd4+cd25+foxp3+ regulatory t cells. *Blood* **2005**, *105*, 4743–4748. [CrossRef]
38. Wang, G.Y.; Zhang, Q.; Yang, Y.; Chen, W.J.; Liu, W.; Jiang, N.; Chen, G.H. Rapamycin combined with allogenic immature dendritic cells selectively expands cd4+cd25+foxp3+ regulatory t cells in rats. *Hepatobiliary Pancreat. Dis. Int.* **2012**, *11*, 203–208. [CrossRef]
39. Ikeda, K.; Kinoshita, M.; Kayama, H.; Nagamori, S.; Kongpracha, P.; Umemoto, E.; Okumura, R.; Kurakawa, T.; Murakami, M.; Mikami, N.; et al. Slc3a2 mediates branched-chain amino-acid-dependent maintenance of regulatory t cells. *Cell Rep.* **2017**, *21*, 1824–1838. [CrossRef]
40. Asanuma, S.; Tanaka, J.; Sugita, J.; Kosugi, M.; Shiratori, S.; Wakasa, K.; Shono, Y.; Shigematsu, A.; Kondo, T.; Kobayashi, T.; et al. Expansion of cd4(+)cd25 (+) regulatory t cells from cord blood cd4(+) cells using the common gamma-chain cytokines (il-2 and il-15) and rapamycin. *Ann. Hematol.* **2011**, *90*, 617–624. [CrossRef]
41. Cybulski, N.; Hall, M.N. Tor complex 2: A signaling pathway of its own. *Trends Biochem. Sci.* **2009**, *34*, 620–627. [CrossRef] [PubMed]
42. Oh, W.J.; Jacinto, E. Mtor complex 2 signaling and functions. *Cell Cycle* **2011**, *10*, 2305–2316. [CrossRef] [PubMed]
43. Gao, Y.; Tang, J.; Chen, W.; Li, Q.; Nie, J.; Lin, F.; Wu, Q.; Chen, Z.; Gao, Z.; Fan, H.; et al. Inflammation negatively regulates foxp3 and regulatory t-cell function via dbc1. *Proc. Natl. Acad. Sci. USA* **2015**, *112*, E3246–E3254. [CrossRef] [PubMed]
44. Kishore, M.; Cheung, K.C.P.; Fu, H.; Bonacina, F.; Wang, G.; Coe, D.; Ward, E.J.; Colamatteo, A.; Jangani, M.; Baragetti, A.; et al. Regulatory t cell migration is dependent on glucokinase-mediated glycolysis. *Immunity* **2017**, *47*, 875–889.e10. [CrossRef] [PubMed]
45. Shi, L.Z.; Wang, R.; Huang, G.; Vogel, P.; Neale, G.; Green, D.R.; Chi, H. Hif1alpha-dependent glycolytic pathway orchestrates a metabolic checkpoint for the differentiation of th17 and treg cells. *J. Exp. Med.* **2011**, *208*, 1367–1376. [CrossRef] [PubMed]
46. Guan, S.Y.; Leng, R.X.; Tao, J.H.; Li, X.P.; Ye, D.Q.; Olsen, N.; Zheng, S.G.; Pan, H.F. Hypoxia-inducible factor-1alpha: A promising therapeutic target for autoimmune diseases. *Expert Opin. Ther. Targets* **2017**, *21*, 715–723. [CrossRef] [PubMed]
47. Dang, E.V.; Barbi, J.; Yang, H.Y.; Jinasena, D.; Yu, H.; Zheng, Y.; Bordman, Z.; Fu, J.; Kim, Y.; Yen, H.R.; et al. Control of t(h)17/t(reg) balance by hypoxia-inducible factor 1. *Cell* **2011**, *146*, 772–784. [CrossRef]
48. Kong, N.; Lan, Q.; Chen, M.; Zheng, T.; Su, W.; Wang, J.; Yang, Z.; Park, R.; Dagliyan, G.; Conti, P.S.; et al. Induced t regulatory cells suppress osteoclastogenesis and bone erosion in collagen-induced arthritis better than natural t regulatory cells. *Ann. Rheum. Dis.* **2012**, *71*, 1567–1572. [CrossRef]
49. Xu, L.; Kitani, A.; Fuss, I.; Strober, W. Cutting edge: Regulatory t cells induce cd4+cd25-foxp3- t cells or are self-induced to become th17 cells in the absence of exogenous tgf-beta. *J. Immunol.* **2007**, *178*, 6725–6729. [CrossRef]
50. Deng, W.; Ren, Y.; Feng, X.; Yao, G.; Chen, W.; Sun, Y.; Wang, H.; Gao, X.; Sun, L. Hypoxia inducible factor-1 alpha promotes mesangial cell proliferation in lupus nephritis. *Am. J. Nephrol.* **2014**, *40*, 507–515. [CrossRef]
51. Hu, F.; Shi, L.; Mu, R.; Zhu, J.; Li, Y.; Ma, X.; Li, C.; Jia, R.; Yang, D.; Li, Y.; et al. Hypoxia-inducible factor-1alpha and interleukin 33 form a regulatory circuit to perpetuate the inflammation in rheumatoid arthritis. *PLoS ONE* **2013**, *8*, e72650.

52. Giatromanolaki, A.; Sivridis, E.; Maltezos, E.; Athanassou, N.; Papazoglou, D.; Gatter, K.C.; Harris, A.L.; Koukourakis, M.I. Upregulated hypoxia inducible factor-1alpha and -2alpha pathway in rheumatoid arthritis and osteoarthritis. *Arthritis Res. Ther.* **2003**, *5*, 193–201. [CrossRef] [PubMed]
53. Loukovaara, S.; Koivunen, P.; Ingles, M.; Escobar, J.; Vento, M.; Andersson, S. Elevated protein carbonyl and hif-1alpha levels in eyes with proliferative diabetic retinopathy. *Acta Ophthalmol.* **2014**, *92*, 323–327. [CrossRef] [PubMed]
54. Graumann, U.; Reynolds, R.; Steck, A.J.; Schaeren-Wiemers, N. Molecular changes in normal appearing white matter in multiple sclerosis are characteristic of neuroprotective mechanisms against hypoxic insult. *Brain Pathol.* **2003**, *13*, 554–573. [CrossRef] [PubMed]
55. Vasilopoulos, Y.; Sourli, F.; Zafiriou, E.; Klimi, E.; Ioannou, M.; Mamuris, Z.; Simos, G.; Koukoulis, G.; Roussaki-Schulze, A. High serum levels of hif-1alpha in psoriatic patients correlate with an over-expression of il-6. *Cytokine* **2013**, *62*, 38–39. [CrossRef]
56. Mimouna, S.; Goncalves, D.; Barnich, N.; Darfeuille-Michaud, A.; Hofman, P.; Vouret-Craviari, V. Crohn disease-associated escherichia coli promote gastrointestinal inflammatory disorders by activation of hif-dependent responses. *Gut Microbes* **2011**, *2*, 335–346. [CrossRef] [PubMed]
57. Bai, A.; Ma, A.G.; Yong, M.; Weiss, C.R.; Ma, Y.; Guan, Q.; Bernstein, C.N.; Peng, Z. Ampk agonist downregulates innate and adaptive immune responses in tnbs-induced murine acute and relapsing colitis. *Biochem. Pharmacol.* **2010**, *80*, 1708–1717. [CrossRef]
58. Gualdoni, G.A.; Mayer, K.A.; Goschl, L.; Boucheron, N.; Ellmeier, W.; Zlabinger, G.J. The amp analog aicar modulates the treg/th17 axis through enhancement of fatty acid oxidation. *FASEB J.* **2016**, *30*, 3800–3809. [CrossRef]
59. Park, M.J.; Lee, S.Y.; Moon, S.J.; Son, H.J.; Lee, S.H.; Kim, E.K.; Byun, J.K.; Shin, D.Y.; Park, S.H.; Yang, C.W.; et al. Metformin attenuates graft-versus-host disease via restricting mammalian target of rapamycin/signal transducer and activator of transcription 3 and promoting adenosine monophosphate-activated protein kinase-autophagy for the balance between t helper 17 and tregs. *Transl. Res.* **2016**, *173*, 115–130.
60. Tamas, P.; Hawley, S.A.; Clarke, R.G.; Mustard, K.J.; Green, K.; Hardie, D.G.; Cantrell, D.A. Regulation of the energy sensor amp-activated protein kinase by antigen receptor and ca2+ in t lymphocytes. *J. Exp. Med.* **2006**, *203*, 1665–1670. [CrossRef]
61. Barnes, K.; Ingram, J.C.; Porras, O.H.; Barros, L.F.; Hudson, E.R.; Fryer, L.G.; Foufelle, F.; Carling, D.; Hardie, D.G.; Baldwin, S.A. Activation of glut1 by metabolic and osmotic stress: Potential involvement of amp-activated protein kinase (ampk). *J. Cell Sci.* **2002**, *115*, 2433–2442. [PubMed]
62. Sun, Y.; Tian, T.; Gao, J.; Liu, X.; Hou, H.; Cao, R.; Li, B.; Quan, M.; Guo, L. Metformin ameliorates the development of experimental autoimmune encephalomyelitis by regulating t helper 17 and regulatory t cells in mice. *J. Neuroimmunol.* **2016**, *292*, 58–67. [CrossRef] [PubMed]
63. Lee, S.Y.; Lee, S.H.; Yang, E.J.; Kim, E.K.; Kim, J.K.; Shin, D.Y.; Cho, M.L. Metformin ameliorates inflammatory bowel disease by suppression of the stat3 signaling pathway and regulation of the between th17/treg balance. *PLoS ONE* **2015**, *10*, e0135858. [CrossRef] [PubMed]
64. Waickman, A.T.; Powell, J.D. Mtor, metabolism, and the regulation of t-cell differentiation and function. *Immunol. Rev.* **2012**, *249*, 43–58. [CrossRef] [PubMed]
65. Yang, K.; Blanco, D.B.; Neale, G.; Vogel, P.; Avila, J.; Clish, C.B.; Wu, C.; Shrestha, S.; Rankin, S.; Long, L.; et al. Homeostatic control of metabolic and functional fitness of treg cells by lkb1 signalling. *Nature* **2017**, *548*, 602–606. [CrossRef] [PubMed]
66. He, N.; Fan, W.; Henriquez, B.; Yu, R.T.; Atkins, A.R.; Liddle, C.; Zheng, Y.; Downes, M.; Evans, R.M. Metabolic control of regulatory t cell (treg) survival and function by lkb1. *Proc. Natl. Acad. Sci. USA* **2017**, *114*, 12542–12547. [CrossRef]
67. Farooqi, I.S.; Matarese, G.; Lord, G.M.; Keogh, J.M.; Lawrence, E.; Agwu, C.; Sanna, V.; Jebb, S.A.; Perna, F.; Fontana, S.; et al. Beneficial effects of leptin on obesity, t cell hyporesponsiveness, and neuroendocrine/metabolic dysfunction of human congenital leptin deficiency. *J. Clin. Investig.* **2002**, *110*, 1093–1103. [CrossRef]
68. Matarese, G.; Procaccini, C.; De Rosa, V.; Horvath, T.L.; La Cava, A. Regulatory t cells in obesity: The leptin connection. *Trends Mol. Med.* **2010**, *16*, 247–256. [CrossRef]

69. De Rosa, V.; Procaccini, C.; La Cava, A.; Chieffi, P.; Nicoletti, G.F.; Fontana, S.; Zappacosta, S.; Matarese, G. Leptin neutralization interferes with pathogenic t cell autoreactivity in autoimmune encephalomyelitis. *J. Clin. Investig.* **2006**, *116*, 447–455. [CrossRef]
70. Matarese, G.; Di Giacomo, A.; Sanna, V.; Lord, G.M.; Howard, J.K.; Di Tuoro, A.; Bloom, S.R.; Lechler, R.I.; Zappacosta, S.; Fontana, S. Requirement for leptin in the induction and progression of autoimmune encephalomyelitis. *J. Immunol.* **2001**, *166*, 5909–5916. [CrossRef]
71. Wang, X.; Qiao, Y.; Yang, L.; Song, S.; Han, Y.; Tian, Y.; Ding, M.; Jin, H.; Shao, F.; Liu, A. Leptin levels in patients with systemic lupus erythematosus inversely correlate with regulatory t cell frequency. *Lupus* **2017**, *26*, 1401–1406. [CrossRef] [PubMed]
72. De Rosa, V.; Procaccini, C.; Cali, G.; Pirozzi, G.; Fontana, S.; Zappacosta, S.; La Cava, A.; Matarese, G. A key role of leptin in the control of regulatory t cell proliferation. *Immunity* **2007**, *26*, 241–255. [CrossRef] [PubMed]
73. Dreyer, C.; Krey, G.; Keller, H.; Givel, F.; Helftenbein, G.; Wahli, W. Control of the peroxisomal beta-oxidation pathway by a novel family of nuclear hormone receptors. *Cell* **1992**, *68*, 879–887. [CrossRef]
74. Kliewer, S.A.; Umesono, K.; Noonan, D.J.; Heyman, R.A.; Evans, R.M. Convergence of 9-cis retinoic acid and peroxisome proliferator signalling pathways through heterodimer formation of their receptors. *Nature* **1992**, *358*, 771–774. [CrossRef] [PubMed]
75. Choi, J.M.; Bothwell, A.L. The nuclear receptor ppars as important regulators of t-cell functions and autoimmune diseases. *Mol. Cells* **2012**, *33*, 217–222. [CrossRef] [PubMed]
76. Pawlak, M.; Lefebvre, P.; Staels, B. Molecular mechanism of pparalpha action and its impact on lipid metabolism, inflammation and fibrosis in non-alcoholic fatty liver disease. *J. Hepatol.* **2015**, *62*, 720–733. [CrossRef]
77. Zhu, Y.; Ni, Y.; Liu, R.; Hou, M.; Yang, B.; Song, J.; Sun, H.; Xu, Z.; Ji, M. Ppar-gamma agonist alleviates liver and spleen pathology via inducing treg cells during schistosoma japonicum infection. *J. Immunol. Res.* **2018**, *2018*, 6398078. [CrossRef]
78. Cipolletta, D.; Feuerer, M.; Li, A.; Kamei, N.; Lee, J.; Shoelson, S.E.; Benoist, C.; Mathis, D. Ppar-gamma is a major driver of the accumulation and phenotype of adipose tissue treg cells. *Nature* **2012**, *486*, 549–553. [CrossRef]
79. Hontecillas, R.; Bassaganya-Riera, J. Peroxisome proliferator-activated receptor gamma is required for regulatory cd4+ t cell-mediated protection against colitis. *J. Immunol.* **2007**, *178*, 2940–2949. [CrossRef]
80. Cipolletta, D. Adipose tissue-resident regulatory t cells: Phenotypic specialization, functions and therapeutic potential. *Immunology* **2014**, *142*, 517–525. [CrossRef]
81. Bapat, S.P.; Myoung Suh, J.; Fang, S.; Liu, S.; Zhang, Y.; Cheng, A.; Zhou, C.; Liang, Y.; LeBlanc, M.; Liddle, C.; et al. Depletion of fat-resident treg cells prevents age-associated insulin resistance. *Nature* **2015**, *528*, 137–141. [CrossRef] [PubMed]
82. Stockinger, B.; Di Meglio, P.; Gialitakis, M.; Duarte, J.H. The aryl hydrocarbon receptor: Multitasking in the immune system. *Annu. Rev. Immunol.* **2014**, *32*, 403–432. [CrossRef] [PubMed]
83. Quintana, F.J.; Basso, A.S.; Iglesias, A.H.; Korn, T.; Farez, M.F.; Bettelli, E.; Caccamo, M.; Oukka, M.; Weiner, H.L. Control of t(reg) and t(h)17 cell differentiation by the aryl hydrocarbon receptor. *Nature* **2008**, *453*, 65–71. [CrossRef] [PubMed]
84. Smith, K.A. Interleukin-2: Inception, impact, and implications. *Science* **1988**, *240*, 1169–1176. [CrossRef] [PubMed]
85. Sakaguchi, S. Naturally arising foxp3-expressing cd25+cd4+ regulatory t cells in immunological tolerance to self and non-self. *Nat. Immunol.* **2005**, *6*, 345–352. [CrossRef] [PubMed]
86. Mizui, M.; Tsokos, G.C. Low-dose il-2 in the treatment of lupus. *Curr. Rheumatol. Rep.* **2016**, *18*, 68. [CrossRef] [PubMed]
87. Von Spee-Mayer, C.; Siegert, E.; Abdirama, D.; Rose, A.; Klaus, A.; Alexander, T.; Enghard, P.; Sawitzki, B.; Hiepe, F.; Radbruch, A.; et al. Low-dose interleukin-2 selectively corrects regulatory t cell defects in patients with systemic lupus erythematosus. *Ann. Rheum. Dis.* **2016**, *75*, 1407–1415. [CrossRef]
88. Rosenzwajg, M.; Churlaud, G.; Mallone, R.; Six, A.; Derian, N.; Chaara, W.; Lorenzon, R.; Long, S.A.; Buckner, J.H.; Afonso, G.; et al. Low-dose interleukin-2 fosters a dose-dependent regulatory t cell tuned milieu in t1d patients. *J. Autoimmun.* **2015**, *58*, 48–58. [CrossRef]

89. Koreth, J.; Kim, H.T.; Jones, K.T.; Lange, P.B.; Reynolds, C.G.; Chammas, M.J.; Dusenbury, K.; Whangbo, J.; Nikiforow, S.; Alyea, E.P., 3rd; et al. Efficacy, durability, and response predictors of low-dose interleukin-2 therapy for chronic graft-versus-host disease. *Blood* **2016**, *128*, 130–137. [CrossRef]
90. Matsuoka, K.; Koreth, J.; Kim, H.T.; Bascug, G.; McDonough, S.; Kawano, Y.; Murase, K.; Cutler, C.; Ho, V.T.; Alyea, E.P.; et al. Low-dose interleukin-2 therapy restores regulatory t cell homeostasis in patients with chronic graft-versus-host disease. *Sci. Transl. Med.* **2013**, *5*. [CrossRef]

Review

Targeting mTOR in Acute Lymphoblastic Leukemia

Carolina Simioni [1,*], Alberto M. Martelli [2], Giorgio Zauli [3], Elisabetta Melloni [3] and Luca M. Neri [3,4,*]

1 Department of Medical Sciences, University of Ferrara, 44121 Ferrara, Italy
2 Department of Biomedical and Neuromotor Sciences, University of Bologna, 40126 Bologna, Italy; alberto.martelli@unibo.it
3 Department of Morphology, Surgery and Experimental Medicine, University of Ferrara, 44121 Ferrara, Italy; giorgio.zauli@unife.it (G.Z.); elisabetta.melloni@unife.it (E.M.)
4 LTTA-Electron Microscopy Center, University of Ferrara, 44121 Ferrara, Italy
* Correspondence: carolina.simioni@unife.it (C.S.); luca.neri@unife.it (L.M.N.); Tel.: +39-0532-455844 (C.S.); +39-0532-455940 (L.M.N.)

Received: 15 January 2019; Accepted: 16 February 2019; Published: 21 February 2019

Abstract: Acute Lymphoblastic Leukemia (ALL) is an aggressive hematologic disorder and constitutes approximately 25% of cancer diagnoses among children and teenagers. Pediatric patients have a favourable prognosis, with 5-years overall survival rates near 90%, while adult ALL still correlates with poorer survival. However, during the past few decades, the therapeutic outcome of adult ALL was significantly ameliorated, mainly due to intensive pediatric-based protocols of chemotherapy. Mammalian (or mechanistic) target of rapamycin (mTOR) is a conserved serine/threonine kinase belonging to the phosphatidylinositol 3-kinase (PI3K)-related kinase family (PIKK) and resides in two distinct signalling complexes named mTORC1, involved in mRNA translation and protein synthesis and mTORC2 that controls cell survival and migration. Moreover, both complexes are remarkably involved in metabolism regulation. Growing evidence reports that mTOR dysregulation is related to metastatic potential, cell proliferation and angiogenesis and given that PI3K/Akt/mTOR network activation is often associated with poor prognosis and chemoresistance in ALL, there is a constant need to discover novel inhibitors for ALL treatment. Here, the current knowledge of mTOR signalling and the development of anti-mTOR compounds are documented, reporting the most relevant results from both preclinical and clinical studies in ALL that have contributed significantly into their efficacy or failure.

Keywords: Acute Lymphoblastic leukemia; targeted therapy; mTOR; metabolism; cell signalling

1. Introduction

Aberrant intracellular signalling pathways and inadequate continuous activation of cellular networks commonly result in abnormal growth and survival of malignant cells. The PI3K/protein kinase B (Akt)/mTOR network initiates and controls multiple cellular activities, including mRNA translation, cell cycle progression, gene transcription, inhibition of apoptosis and autophagy, as well as metabolism [1–5]. Constitutive activation of this pathway not only promotes uncontrolled production of malignant cells but also induces chemotherapy resistance mechanisms, also in leukemias. ALL is an aggressive malignancy of lymphoid progenitor cells in both pediatric and adult patients. In adults, 75% of cases develop from precursors of the B-cell lineage, the others consisting of malignant T-cell precursors [5–10]. T-ALL is also found in a range of 15% to 20% in children, affecting boys more than girls. Modern genomic approaches have identified a number of recurrent mutations that can be grouped into several different signalling pathways, including Notch, Jak/Stat, MAPK and PI3K/Akt/mTOR. Phosphatase and tensin homolog (PTEN), which acts as a tumour suppressor gene, represents one of the main negative regulator of PI3K/Akt/mTOR network. PTEN is the key regulator of phosphatidylinositol (3,4,5)-trisphosphate (PIP3) dephosphorylation into phosphatidylinositol (4,5)-bisphosphate (PIP2), thus blunting PI3K activity. In human T-ALL, PTEN is often mutated or deleted, leading to the upregulation of PI3K/Akt/mTOR, in combination with additional genetic anomalies [11,12]. Therefore, targeting the PI3K/Akt/mTOR signalling network has been investigated extensively in preclinical models of ALL, with initial studies focused on mTOR inhibition, demonstrating significant efficacy for mTOR drugs used as single inhibitors and synergistic effects in association with conventional chemotherapy [13].

It should be highlighted that, in addition to the standard chemotherapy, other treatment options such as immunotherapy represent today a new pharmacological approach, by targeting ALL surface markers expressed on B lymphoblasts, that are, CD19, CD20 or CD22 [14]. One immunotherapy strategy is represented by the bispecific T-cell engager (BiTE) antibodies, that bind the surface antigens on two different target cells, generating a physical link of a tumour cell to a T cell: from one side they can recognize the malignant B-cells through the CD19 and from the other side they activate T-cell receptor (TCR) through the interaction with the CD3 receptor on T-lymphocytes [15,16]. Blinatumomab is a first-in-class BiTE antibody and it is a bispecific CD19-directed CD3 T-cell mAb that has induced durable responses in patients with B-cell malignancies [17]. Blinatumomab has demonstrated important response rates in minimal residual disease (MRD) positive and relapsed or refractory B-ALL, both in adults and in children [16]. Another immunotherapeutic strategy in relapsed/refractory CD22^{+} ALL is represented by Inotuzumab ozogamicin, a novel mAb against CD22 conjugated to the toxin calicheamicin [18]. Another promising new therapy is the adoptive immunotherapy using chimeric antigen receptors (CARs) modified T cells, developed in recent years. CARs are artificial engineered receptors that can target specific cancer cell surface antigens, activates T cells and, moreover, enhances T-cell immune function [19,20]. The first constructs consisted of CAR T cells targeting CD19 marker and today different other antigens are under development. It has to be underlined that a CD19-directed genetically modified autologous T-cell immunotherapy, Kymriah (Tisagenlecleucel), has already been approved by FDA for patients up to 25 years of age with relapsed or refractory B-cell ALL [21]. In pediatric patients and young adults, treatment consisting of fludarabine and cyclophosphamide followed by a single infusion of Kymriah induced a significant (63%) Complete Remission (CR), negative for MRD with an acceptable benefit–risk profile for this patient population (see also www.clinicaltrials.gov: NCT02435849).

However these immunotherapies are not considered completely curative and this is due to the fact that deadly relapses are common (median overall survival is <8 months with Blinatumomab and Inotuzumab ozogamicin and ~50% of CAR-T relapse within a year). Thus, novel approaches and further studies are needed.

Concerning mTOR inhibition, Rapamycin (Sirolimus) was first discovered as a new antifungal agent in soil samples from Rapa Nui, from which it received its name. In 1999 it was approved by FDA to prevent immune rejection of transplanted organs and studies in the budding yeast Saccharomyces cerevisiae revealed that a serine/threonine (Ser/Thr) protein kinase was the mediator of rapamycin toxic effects in yeast (Target of Rapamycin, TOR) [22]. mTOR is a 289 kDa serine/threonine kinase that works as a regulator of cellular progression and metabolic mechanisms in response to nutrients and hormonal signals [23]. mTOR forms with other components two distinct complexes, mTOR complex 1 (mTORC1) and mTOR complex 2 (mTORC2), that differ from each other according to their different functionality [24]. These complexes have in common the catalytic mTOR subunit and other three known complex components, mLST8, DEP domain-containing mTOR-interacting protein (DEPTOR) and Tti1/Tel2 [22,25–27]. mTORC1 is Rapamycin-sensitive and has unique components including regulatory-associated protein of mammalian target of rapamycin (raptor) and the 40 kDa proline-rich Akt substrate (PRAS40); mTORC2 comprises rapamycin insensitive companion of mTOR (rictor), mSIN1 and Proctor 1/2 as its specific components [28–30].

In this review we will summarize the current knowledge of mTOR signalling, the roles of its complexes, the mTOR involvement in the metabolism process, its relevance in ALL diseases and the status of mTOR inhibitors, reporting the most consistent results obtained from both preclinical and clinical analysis in ALL.

2. Activity of mTOR Complex 1

mTORC1 represents a nutrient/energy/redox sensor and is a controller of protein translation, cell metabolism and growth. For protein synthesis cells must have enough energy resources, nutrient support, oxygen abundance and proper growth factors to begin mRNA translation [31,32].

The binding of different growth factors and various cytokines to cell surface tyrosine kinase receptors leads to the PI3K/Akt/mTOR network activation. In particular, a key negative regulator of mTORC1 is TSC1/TSC2, a heterodimer with GTPase activity consisting of tuberous sclerosis 1 (TSC1) and tuberous sclerosis 2 (TSC2). This complex acts as a molecular switch for mTORC1 activity: during high stress situations, mTOR activity is blocked by this heterodimer, while it is again reactivated under favourable circumstances, that are, with growth factor stimulation and high cell growth conditions. The GTP-bound form of Ras homolog enriched in brain (Rheb) directly cooperates with mTORC1, strengthening its kinase activity and inducing proliferation and cell survival. As a Rheb GTPase-activating protein (GAP), TSC1/TSC2 negatively regulates mTORC1 by converting Rheb into its inactive form [33].

Among the upstream signalling networks, Akt and extracellular-signal-regulated kinase (ERK1/2) inactivate TSC1/2 and thus activate mTORC1.

Activated mTORC1 directly phosphorylates eIF4E-binding protein 1 (4E-BP1) and ribosomal S6 kinase (S6K), inducing protein synthesis [34–36]. The phosphorylation of 4E-BP1 impedes its binding to cap-binding protein eIF4E complex and leads to the initiation of cap-dependent translation. S6K controls the translation of several mRNAs that encode for ribosomal proteins and other constituents of the translational machinery such as elongation factors (i.e., the eukaryotic elongation factor 2 or eEF2). mTORC1-dependent anabolic induction is achieved through the phosphorylation of the downstream kinase S6K, as well as 4E-BP1, Lipin1 (lipid synthesis) and ATF4 (nucleotide synthesis). mTORC1 also up-regulates the glycolytic pathway by stimulating the transcriptional activation capability of Hypoxia Inducible Factor 1α (HIFα), a positive regulator of many glycolytic genes [37,38]. Furthermore, mTORC1 regulates the lysosomal function through transcription factor EB (TFEB), which controls the expression of many genes with key roles in lysosomal biogenesis and autophagy mechanism [39]. TFEB-mediated endocytosis induces the assembly of the mTORC1-containing nutrient sensing complex through the formation of endosomes, with further activation of Akt (Akt p-T308) [40].

mTORC1 promotes cell growth by blocking catabolic pathways such as autophagy, that represents the major degradation pathway in eukaryotic cells [41–43]. The inhibition of mTORC1 stimulates autophagy and this is correlated, for example during amino acid starvation condition, with the activation of the mTORC1 direct substrate Unc-51 like autophagy activating kinase (ULK1, including ULK2 isoform) [44]. The phosphorylation of ULK1 is a key signalling mechanism through which starvation-induced autophagy is regulated. Therefore, under starvation, activated phosphatases such as PP2A lead to a rapid dephosphorylation of ULK1, phosphorylated by mTORC1. ULK1 dephosphorylation correlates with an increment of the autophagic process [43]. Alterations in autophagy processes are correlated to different disorders including diabetes, cardiovascular disease, neurodegenerative diseases and cancer [45–47].

3. Activity of mTOR Complex 2

Differently to the role of mTORC1, little is known about the regulatory mechanism of mTORC2. mTORC2 specifically senses growth factors and controls cell survival, metabolism and actin rearrangement [48]. The nutrient-sensing role of mTOR is mainly dependent on mTORC1 [49]. mTORC2 is abnormally overexpressed in several cancer types and this characteristic leads to poor survival [50]. Although initial studies reported mTORC2 as a rapamycin-insensitive complex, there are actually several reports showing that rapamycin is capable of inhibiting mTORC2 upon longer exposure, most likely by negatively affecting the assembly of new mTORC2 complexes and therefore reducing mTORC2 levels that are usually required to maintain Akt/PKB signalling [51]. The different sensitivity to rapamycin may be explained by the fact that, in various cell types, a fraction of mTORC2 assembles with the FKBP12-rapamycin binding site not accessible to the drug. PTEN and FKBP12 expression represent possible modulators of rapamycin-mediated inhibition of Akt/PKB phosphorylation [51].

Liu and co-workers identified a correlation between PI3K growth factor stimulation and mTORC2 activity. SIN1 pleckstrin homology (PH) domain suppresses mTOR kinase domain function, as PIP3 interacts with the PH domain of mSIN1 to repress its inhibitory activity on mTOR, thus leading to mTORC2 activation [52].

mTORC2 substrates include the Ser/Thr cytosolic protein kinase Akt and protein kinase C (PKC), which share the hydrophobic motif at their phosphorylation site [53]. mTORC2 directly phosphorylates Akt at S473 residue and fully activates it [54,55]. Akt phosphorylation inhibits mTORC2 activity and thereby reduces the function of some Akt targets such as FoxO1/2 [56]. mTORC2 also stimulates serum and glucocorticoid-regulated kinase 1 (SGK1), a kinase that contributes to the regulation of glucose and creatine transporters, hormone release, inflammation, growth and apoptosis [57], belongs to the AGC family of protein kinases [54] and is involved in aberrant cell growth, survival and invasiveness. Nevertheless, recent findings have highlighted how mTORC2 could also act as a repressor of chaperone-mediated autophagy [58,59], which is frequently deregulated in numerous age-related disorders [60,61]. Moreover, mTORC2 increases Na^+ transport and regulates cell migration [62,63].

The major functions and downstream targets of mTORC1 and 2 are highlighted in Figure 1.

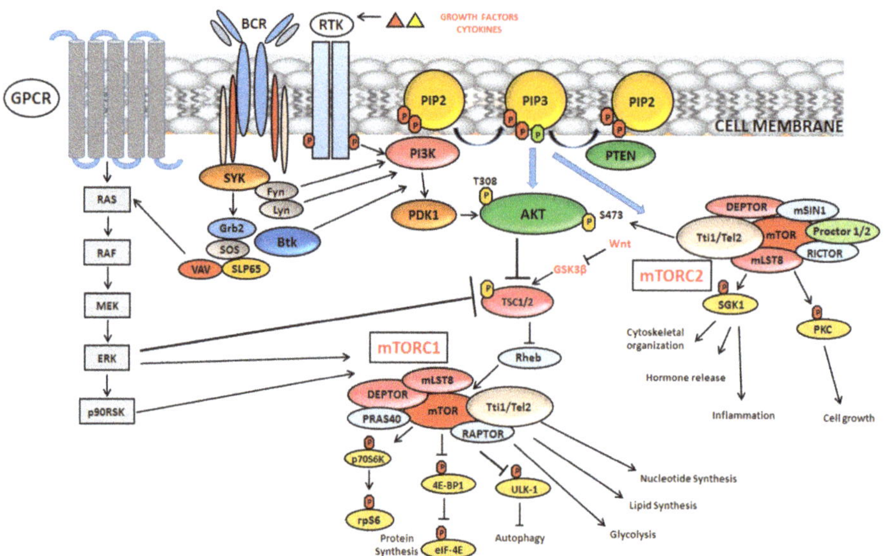

Figure 1. Signalling mechanisms and major functions of mTORC1 and mTORC2.

mTOR, activated by specific growth factors and cytokines, nutrients, cellular stress and oxygen levels, forms two distinct complexes, mTOR complex 1 (mTORC1) and mTOR complex 2 (mTORC2), that differ from each other according to their different activities. A key negative regulator of mTORC1 is TSC1/TSC2, consisting of tuberous sclerosis 1 (TSC1) and tuberous sclerosis 2 (TSC2). Among the upstream signalling networks, Akt and extracellular-signal-regulated kinase (ERK1/2) inactivate TSC1/2 and activate mTORC1. mTORC2 directly phosphorylates Akt at S473 residue leading to its complete activation. In B-cell development, PI3K/Akt/mTOR can also be activated by pre-BCR network, which involves constitutive activity of different kinases such as SYK, Fyn and Lyn. Wnt network, involved in embryonic development, can inhibit mTORC1 by blocking GSK3β expression. GSK3β is a positive modulator of the TSC complex. For other details, see the text.

4. mTOR Involvement in Metabolism

Metabolic modifications represent a hallmark of oncogenesis and tumour progression [64] and there is a growing interest in understanding these alterations associated with cancer cells resistant to therapy. Relapsed cancer cells, that have an aggressive phenotype, display an overexpression of the ATP-binding cassette transporter ABCB1 gene product and a high expression of the efflux pump P-glycoprotein (P-gp) [65]. The metabolic program often used by cancer cells is the aerobic glycolysis, which is characterized by increased glucose flux and production of lactate. Because of aerobic glycolysis tumour cells generate new lipids or amino acids necessary for cell proliferation [66]. Therefore, the metabolic alterations that characterize drug-resistance in cancer cells could represent an attractive therapeutic target. PI3K/Akt/mTOR network is a key regulator of glycometabolic homeostasis [67,68]. The active PI3K/Akt pathway is involved in glucose uptake by upregulating several cell surface glucose transporters (GLUT) expression and by stimulating several glycolytic enzymes, such as hexokinase [68]. mTORC1 activation is induced by different amino acids, particularly leucine and glutamine and also activates feedback mechanisms that further increase nutrient uptake to fuel anabolic reactions [69]. This complex is also involved in the upregulation of the expression of HIF1α, which has a key function in cell energy regulation, lipid and glutamine metabolic flux and promotes glycolysis with the conversion of glucose to pyruvate [70]. Moreover, mTORC1 positively regulates mitochondrial function and metabolism by selectively inducing translation of nucleus-encoded mitochondria-related mRNAs [71]. Upon starvation, mTORC1 activity is inhibited, alleviating the repression of ULK1/2. Active ULK1/2 phosphorylates key enzymes involved in the glycolytic pathway and promotes autophagy ensuring a good level of cellular energy and redox homeostasis [72]. Less is known about mTORC2 in growth factor signalling and metabolism. Recent data indicate that mTORC2 direct association with ribosomes ensures that this complex is active in cells that are growing and undergoing protein synthesis [63,68,73]. The metabolic profile of ALL is yet not well understood. However, it has been reported that in primary human peripheral T-ALL blood samples Glut1 and Hexokinase 1 and 2 (HK1 and HK2) were significantly elevated in comparison to T cells from healthy donors. In T-ALL, oncogenic Notch induced metabolic stress that stimulated 5′ AMP-activated kinase (AMPK). AMPK has growth suppressive functions and may act to block growth also in T-ALL [74]. Kishton et al demonstrated that unlike stimulated T cells, AMPK restrained aerobic glycolysis in T-ALL, also maintaining mitochondrial function to mitigate stress. The reduction of energy demand was modulated through mTORC1 inhibition regulated anabolic growth, resulting in decreased aerobic glycolysis. The metabolic stress and apoptosis modulation by AMPK may thus provide new, potential approaches to treat T-ALL. To date, few studies have analysed the adaptations of the cellular cancer metabolism in a drug-resistance context. The transcriptional profile of several glucocorticoid-resistant T-ALL cells with high expression of genes related to the metabolic pathway was recently analysed. Treatment with rapamycin augmented cell sensibility to the glucocorticoid dexamethasone, indirectly correlating metabolism upregulation and mTOR function [75]. Compared with sensitive ALL cells, a recent study reported the metabolic profile of daunorubicin-resistant T-ALL cells and transcriptomic and metabolomic analysis revealed a higher dependence on glucose and a lower dependence on glutamine and fatty acids. The reduction of glutamate-ammonia ligase (GLUL) expression, the low levels of glutamine metabolism genes ASNS, ASS1, of the transporter SLC1A5 and the reduced level of pantothenic acid may reflect a more general adaptation based on metabolic rewiring that characterizes drug resistance in tumour cells [64]. Moreover, transforming oncogenes, such as RAS or BCR-ABL can impose significant metabolic requirements on glucose and energy supply [76,77].

Given the high presence of genetic lesions of B-lymphoid transcription factors in pre-B ALL cells, some studies analysed whether these transcription factors could restrict glucose and energy supply, thus getting in the way against malignant transformation. By combining ChIP-seq and RNA-seq studies, it was shown that two transcription factors, PAX5 and IKZF1, that are reported to have a key role in normal B-cell development, enforce a state of chronic energy deprivation and a constitutive activation of AMPK. In pre-B ALL patients, inducible wildtype reconstitution of PAX5 and IKZF1 decreased Akt activation, glucose-metabolism effectors and glucose transporters. On the other hand, a strong induced expression of glucose transport inhibitors was shown. NR3C1 (glucocorticoid receptor), TXNIP (glucose feedback sensor) and CNR2 (cannabinoid receptor) were identified as central effectors of B-lymphoid glucose restriction and energy supply. Indeed, the genetic loss-of-function of NR3C1, TXNIP and CNR2 significantly improved glucose uptake and increased ATP-levels [78]. Therefore, B-lymphoid transcription factors can represent metabolic gatekeepers limiting the amount of cellular ATP to levels insufficient to achieve malignant transformation. Moreover, the effectors NR3C1, TXNIP and CNR2 could represent valuable therapeutic targets for the treatment of pre-B ALL.

5. Targeted Therapy: Inhibition of mTOR in ALL

The crucial function of mTOR as a regulator of Akt, its involvement in modulation of other signalling cascades such as NOTCH-1 (principally via-mTORC2) and its ability in monitoring metabolic functions and energy homeostasis in healthy and tumoral cells led to a growing interest in developing targeted and personalized ALL therapies. There is a marked interest in targeting mTOR protein kinase, for cancer therapy, also based on genetic studies showing selective effects on tumour cells following mTOR inactivation [79]. Indeed, some recognized molecular lesions related to adverse clinical prognosis in ALL are involved in mTOR-mediated signalling. Based on these observations, three classes of mTOR inhibitors are included in the scenario of ALL treatment: allosteric inhibitors [(rapamycin and rapalogs, that are, RAD001 (everolimus), CCI-779 (temsirolimus)] mainly targeting mTORC1 [80–82], ATP-competitive dual PI3K/mTOR inhibitors [6,83,84] and mTOR kinase inhibitors (TORKIs) that specifically have as a target mTORC1 and mTORC2 but not PI3K [79,83,85]. Rapamycin and rapalogs belong to the mTOR first generation inhibitors, are the most well documented drugs and have shown antitumor activity in clinical trials, not only for ALL cases but also in non-haematological neoplasms. Rapamycin interacts with the intracellular receptor, FK506-binding protein 12 (FKBP12) and interferes with growth-stimulating cytokine signalling [86]. Together with the immunophilin FKBP12, rapamycin/rapalogs associate with the FKBP12-rapamycin-binding (FRB) domain of mTORC1 [87]. This association results in decreased interactions between mTOR and Raptor with a consequent downregulation of mTORC1 activity [88]. This was reported to occur by steric hindrance through reduction in the size of the mTOR active-site [87]. Rapamycin has been tested alone and in combination with Janus Kinase (JAK), ABL protein inhibitors [84], focal adhesion kinase (FAK) or also with cyclin D3 (CCND3) and CDK4/6 inhibitors [89,90] in xenografts model and in in vitro cancer cell lines, showing relevant synergistic effects and induction of cellular mechanisms as autophagy. Nevertheless, rapamycin has also been reported to have pharmacological limitations. For this reason rapamycin derivatives (so called "rapalogs") for the treatment of ALL have been developed, with minor immunosuppressive activity [86] and greater antitumoral action. Among these rapalogs, RAD001 has been widely investigated in in vivo and in vitro models for its antiproliferative activities. It is more selective for the mTORC1 protein complex, with lower impact on mTORC2 and different trials are underway also in solid tumours like gastric cancer and hepatocellular carcinoma. Numerous studies reported the efficacy of RAD001 in inducing caspase-independent cell death and cell cycle regulation changes [91] or in overcoming resistance in ALL versus several inhibitors like Tirosine Kinase Inhibitors (TKI) [92]. However, limited apoptosis was reported by rapalogs, despite the delayed progression of tumour growth. Therefore, second-generation anti mTOR drugs were developed, to compete with the catalytic site of mTOR blocking the feedback activation of PI3K/Akt network (mTOR kinase inhibitors) and to repress both mTOR complexes, overcoming some of the limitations of rapalogs.

BEZ235 and BGT226, in addition to others that will be described in the next sections, belong to the PI3K/mTOR inhibition class and for both are reported significant apoptotic and anti-leukemic activity in vitro and in vivo [8,93–95]. AZD8055, AZD2014, TAK-228, CC-223 and OSI-027 are examples of TORKIs and entered phase 1/2 clinical trials for the treatment of different solid tumours but not yet for ALL. mTOR inhibition also appears to be effective when combined with conventional chemotherapy in ALL [85,96] and with drugs targeting the epigenetic machinery, inducing apoptosis, as well as increased mitochondria sensitivity to initiate cell death [97] and inhibiting BCL2 protein family leading to increased cytotoxicity [98].

mTOR inhibition either alone or in association with conventional ALL therapies or with targeted drugs for different cellular cascades is able to block distinct mechanisms of cell survival in ALL, providing a strong rationale for the investigation of mTOR inhibition particularly in the setting of resistance to chemotherapeutic drugs. In the next sections the most recent advances in mTOR inhibition in B- and T-ALL are discussed, pointing out the therapeutic importance of this protein kinase in this hematologic malignancy treatment.

5.1. mTOR Inhibition in T-ALL

T-ALL is a neoplasm caused by numerous and relevant genomic lesions that affect the development of T cells [99–101]. In T-ALL, high expression of mTOR was reported to be more frequent in adults than in children [102]. Literature data documented, in several preclinical studies, a good efficacy of allosteric mTOR inhibitors in T-ALL cells, with cytostatic effects [103,104]. Rapamycin and CCI-779 are able to block Interleukin-7 (IL-7)-dependent T-ALL growth [105]. Indeed, the two inhibitors induced T-ALL cell death when cultured in the presence of this interleukin. Moreover, in T-ALL cell lines, rapamycin re-establishes the expression of p14, p15 and p57 genes, that normally act as cell cycle regulators and are reported to be inactivated in adult ALL through promoter methylation [90]. Demethylation of the promoter of the G1/S transition genes mediated these effects, accompanied by a marked decrease of mTOR and p70S6K expression. However, the molecular mechanism concerning this regulation still has to be clarified. Rapamycin and derivatives could be significantly combined with drugs currently employed in the treatment of T-ALL, that is, Doxorubicin (Doxo) [89,105], cyclophosphamide [106] and methotrexate [107]. Pikman et al. documented also a good synergism in the combinations of the CDK4/6 inhibitor LEE-01 (ribociclib) with the rapalog RAD001 and with glucocorticoids (GCs), currently used in the therapeutic protocols in the treatment of T- and B-ALL. The efficacy of the combination rather than the single agent was shown both in vitro and in T-ALL mouse models [89]. Other relevant allosteric mTOR inhibitors combinations to be cited are those with inhibitors of NOTCH1 signalling network. This evolutionally conserved signalling network, with key roles in modulation of haematopoiesis, cell growth, apoptosis and angiogenesis, is commonly dysregulated in T-ALL representing the most common abnormality in this subtype. NOTCH1 can activate PI3K/AKT/mTOR network at multiple levels, regulating cell size, glucose accumulation and glycolysis during T-cell development [108]. Consequently, inhibition of NOTCH1 correlates with suppression of mTOR, highlighting the close interconnection between the two signalling cascades. Different PI3K upstream signalling receptors, such as the interleukin 7 receptor α chain (IL7RA), are upregulated by NOTCH1 signalling in T-cell progenitors. The oncogene MYC, prominent target gene in T-cell transformation, is able to revert the inhibitory effects of blocking NOTCH1 on the mTOR network [108–110]. The dual inhibitor PI3K/mTOR inhibitor PKI-587 (Gedatolisib) exhibited a significant inhibitory effect on T-ALL leukemia cells and in T-ALL patients with poor prognosis. In T-ALL cells PKI-587 blocked proliferation and colony formation and, in immune-deficient mouse models, delayed tumour progression, enhancing the survival rate. PKI-587 was also particularly effective in CRLF2/JAK-mutant models with a 92.2% leukemia proliferation reduction versus vehicle controls and with significantly prolonged mice survival [83]. Preclinical studies of the dual PI3K/mTOR inhibitor BEZ235 showed anti-proliferative activity in ALL cell lines [111,112]. In particular, this drug induced anti-leukemic activity when associated with glucocorticoids in vitro and in vivo models [112,113]. In PTEN null cells, BEZ235 controlled GC-resistance by increasing the level of the proapoptotic BIM protein, inducing a marked apoptosis [112,114]. The pronounced antiproliferative effects of BEZ235 have also been observed in Jurkat and MOLT4 T-ALL cells, when administered with cytarabine (AraC), Doxo or glucocorticoids [111]. TORKIs interfere only with the mTOR catalytic domain [54]. The levels and activation status of regulators of mTORC1 were recently examined in γ secretase inhibitors (GSI)-resistant T-ALL cells. These inhibitors are reported to suppress Aβ peptide and to suppress the Notch signalling pathway [115]. The combination of the mTOR inhibitor AZD8055 with the BH3-mimetic, ABT-263, induced a decreased phosphorylation of 4EBP1 and S6, also lowering MCL-1 expression and inducing tumour regression in vivo [116]. In T-ALL Jurkat cells, the mTOR kinase inhibitor OSI-027 induced c-Myc reduction and activation of the PUMA BCL2 family member. At the same time, the mTORC2 activity inhibition, with 4EBP1 phosphorylation, resulted in NF κB–mediated expression of the early growth response 1 (EGR1) gene, which encodes for the proapoptotic protein BIM. Therefore, different signalling networks are involved in T-ALL apoptosis, after OSI-027 treatment [117]. Table 1 reports a summary of the main mTOR inhibitors in T-ALL models. However, there are still few data concerning the efficacy of this class of inhibitors for T-ALL, while some Phase I clinical studies conducted mostly on solid tumours have been released [118]. Therefore, further studies are needed to predict valuable therapeutic protocols with TORKIs in T-ALL models.

5.2. *mTOR Inhibition in B-ALL*

B-ALL is characterized by the uncontrolled proliferation of B-cell precursors [119] and is further classified by the differentiation status as pro-B, common, precursor B (pre-B) and mature B-cell ALL. Pro-B ALL is an ALL unfavourable subset in childhood and adults and lacks the B cell marker of therapeutic resistance CD10 [120]. Pre-B ALL represents the most common type in adults, with cells that characteristically co-express CD10 and CD19 surface markers [121], while mature B cell ALL is sometimes called Burkitt type ALL because it shares similar characteristics to the Burkitt lymphoma [122]. Despite considerable progress in treatment protocols, B-ALL displays a poor prognosis in about 15–20% of pediatric cases and about 60% of adult patients. Factors for higher risk of relapse in adults include the Philadelphia chromosome alteration (Ph) formed upon the t(9;22) reciprocal chromosomal translocation, with consequent formation of a Bcr-Abl chimera gene [107]. The incidence of Ph$^+$ B-ALL increases with age and occurs up to 50% of B-ALL diagnosed in subjects over 50 years old [85]. Philadelphia (Ph)-like ALL is a recently characterized subtype. This subgroup has been reported to have a high expression of cytokine receptors and signalling tyrosine kinases, resulting in kinase activation through stimulation of ABL and JAK/STAT pathways. This subtype is very aggressive and the gene expression profile is reported similar to that of Ph$^+$ B-ALL, although the Bcr-Abl fusion protein is lacking. For this aspect, the development of sensitive, cost-effective and commercially available diagnostic approaches are needed in order to better identify this type of patients [123]. Many Small Molecule Inhibitors (SMI) have been or are in clinical development for both T- and B-ALL subtype [10,124–127]. Indeed, hyperactivation of PI3K/Akt/mTOR network is usually detected also in B-ALL subtype [128] and correlates with poor prognosis and drug resistance both in adults and in pediatric B-ALL patients [129,130]. In pre-B ALL the combination of RAD001 with the Akt inhibitor MK-2206 significantly exerts anti-leukemic activity, with increased apoptosis and autophagy induction [5]. For this ALL subtype and in particular for Ph$^+$ B-ALL, mTOR inhibitors are usually given in combination with other signalling cascade drugs or in association with TKI, since the antiproliferative activity of a single mTOR inhibitor is not often effective in overcoming drug-resistance.

5.3. *mTOR Activity in B-ALL*

High basal levels of Akt and mTOR activation in B cell leukemias and lymphomas have been reported in different works [10,131,132]. Recent studies in immune cells highlighted that mTOR not only couples nutrient availability with cell proliferation but also controls cell differentiation and activation-induced responses in B and T lymphocytes, natural killer cells, neutrophils, macrophages and other cellular types [132]. By assessing the total expression of mTORC1 signalling proteins Raptor and mTOR, as well as the phosphorylation of mTORC1 downstream S6 ribosomal (S6R) and 4EBP1 proteins, it has been found that mTORC1 pathway is predominantly activated during the pro-B, large and small pre-B cells stages and to a lesser extent in resting immature and mature B cells in the bone marrow. These results are consistent with the expression of the IL-7 receptor (IL-7R) during the pro-B and pre-B cell processes and the activation of mTORC1 downstream of IL7R. The importance of mTOR in cell maturation and differentiation has been documented in a study involving an hypomorphic mouse, characterized by a neomycin cassette insertion in an intron that partially disrupted mTOR transcription. A partial block in B-cell development in the bone marrow was detected with a reduction of pro-B, pre-B and immature B cell populations, simultaneously with a reduced mature B cell populations in the spleen. mTORC1 signalling is therefore specifically required in early B cell development [133,134]. mTORC2 signalling is also important in B cell development and function. Indeed, Rictor knockout mice reported a decrease in the quantity of mature B cells in the peripheral blood and in the spleen and impaired early B cell development in the bone marrow. Rictor deficient B cells exhibited an aberrant increase in FoxO1 and recombination-activating gene 1 (Rag-1) and rapamycin aggravated the Rictor deletion-induced defect in B cells via the inhibition of mTORC1 activity [135]. The mTOR hyperactivation in B-ALL is strictly dependent on oncogenic drivers, such as BCR-ABL1 fusion gene, on kinases commonly found aberrant in B-ALL such as cytokine receptor-like factor 2 (CRLF2) or JAKs, on hyperactivated growth receptors as colony-stimulating factor 1 receptor (CSF1R) and on overexpressed fms-related tyrosine kinase 3 (FLT3) and IL7R. Besides BCR-ABL1 fusion gene, the ETV6/RUNX1 (E/R) mutation is frequently found in childhood B-ALL and literature reported the importance of using dual PI3K/mTOR drugs for a marked inhibition of cells harbouring E/R mutation and a decreased cell resistance to glucocorticoids [136]. Activating mutations in upstream kinases such as protein kinase C delta (PKCδ) or the cytoplasmic protein spleen tyrosine kinase (SYC), are the most common causes of mTOR hyperactivity. SYK, which is downstream of pre-BCR signalling, promotes the activation of PI3K/Akt pathway [137,138]. Concerning Ph^- B-ALL cases, literature reported that the upregulation of the PI3K/Akt/mTOR network could be dependent on constitutively active pre-BCR signalling that characterizes approximately 13% of Ph^- B-ALL cases [139]. Rapamycin, RAD001 and CCI-779 have all been tested in various clinical trials for ALL in combination with multi-agent chemotherapy, with promising results including CR of relapsed childhood ALL [140]. However, a disadvantage of rapalog therapy is the upregulation of the pro-survival/proliferation PI3K pathway, which occurs through loss of mTORC1 negative feedback on PI3K and on mTORC2 and the incomplete suppression of the mTORC1 substrate 4E-BP1 [140]. Moreover, rapamycin and rapalogs suppress p70S6K1 phosphorylation and switch off the Insulin receptor substrate (IRS)-dependent negative feedback mechanism that prevents aberrant activation of PI3K/Akt network in response to insulin/IGF-1, leading to Akt/mTORC1 activity up-regulation [54]. Treatments with TORKIs more effectively block mTORC1 substrate phosphorylation relative to rapalogs and also inhibit mTORC2 activity, thus attenuating the Akt pathway and reducing the unwanted upregulation of PI3K pathway. Different pharmacological combinations involving mTOR inhibitors, dual PI3K/mTOR inhibitors and other drugs are used to overcome possible other resistance mechanisms such as the upregulation of the Ras/MAPK/ERK pathway or Receptor Tyrosine Kinase (RTK) overexpression. Targeting BCL-2 proteins represents a direct approach for apoptosis subsequently to mTOR inhibition, therefore modulating the anti-apoptotic components (BCL-2 or BCL-XL) and proapoptotic sensors (BAD or PUMA), activators (BIM) or effectors (BAX) [140]. Analysis of structurally distinct TORKIs in B-ALL reported that mTOR inhibition was capable to induce apoptosis when compared to rapamycin [141,142]. The main challenge will be to achieve the most advantageous drug combinations targeting multiple key survival pathways, at the same time selective for cancer cells but with little or no side toxicity that at present constitutes a major concern.

5.3.1. Targeting mTOR in Ph+ B-ALL and in Ph-like B-ALL

The discovery of TKIs has led to significant improvement in the treatment of Chronic Myeloid Leukemia (CML) and Ph^+ ALL [143–146]. However, despite the implementation of TKI for the treatment of Ph^+ B-ALL, survival outcomes still remain poor compared to Ph^-B-ALL [124]. The combination of Imatinib (IM, first-generation TKI) with standard chemotherapy or with allogeneic hematopoietic stem cell transplantation (HSCT) has significantly ameliorated the survival of Ph+ ALL [147–149]. More recently, second- or third-generation TKI (dasatinib, nilotinib, bosutinib and ponatinib) have been used as first-line treatment in Ph^+ B-ALL with positive outcomes [150,151]. In Ph^+ B-ALL, the BCR-ABL fusion gene directly activates the mTOR network, that could represent a mechanism of disease resistance to TKI therapy [152]. More than 50 types of mutations in the BCR–ABL fusion gene have been identified including Y253H, E255K, M351T, G250E and T315I [146]. Targeting the signalling pathway downstream from BCR-ABL, rapamycin could circumvent imatinib resistance in cells carrying the T315I mutation. This mutation usually confers resistance to all first-and second-generation TKIs, except to the third-generation TKI Ponatinib (AP24534), that represents the treatment of choice for CML [146,153] and Ph^+ ALL [154,155]. Ponatinib has also potent activity against FLT3, that confers resistance to imatinib, nilotinib and dasatinib [156]. The efficacy of ponatinib versus imatinib is actually in clinical studies for Ph^+ ALL, besides another study focused on the effects of ponatinib with the monoclonal antibody Blinatumomab in Ph^+ and Ph^- ALL (see www.clinicaltrials.gov: NCT03263572). Martinelli et al. published a study revealing a promising activity of ponatinib in patients affected by CML and positive for T315I [157]. Its efficacy is better documented in association with other molecules, such as blinatumomab and the WEE1 inhibitor AZD-1775, in patients with relapsed/refractory Ph^+ disease and in T-ALL cell lines [158–160]. Imatinib could cause an abnormal activation of the mTOR pathway, leading to treatment resistance. The addition of rapamycin to imatinib mesylate overcame this effect in Ph^+ B-ALL and induced apoptosis [161]. Moreover, rapamycin could potentiate the proliferation inhibition induced by daunorubicin in Ph^+ B-ALL cells and primary samples and at the same time eliminated the abnormal effect of daunorubicin to aberrantly upregulate mTORC1 signalling [162]. Therefore, there is a rationale also in using mTOR inhibitors for this ALL subtype and this may be a mechanism of improving outcomes in Ph^+ B-ALL. The importance and efficacy of the co-treatment of allosteric mTOR inhibitors with conventional chemotherapy or with TKI therapy has also been highlighted by Kuwatsuka et al., whose study demonstrated that RAD001 overcame resistance to imatinib by targeting in vitro and in vivo a mostly quiescent Ph^+ B-ALL cell subset ($CD34^+$/$CD38^-$) [75]. RAD001/imatinib co-treatment induced in vitro apoptosis of $CD34^+$/$CD38^-$ cells more selectively than RAD001 alone and the treatment was more effective in reducing Mcl-1 expression than either drug alone. Co-treatment with RAD001 can therefore overcome resistance to imatinib in Ph^+ B-ALL leukemic stem cells (LSCs), introducing more effective therapeutic treatments aimed to lower the number of patients who relapse after TKI treatment. PI-103, BEZ235 and PKI-587 (Gedatolisib) have also been employed in pre-clinical models of B-ALL subtypes [10]. PI-103 was more effective than rapamycin in suppressing proliferation of Ph^+ B-ALL leukemia cells treated with imatinib [148], both in mouse pre-B ALL and human $CD19^+$ $CD34^+$ Ph^+ ALL cells. BEZ235 was reported to induce apoptosis in Ph^+ B-ALL nilotinib-resistant cells, leading to a marked downregulation of the anti-apoptotic MDM2 protein (or human homolog of the murine double minute-2) [163]. As for the dual PI3K/mTOR inhibitor PKI-587 (Gedatolisib), it displayed antitumoral activity in childhood B-ALL patient-derived xenograft models having various Ph-like genomic alterations [84]. Regarding TORKIs, actually only PP-242 and MLN0128 (known also as INK128) have been tested in Ph^+ ALL models. PP242 displayed more significant cytotoxic effects and a more complete inhibition of mTORC1 in combination with Imatinib in Ph+ SUP-B15 cells, with a marked up-regulation of the apoptosis associated proteins (Bax and cleaved caspase-3) [150]. In preclinical models of paediatric and adult Ph^+ B-ALL but also in Ph^- B-ALL, MLN0128 suppressed proliferation and increased the efficacy of the second generation TKI Dasatinib, supporting the hypothesis for potential clinical analysis of this TORKI. Moreover, in in vivo models, this inhibitor displayed a low toxicity [164]. Further studies are needed to highlight the importance of TORKIs for a more precise and personalized treatment in Ph+ B-ALL models, with protocols that also involve more than one inhibitor targeting different signalling cascades. Beagle et al. demonstrated, for example, the benefit of using histone deacetylases inhibitors (HDACi) in combination with TORKIs. Indeed, in Ph^+ and Ph^- B-ALL and in primary pediatric B-ALL, the cytotoxic role of TORKIs can be augmented by the HDACi vorinostat or panobinostat, with a resulting increased expression of pro-death genes and transcription factors [97]. The high risk pediatric Ph-like B-ALL cohort, that suffers high rates of relapse and mortality, frequently displays a panel of genetic rearrangements in the cytokine receptor like factor 2 (CLRF2), JAK 1/2/3, IL-7R, FLT3 or platelet-derived growth factor receptor-β (PDGFRB) [165]. All of these aberrations have the potential to modulate the PI3K/Akt/mTOR network leading to its aberrant activation. It has been shown that, in xenograft models with and without CRLF2 and JAK genetic lesions, rapamycin reduced leukemia blasts, prolonging also survival [166]. Gotesman et al. further gave relevance to mTORC1 inhibition combining TORKIs with dasatinib in ABL-rearranged Ph-like B-ALL. The combination was more effective than TKI alone against patient-derived Ph-like B-ALL cells, suggesting a rationale for clinical testing of TKI associations with TORKIs in pediatric and adults Ph-like B-ALL patients and new therapeutic strategies in this poor prognosis subtype of B-ALL [143].

5.3.2. Targeting mTOR in Ph^- B-ALL

Evidence suggests that B-cell receptor (BCR) plays an important role not only in Ph^+ B-ALL but also in Ph^- B-ALL, where several molecules, such as IL-7, modulate survival and cell death mechanisms [167]. Indeed, the precursor-B-cell receptor (pre-BCR) activation depends on different signals that are required to initiate several aberrant cellular processes in pre-B cells, such as abnormal proliferation. Prior to become a functional mature B cell, B cell progenitors must successfully proceed through several checkpoints that ensure in mature B cells the expression of functional immunoglobulin receptors capable of recognizing a wide-array of antigens [132]. Pre-BCR receptor ensures pre-B cells differentiation into mature B-cells. Therefore, pre-BCR acts as a checkpoint in B-cell development and is involved in the recombination of light chain gene IgL through the termination of surrogate light chains (SLC) expression [168]. Together with BCR, mature B cells development is strictly correlated to the activation of the receptor for the tumour necrosis factor (TNF) family cytokine, BAFF, that signals mainly through NF-κB pathway [169]. In Ph^+ALL, the oncogenic fusion BCR-ABL stimulates some pre-BCR downstream effectors, such as Bruton's tyrosine kinase (BTK), the transcription regulator protein BACH2 and B Cell CLL/Lymphoma 6 (BCL6) [138,170]. In Ph-B-ALL the number of cases displaying high expression of BCL6 protein and a constitutively active BCR signalling is around 13% [171]. Several studies reported the efficacy of different pharmacological combinations in inducing downregulation of PI3K/Akt signalling, as an arm of BCR pathway. In $Pre\text{-}BCR^+$ ALL cells the BTK inhibitor ibrutinib induced the suppression of some pre-BCR signalling negative regulators, inhibited the phosphorylation of PI3K/Akt network and its substrates, reduced BCL6 levels and synergized with glucocorticoids. Moreover, it induced apoptosis and prolonged survival in $Pre\text{-}BCR^+$ ALL mouse models [172]. On a panel of B-ALL, the Akt inhibition and the reduction of phosphorylation of its downstream target glycogen synthase kinase 3β (GSK3β) have been detected after treatment for 48h with the PKCβ selective inhibitor enzastaurin (ENZ). PKCβ represents a key mediator of BCR and pre-BCR signalling. The reduction of activated GSK3β correlated with an abnormal accumulation of β-catenin [173]. β-catenin, that belongs to the Wnt/β-catenin signalling, is involved in stem cell abilities to self-renew and is implicated in growth and drug resistance of B-ALL cells. Wnt was reported to inhibit mTORC1 by inhibiting GSK3β, positive regulator of the TSC complex [132]. In B-cell precursor (BCP) ALL with the TCF3-PBX1 (E2A-PBX1) gene fusion, the PI3K delta (p110δ) inhibitor idelalisib represents a promising pharmacological approach for this subtype, among a panel of 302 investigational and approved anti-neoplastic drugs. The idelalisib insensitive 697 BCP-ALL cell line harbours an activating NRAS mutation, which may cause resistance to p110δ inhibition [174]. Also an aberrant expression of C-X-C chemokine receptor type 4 (CXCR4), may influence additional drug sensitivity of this cell subtype. Concerning mTOR inhibition, Rapamycin, CCI-779 and RAD001 have been used also in pre-clinical models of Ph^- B-ALL. Rapamycin induced apoptosis and exerted anti leukemic effects in the pre-B ALL cell line T309 and the treatment, in vivo, with the mTOR inhibitor in transgenic mice displayed a reduction in nodal masses and a prolonged survival. Interestingly, the inhibitory effects of rapamycin could be reversed by IL-7. This suggests an important role of this cytokine in the control of mTOR activity in B-ALL cells [175]. Moreover, rapamycin has been reported to be synergistic with focal adhesion kinase (FAK) down-regulation in REH cells with significant down-regulation of cell growth, cell cycle and apoptosis [7]. CCI-779 significantly decreased survival and induced apoptosis of lymphoblasts from Ph^- B-ALL adult patients co-cultured with bone marrow stromal cells. This drug was also effective in vivo in a NOD/SCID xenograft model, where inhibition of mTORC1 showed a significant reduction in peripheral-blood blasts and splenomegaly [81]. Also RAD001 has proven its efficacy, especially in models of Ph^- pediatric B-ALL. RAD001 is able to synergize with conventional chemotherapy (i.e., vincristine) or novel agents (i.e., bortezomib) both in vitro and in vivo, with increased caspase-dependent but p53-independent cell killing [176,177]. In further several studies it has also been shown that this inhibitor induces autophagy [178] and caspase-independent programmed cell death [179]. Our group recently reported the synergistic effect of RAD001, in both Ph^- B-ALL cell lines and patient samples, with MK-2206, a specific, potent and orally bioavailable allosteric Akt inhibitor that targets both its catalytic and PH domains [180]. BEZ235 has also shown to exert anti-proliferative activity in Ph^- B-ALL cells [95], as well as the PI3K/mTOR inhibitor BTG226, that showed a more powerful effect than BEZ235 [95]. Both drugs, however, inhibited the proliferation of Ph^- B-ALL cell lines with a three log greater potency than RAD001 alone. It has also been recently reported that BEZ235 synergizes with the Bcl-2 inhibitor, GX15-070 (Obatoclax), in Ph^+ and Ph^- primary B-ALL cells [181], representing a potent approach to counteract growth and survival of ALL cells. Regarding TORKIs, our group has recently documented that Torin-2, displayed cytotoxicity to a panel of Ph^- B-ALL cell lines and it was found that the drug as a single agent was able to suppress feedback activation of PI3K/Akt, whereas RAD001 needed the addition of MK-2206 to show the same effect [142]. Table 2 reports a summary of the main mTOR inhibitors in B-ALL models. An effective target inhibition coupled with safety characterization of the targeted drugs could help to better identify therapeutic responses with limited adverse side effects.

Table 1. mTOR inhibitors used alone or in association in T-ALL.

Drug	Drug Target	Reported Synergism	Clinical Trials	Reference(s)
Rapamycin (Sirolimus)	mTORC1	• Doxorubicin • Janus kinase inhibitor • ABL protein inhibitors • Focal Adhesion Kinase (FAK) inhibitor • Cyclin D3 and CDK4/6 inhibitors • Cyclophosphamide • Methotrexate • γ-secretase inhibitors • CCI-779	NCT00968253 NCT01184885	[80,82,84,86–90,96,105–107]
CCI-779 (Temsirolimus)	mTORC1	Doxorubicin	NCT01614197 NCT01403415	[80–82,96,105,142,182]
RAD001 (Everolimus)	mTORC1	• LEE-01 • Glucocorticoids	NCT03328104	[80,82,89,91,92,96,97,182]
PKI-587 (Gedatolisib)	PI3K/mTOR	-	-	[6,83,84]
BEZ235	PI3K/mTOR	• cytarabine (AraC) • Doxorubicin • Dexamethasone	-	[6,84,111–114]
AZD8055	mTORC1/mTORC2	PP-242	-	[79,85,115]
OSI-027	mTORC1/mTORC2	-	-	[79,85,116]

Table 2. mTOR inhibitors used alone or in association in B-ALL.

Drug/Cells	Drug Target	Reported Synergism	Clinical Trials	Reference(s)
Rapamycin (Sirolimus) (Ph^+ B-ALL, Ph^- B-ALL, Ph-like B-ALL)	mTORC1	• Imatinib mesylate • Daunorubicin • Focal adhesion kinase (FAK) inhibitor • Methotrexate • 6-mercaptopurine	NCT01184885 NCT00792948	[80,82,154,155,171]
CCI-779 (Temsirolimus) (Ph^- B-ALL)	mTORC1	-	NCT01614197 NCT01403415	[80–82,96,144,182]
RAD001 (Everolimus) (Ph^+ B-ALL, Ph^- B-ALL)	mTORC1	• Vincristine • Bortezomib • MK2206 • LEE-01 • (CDK4/6 inhibitor) • Glucocorticoids	NCT01523977	[5,75,80,82,89,96,176–181,183]
PI-103 (Ph^+ B-ALL)	PI3K/mTOR	Imatinib	-	[6,10,77,127,148]
PKI-587 (Gedatolisib) (Ph^+ B-ALL, Ph^- B-ALL, Ph-like B-ALL)	PI3K/mTOR	-	-	[6,76,77,136]
BEZ235 (Ph^+ B-ALL, Ph^- B-ALL)	PI3K/mTOR	• Nilotinib • GX15-070 • Methotrexate • 6-mercaptopurine	-	[6,10,84,95,162,181]
BGT226 (Ph^- B-ALL)	PI3K/mTOR	-	-	[8,100–102]
PP-242 (Ph^+ B-ALL)	mTORC1/mTORC2	-	-	[80,85,150]
Torin-2 (B-pre-ALL)	mTORC1/mTORC2	MK2206	-	[80,85,180]
MLN0128 (Ph^+ B-ALL, Ph^- B-ALL)	mTORC1/mTORC2	Dasatinib	-	[80,85,164]

6. Clinical Trials

Actually some clinical trials are performed in both T- and B-ALL, some of them in completed status with results, others in recruiting phase. Among the different clinical trials, a Phase I/II study reported the results of RAD001 or CCI-779 in combination with chemotherapy for treatment of ALL patients with relapse episodes. The combination of RAD001 with Hyper-CVAD (Hyperfractionated Cyclophosphamide, Vincristine, Doxo and Dexamethasone) high-intensity chemotherapy in B-lineage or T-lineage acute leukemia patients significantly inhibited the phosphorylation of S6RP. Of note, the combination of Hyper-CVAD and RAD001 did not induce a relevant increased toxicity, compared with Hyper-CVAD alone (see www.clinicaltrials.gov: NCT00968253) [96]. In an early Phase I pilot study also rapamycin was given in combination with Hyper-CVAD in adults with B- and T-ALL and other aggressive lymphoid malignancies, with the aim to assess the feasibility, safety and toxicity of the drug combination (see www.clinicaltrials.gov: NCT01614197). Rapamycin efficacy is also under evaluation in combination chemotherapy with or without Donor Stem Cell Transplant in adult patients affected by Ph^{+} B-ALL (see www.clinicaltrials.gov: NCT00792948). The toxicity of CCI-779 in combination with dexamethasone, cyclophosphamide and etoposide is also under evaluation in children with relapsed ALL and in patients affected by lymphoma. In this clinical trial (see www.clinicaltrials.gov: NCT01614197) different analytical tests aimed on dose limiting toxicity (DLT) measurement or on rate of remission assessment will be performed. MRD and in general CCI-779 effect on glucocorticoid resistance and on the mTOR inhibition will be also evaluated. In Another Phase I trial CCI-779 was tested in combination with UKALL R3 re-induction protocol. Sixteen T- or B-ALL patients participated in the study. Unfortunately, the addition of CCI-779 to reinduction chemotherapy resulted in consistent toxicity, without disease reduction. (see www.clinicaltrials.gov: NCT01403415) [182]. In a third Phase I trial study, the activity of RAD001 was evaluated in combination with different drugs and conventional chemotherapy in childhood ALL having a first relapse at bone marrow level (see www.clinicaltrials.gov: NCT01523977) [183]. More than 20 patients were involved in the study. The drug combination was well tolerated and correlated with low level of MRD re-induction, thus enhancing further evaluations in the clinic for this combination efficacy. Further trials are ongoing for the evaluation of rapalogs efficacy (see for example www.clinicaltrials.gov: NCT03328104). BEZ235 is also under clinical evaluation, especially in patients with relapsed ALL. This study is currently in active status and the main objective is the identification of the patient MTD and BEZ235 DLT when administered twice daily. Moreover, changes in PI3K/Akt/mTOR network molecules in bone marrow are being analysed (see for example www.clinicaltrials.gov: NCT01756118). The selection of effective chemotherapy drugs that, in combination with targeted mTOR SMI, could be quite well tolerated with minimal toxicity is a priority for ALL treatment, with the aim to define accurate pharmacological protocols and to optimize the drug doses to alleviate adverse effects.

7. Conclusions

The mTOR signalling pathway is physiologically upregulated in different cellular mechanisms and can be aberrantly activated in human pathologies and in particular in tumours. The key role of mTOR in the early stages of leukemia and cell drug resistance has been documented in many scientific works [6,24,184].

The field of mTOR targeted therapies has reported a rapid progression over the past few decades and several mTOR network inhibitors have been analysed, with their own inhibitory activities and profiles on their levels of toxicity. However, the advances in knowledge on mTOR drugs is unfortunately still quite limited. Therefore, the challenge ahead is to understand the most suitable inhibitor that could exert a considerable efficacy with minor toxicity in each patient affected by cancer, in this case by ALL. Modern techniques such as kinase activity profiling [185] or next-generation sequencing analysis [186] will help to enter more specifically in the molecular aspects of signal transduction, pointing out the most appropriate druggable molecules or gene mutations and therefore to define the percentage of ALL patients that can benefit from drug-sensitiveness or drug-resistance.

A better overview of the inhibitory and anti-proliferative effects of targeted inhibitors and their role on the complexity and diversity of the immune context of the tumour microenvironment could further improve therapy response [187].

The advances in the comprehension of the biological activities and of the impact that mTOR performs in ALL could give a significant contribution to the advancement of new therapeutic strategies, aimed to inhibit mTOR, for ameliorating ALL patient outcomes.

Funding: This work was supported by current research funds Fondo di Ateneo per la Ricerca Scientifica (FAR) and Fondo per l'Incentivazione alla Ricerca (FIR) (G.Z. and L.M.N), Fondazione Di Bella Onlus (L.M.N.).

Conflicts of Interest: The authors declare no conflict of interest.

References

1. Wullschleger, S.; Loewith, R.; Hall, M.N. TOR signaling in growth and metabolism. *Cell* **2006**, *124*, 471–484. [CrossRef] [PubMed]
2. Carneiro, B.A.; Kaplan, J.B.; Altman, J.K.; Giles, F.J.; Platanias, L.C. Targeting mTOR signaling pathways and related negative feedback loops for the treatment of acute myeloid leukemia. *Cancer Biol. Ther.* **2015**, *16*, 648–656. [CrossRef] [PubMed]
3. Gentzler, R.D.; Altman, J.K.; Platanias, L.C. An overview of the mTOR pathway as a target in cancer therapy. *Expert Opin. Ther. Targets* **2012**, *16*, 481–489. [CrossRef] [PubMed]
4. Evangelisti, C.; Chiarini, F.; Lonetti, A.; Buontempo, F.; Bressanin, D.; Cappellini, A.; Orsini, E.; McCubrey, J.A.; Martelli, A.M. Therapeutic potential of targeting mTOR in T-cell acute lymphoblastic leukemia (review). *Int. J. Oncol.* **2014**, *45*, 909–918. [CrossRef] [PubMed]
5. Neri, L.M.; Cani, A.; Martelli, A.M.; Simioni, C.; Junghanss, C.; Tabellini, G.; Ricci, F.; Tazzari, P.L.; Pagliaro, P.; McCubrey, J.A.; et al. Targeting the PI3K/Akt/mTOR signaling pathway in B-precursor acute lymphoblastic leukemia and its therapeutic potential. *Leukemia* **2014**, *28*, 739–748. [CrossRef] [PubMed]
6. Martelli, A.M.; Chiarini, F.; Evangelisti, C.; Cappellini, A.; Buontempo, F.; Bressanin, D.; Fini, M.; McCubrey, J.A. Two hits are better than one: Targeting both phosphatidylinositol 3-kinase and mammalian target of rapamycin as a therapeutic strategy for acute leukemia treatment. *Oncotarget* **2012**, *3*, 371–394. [CrossRef] [PubMed]
7. Shi, P.J.; Xu, L.H.; Lin, K.Y.; Weng, W.J.; Fang, J.P. Synergism between the mTOR inhibitor rapamycin and FAK down-regulation in the treatment of acute lymphoblastic leukemia. *J. Hematol. Oncol.* **2016**, *9*, 12. [CrossRef]
8. Simioni, C.; Ultimo, S.; Martelli, A.M.; Zauli, G.; Milani, D.; McCubrey, J.A.; Capitani, S.; Neri, L.M. Synergistic effects of selective inhibitors targeting the PI3K/AKT/mTOR pathway or NUP214-ABL1 fusion protein in human Acute Lymphoblastic Leukemia. *Oncotarget* **2016**, *7*, 79842. [CrossRef]
9. Terwilliger, T.; Abdul-Hay, M. Acute lymphoblastic leukemia: A comprehensive review and 2017 update. *Blood Cancer J.* **2017**, *7*, e577. [CrossRef]
10. Simioni, C.; Martelli, A.M.; Zauli, G.; Vitale, M.; McCubrey, J.A.; Capitani, S.; Neri, L.M. Targeting the phosphatidylinositol 3-kinase/Akt/mechanistic target of rapamycin signaling pathway in B-lineage acute lymphoblastic leukemia: An update. *J. Cell Physiol.* **2018**, *233*, 6440–6454. [CrossRef]
11. Evangelisti, C.; Cappellini, A.; Oliveira, M.; Fragoso, R.; Barata, J.T.; Bertaina, A.; Locatelli, F.; Simioni, C.; Neri, L.M.; Chiarini, F.; et al. Phosphatidylinositol 3-kinase inhibition potentiates glucocorticoid response in B-cell acute lymphoblastic leukemia. *J. Cell Physiol.* **2018**, *233*, 1796–1811. [CrossRef]
12. Mendes, R.D.; Cante-Barrett, K.; Pieters, R.; Meijerink, J.P. The relevance of PTEN-AKT in relation to NOTCH1-directed treatment strategies in T-cell acute lymphoblastic leukemia. *Haematologica* **2016**, *101*, 1010–1017. [CrossRef]
13. Tasian, S.K.; Teachey, D.T.; Rheingold, S.R. Targeting the PI3K/mTOR Pathway in Pediatric Hematologic Malignancies. *Front. Oncol.* **2014**, *4*, 108. [CrossRef] [PubMed]
14. Wei, G.; Wang, J.; Huang, H.; Zhao, Y. Novel immunotherapies for adult patients with B-lineage acute lymphoblastic leukemia. *Int. J. Hematol.* **2017**, *10*, 150. [CrossRef]
15. Portell, C.A.; Advani, A.S. Novel targeted therapies in acute lymphoblastic leukemia. *Leuk. Lymphoma* **2014**, *55*, 737–748. [CrossRef]

16. Ribera, J.M. Efficacy and safety of bispecific T-cell engager blinatumomab and the potential to improve leukemia-free survival in B-cell acute lymphoblastic leukemia. *Expert Rev. Hematol.* **2017**, *10*, 1057–1067. [CrossRef] [PubMed]
17. Frankel, S.R.; Baeuerle, P.A. Targeting T cells to tumor cells using bispecific antibodies. *Curr. Opin. Chem. Biol.* **2013**, *17*, 385–392. [CrossRef]
18. Tvito, A.; Rowe, J.M. Inotuzumab ozogamicin for the treatment of acute lymphoblastic leukemia. *Exp. Opin. Biol. Ther.* **2017**, *17*, 1557–1564. [CrossRef] [PubMed]
19. Leyfman, Y. Chimeric antigen receptors: Unleashing a new age of anti-cancer therapy. *Cancer Cell Int.* **2018**, *18*, 182. [CrossRef] [PubMed]
20. Qasim, W. Allogeneic CAR T cell therapies for leukemia-R1. *Am. J. Hematol* **2019**. [CrossRef]
21. O'Leary, M.C.; Lu, X.; Huang, Y.; Lin, X.; Mahmood, I.; Przepiorka, D.; Gavin, D.; Lee, S.; Liu, K.; George, B.; et al. FDA Approval Summary: Tisagenlecleucel for Treatment of Patients with Relapsed or Refractory B-Cell Precursor Acute Lymphoblastic Leukemia. *Clin. Cancer Res.* **2018**. [CrossRef]
22. Wang, X.; Chu, Y.; Wang, W.; Yuan, W. mTORC signaling in hematopoiesis. *Int. J. Hematol.* **2016**, *103*, 510–518. [CrossRef]
23. Kaur, A.; Sharma, S. Mammalian target of rapamycin (mTOR) as a potential therapeutic target in various diseases. *Inflammopharmacology* **2017**, *25*, 293–312. [CrossRef]
24. Mirabilii, S.; Ricciardi, M.R.; Piedimonte, M.; Gianfelici, V.; Bianchi, M.P.; Tafuri, A. Biological Aspects of mTOR in Leukemia. *Int. J. Mol. Sci.* **2018**, *19*, 2396. [CrossRef]
25. Conciatori, F.; Ciuffreda, L.; Bazzichetto, C.; Falcone, I.; Pilotto, S.; Bria, E.; Cognetti, F.; Milella, M. mTOR Cross-Talk in Cancer and Potential for Combination Therapy. *Cancers* **2018**, *10*, 23. [CrossRef]
26. Harachi, M.; Masui, K.; Okamura, Y.; Tsukui, R.; Mischel, P.S.; Shibata, N. mTOR Complexes as a Nutrient Sensor for Driving Cancer Progression. *Int. J. Mol. Sci.* **2018**, *19*, 3267. [CrossRef]
27. Kim, L.C.; Cook, R.S.; Chen, J. mTORC1 and mTORC2 in cancer and the tumor microenvironment. *Oncogene* **2017**, *36*, 2191–2201. [CrossRef]
28. Hara, K.; Maruki, Y.; Long, X.; Yoshino, K.; Oshiro, N.; Hidayat, S.; Tokunaga, C.; Avruch, J.; Yonezawa, K. Raptor, a binding partner of target of rapamycin (TOR), mediates TOR action. *Cell* **2002**, *110*, 177–189. [CrossRef]
29. Jacinto, E.; Facchinetti, V.; Liu, D.; Soto, N.; Wei, S.; Jung, S.Y.; Huang, Q.; Qin, J.; Su, B. SIN1/MIP1 maintains rictor-mTOR complex integrity and regulates Akt phosphorylation and substrate specificity. *Cell* **2006**, *127*, 125–137. [CrossRef]
30. Sancak, Y.; Thoreen, C.C.; Peterson, T.R.; Lindquist, R.A.; Kang, S.A.; Spooner, E.; Carr, S.A.; Sabatini, D.M. PRAS40 is an insulin-regulated inhibitor of the mTORC1 protein kinase. *Mol. Cell* **2007**, *25*, 903–915. [CrossRef]
31. Rabanal-Ruiz, Y.; Otten, E.G.; Korolchuk, V.I. mTORC1 as the main gateway to autophagy. *Essays Biochem.* **2017**, *61*, 565–584. [CrossRef]
32. Zhang, Y.; Manning, B.D. mTORC1 signaling activates NRF1 to increase cellular proteasome levels. *Cell Cycle* **2015**, *14*, 2011–2017. [CrossRef]
33. Laplante, M.; Sabatini, D.M. mTOR signaling in growth control and disease. *Cell* **2012**, *149*, 274–293. [CrossRef]
34. Ma, X.M.; Blenis, J. Molecular mechanisms of mTOR-mediated translational control. *Nat. Rev. Mol. Cell Biol.* **2009**, *10*, 307–318. [CrossRef]
35. Nojima, H.; Tokunaga, C.; Eguchi, S.; Oshiro, N.; Hidayat, S.; Yoshino, K.; Hara, K.; Tanaka, N.; Avruch, J.; Yonezawa, K. The mammalian target of rapamycin (mTOR) partner, raptor, binds the mTOR substrates p70 S6 kinase and 4E-BP1 through their TOR signaling (TOS) motif. *J. Biol. Chem.* **2003**, *278*, 15461–15464. [CrossRef]
36. Schalm, S.S.; Fingar, D.C.; Sabatini, D.M.; Blenis, J. TOS motif-mediated raptor binding regulates 4E-BP1 multisite phosphorylation and function. *Curr. Biol.* **2003**, *13*, 797–806. [CrossRef]
37. Laplante, M.; Sabatini, D.M. Regulation of mTORC1 and its impact on gene expression at a glance. *J. Cell Sci.* **2013**, *126*, 1713–1719. [CrossRef]
38. Toschi, A.; Lee, E.; Gadir, N.; Ohh, M.; Foster, D.A. Differential dependence of hypoxia-inducible factors 1 alpha and 2 alpha on mTORC1 and mTORC2. *J. Biol. Chem.* **2008**, *283*, 34495–34499. [CrossRef]

39. Settembre, C.; Zoncu, R.; Medina, D.L.; Vetrini, F.; Erdin, S.; Huynh, T.; Ferron, M.; Karsenty, G.; Vellard, M.C.; Facchinetti, V.; et al. A lysosome-to-nucleus signalling mechanism senses and regulates the lysosome via mTOR and TFEB. *EMBO J.* **2012**, *31*, 1095–1108. [CrossRef]
40. Nnah, I.C.; Wang, B.; Saqcena, C.; Weber, G.F.; Bonder, E.M.; Bagley, D.; De Cegli, R.; Napolitano, G.; Medina, D.L.; Ballabio, A.; et al. TFEB-driven endocytosis coordinates MTORC1 signaling and autophagy. *Autophagy* **2019**, *15*, 151–164. [CrossRef]
41. Hong, Z.; Pedersen, N.M.; Wang, L.; Torgersen, M.L.; Stenmark, H.; Raiborg, C. PtdIns3P controls mTORC1 signaling through lysosomal positioning. *J. Cell Biol.* **2017**, *216*, 4217–4233. [CrossRef]
42. Klionsky, D.J.; Abdelmohsen, K.; Abe, A.; Abedin, M.J.; Abeliovich, H.; Acevedo Arozena, A.; Adachi, H.; Adams, C.M.; Adams, P.D.; Adeli, K.; et al. Guidelines for the use and interpretation of assays for monitoring autophagy (3rd edition). *Autophagy* **2016**, *12*, 1–222. [CrossRef]
43. Ma, X.; Zhang, S.; He, L.; Rong, Y.; Brier, L.W.; Sun, Q.; Liu, R.; Fan, W.; Chen, S.; Yue, Z.; et al. MTORC1-mediated NRBF2 phosphorylation functions as a switch for the class III PtdIns3K and autophagy. *Autophagy* **2017**, *13*, 592–607. [CrossRef]
44. Wong, P.M.; Feng, Y.; Wang, J.; Shi, R.; Jiang, X. Regulation of autophagy by coordinated action of mTORC1 and protein phosphatase 2A. *Nat. Commun.* **2015**, *6*, 8048. [CrossRef]
45. Codogno, P.; Meijer, A.J. Autophagy and signaling: Their role in cell survival and cell death. *Cell Death Differ.* **2005**, *12* (Suppl. 2), 1509–1518. [CrossRef]
46. Klionsky, D.J.; Meijer, A.J.; Codogno, P. Autophagy and p70S6 kinase. *Autophagy* **2005**, *1*, 59–60. [CrossRef]
47. Paquette, M.; El-Houjeiri, L.; Pause, A. mTOR Pathways in Cancer and Autophagy. *Cancers* **2018**, *10*, 18. [CrossRef]
48. Zou, Z.; Chen, J.; Yang, J.; Bai, X. Targeted Inhibition of Rictor/mTORC2 in Cancer Treatment: A New Era after Rapamycin. *Curr. Cancer Drug Targets* **2016**, *16*, 288–304. [CrossRef]
49. Hoshii, T.; Matsuda, S.; Hirao, A. Pleiotropic roles of mTOR complexes in haemato-lymphopoiesis and leukemogenesis. *J. Biochem.* **2014**, *156*, 73–83. [CrossRef]
50. Gkountakos, A.; Pilotto, S.; Mafficini, A.; Vicentini, C.; Simbolo, M.; Milella, M.; Tortora, G.; Scarpa, A.; Bria, E.; Corbo, V. Unmasking the impact of Rictor in cancer: Novel insights of mTORC2 complex. *Carcinogenesis* **2018**, *39*, 971–980. [CrossRef]
51. Sarbassov, D.D.; Ali, S.M.; Sengupta, S.; Sheen, J.H.; Hsu, P.P.; Bagley, A.F.; Markhard, A.L.; Sabatini, D.M. Prolonged rapamycin treatment inhibits mTORC2 assembly and Akt/PKB. *Mol. Cell* **2006**, *22*, 159–168. [CrossRef] [PubMed]
52. Liu, P.; Gan, W.; Chin, Y.R.; Ogura, K.; Guo, J.; Zhang, J.; Wang, B.; Blenis, J.; Cantley, L.C.; Toker, A.; et al. PtdIns(3,4,5)P3-Dependent Activation of the mTORC2 Kinase Complex. *Cancer Discov.* **2015**, *5*, 1194–1209. [CrossRef] [PubMed]
53. Moore, S.F.; Hunter, R.W.; Hers, I. mTORC2 protein complex-mediated Akt (Protein Kinase B) Serine 473 Phosphorylation is not required for Akt1 activity in human platelets [corrected]. *J. Biol. Chem.* **2011**, *286*, 24553–24560. [CrossRef] [PubMed]
54. Martelli, A.M.; Buontempo, F.; McCubrey, J.A. Drug discovery targeting the mTOR pathway. *Clin. Sci.* **2018**, *132*, 543–568. [CrossRef] [PubMed]
55. Zhang, Q.; Wan, M.; Shi, J.; Horita, D.A.; Miller, L.D.; Kute, T.E.; Kridel, S.J.; Kulik, G.; Sui, G. Yin Yang 1 promotes mTORC2-mediated AKT phosphorylation. *J. Mol. Cell Biol.* **2016**, *8*, 232–243. [CrossRef] [PubMed]
56. Zeng, H.; Cohen, S.; Guy, C.; Shrestha, S.; Neale, G.; Brown, S.A.; Cloer, C.; Kishton, R.J.; Gao, X.; Youngblood, B.; et al. mTORC1 and mTORC2 Kinase Signaling and Glucose Metabolism Drive Follicular Helper T Cell Differentiation. *Immunity* **2016**, *45*, 540–554. [CrossRef] [PubMed]
57. Lang, F.; Strutz-Seebohm, N.; Seebohm, G.; Lang, U.E. Significance of SGK1 in the regulation of neuronal function. *J. Physiol.* **2010**, *588*, 3349–3354. [CrossRef] [PubMed]
58. Arias, E.; Koga, H.; Diaz, A.; Mocholi, E.; Patel, B.; Cuervo, A.M. Lysosomal mTORC2/PHLPP1/Akt Regulate Chaperone-Mediated Autophagy. *Mol. Cell* **2015**, *59*, 270–284. [CrossRef] [PubMed]
59. Shi, D.; Niu, P.; Heng, X.; Chen, L.; Zhu, Y.; Zhou, J. Autophagy induced by cardamonin is associated with mTORC1 inhibition in SKOV3 cells. *Pharmacol. Rep.* **2018**, *70*, 908–916. [CrossRef] [PubMed]
60. Moreno-Blas, D.; Gorostieta-Salas, E.; Castro-Obregon, S. Connecting chaperone-mediated autophagy dysfunction to cellular senescence. *Ageing Res. Rev.* **2018**, *41*, 34–41. [CrossRef] [PubMed]

61. Lee, H.K.; Kwon, B.; Lemere, C.A.; de la Monte, S.; Itamura, K.; Ha, A.Y.; Querfurth, H.W. mTORC2 (Rictor) in Alzheimer's Disease and Reversal of Amyloid-beta Expression-Induced Insulin Resistance and Toxicity in Rat Primary Cortical Neurons. *J. Alzheimer's Dis.* **2017**, *56*, 1015–1036. [CrossRef] [PubMed]
62. Guri, Y.; Colombi, M.; Dazert, E.; Hindupur, S.K.; Roszik, J.; Moes, S.; Jenoe, P.; Heim, M.H.; Riezman, I.; Riezman, H.; et al. mTORC2 Promotes Tumorigenesis via Lipid Synthesis. *Cancer Cell* **2017**, *32*, 807–823. [CrossRef] [PubMed]
63. Oh, W.J.; Jacinto, E. mTOR complex 2 signaling and functions. *Cell Cycle* **2011**, *10*, 2305–2316. [CrossRef]
64. Staubert, C.; Bhuiyan, H.; Lindahl, A.; Broom, O.J.; Zhu, Y.; Islam, S.; Linnarsson, S.; Lehtio, J.; Nordstrom, A. Rewired metabolism in drug-resistant leukemia cells: A metabolic switch hallmarked by reduced dependence on exogenous glutamine. *J. Biol. Chem.* **2015**, *290*, 8348–8359. [CrossRef] [PubMed]
65. Styczynski, J.; Wysocki, M.; Debski, R.; Czyzewski, K.; Kolodziej, B.; Rafinska, B.; Kubicka, M.; Koltan, S.; Koltan, A.; Pogorzala, M.; et al. Predictive value of multidrug resistance proteins and cellular drug resistance in childhood relapsed acute lymphoblastic leukemia. *J. Cancer Res. Clin. Oncol.* **2007**, *133*, 875–893. [CrossRef] [PubMed]
66. Vander Heiden, M.G.; Cantley, L.C.; Thompson, C.B. Understanding the Warburg effect: The metabolic requirements of cell proliferation. *Science* **2009**, *324*, 1029–1033. [CrossRef] [PubMed]
67. Simabuco, F.M.; Morale, M.G.; Pavan, I.C.B.; Morelli, A.P.; Silva, F.R.; Tamura, R.E. p53 and metabolism: From mechanism to therapeutics. *Oncotarget* **2018**, *9*, 23780–23823. [CrossRef] [PubMed]
68. Ward, P.S.; Thompson, C.B. Signaling in control of cell growth and metabolism. *Cold Spring Harb. Perspect. Biol.* **2012**, *4*, a006783. [CrossRef] [PubMed]
69. Schriever, S.C.; Deutsch, M.J.; Adamski, J.; Roscher, A.A.; Ensenauer, R. Cellular signaling of amino acids towards mTORC1 activation in impaired human leucine catabolism. *J. Nutr. Biochem.* **2013**, *24*, 824–831. [CrossRef] [PubMed]
70. Altomare, D.A.; Khaled, A.R. Homeostasis and the importance for a balance between AKT/mTOR activity and intracellular signaling. *Curr. Med. Chem.* **2012**, *19*, 3748–3762. [CrossRef] [PubMed]
71. Morita, M.; Gravel, S.P.; Chenard, V.; Sikstrom, K.; Zheng, L.; Alain, T.; Gandin, V.; Avizonis, D.; Arguello, M.; Zakaria, C.; et al. mTORC1 controls mitochondrial activity and biogenesis through 4E-BP-dependent translational regulation. *Cell Metab.* **2013**, *18*, 698–711. [CrossRef]
72. Li, T.Y.; Sun, Y.; Liang, Y.; Liu, Q.; Shi, Y.; Zhang, C.S.; Zhang, C.; Song, L.; Zhang, P.; Zhang, X.; et al. ULK1/2 Constitute a Bifurcate Node Controlling Glucose Metabolic Fluxes in Addition to Autophagy. *Mol. Cell* **2016**, *62*, 359–370. [CrossRef] [PubMed]
73. Zinzalla, V.; Stracka, D.; Oppliger, W.; Hall, M.N. Activation of mTORC2 by association with the ribosome. *Cell* **2011**, *144*, 757–768. [CrossRef] [PubMed]
74. Mavrakis, K.J.; Wolfe, A.L.; Oricchio, E.; Palomero, T.; de Keersmaecker, K.; McJunkin, K.; Zuber, J.; James, T.; Khan, A.A.; Leslie, C.S.; et al. Genome-wide RNA-mediated interference screen identifies miR-19 targets in Notch-induced T-cell acute lymphoblastic leukaemia. *Nat. Cell Biol.* **2010**, *12*, 372–379. [CrossRef] [PubMed]
75. Beesley, A.H.; Firth, M.J.; Ford, J.; Weller, R.E.; Freitas, J.R.; Perera, K.U.; Kees, U.R. Glucocorticoid resistance in T-lineage acute lymphoblastic leukaemia is associated with a proliferative metabolism. *Br. J. Cancer* **2009**, *100*, 1926–1936. [CrossRef] [PubMed]
76. Liu, T.; Kishton, R.J.; Macintyre, A.N.; Gerriets, V.A.; Xiang, H.; Liu, X.; Abel, E.D.; Rizzieri, D.; Locasale, J.W.; Rathmell, J.C. Glucose transporter 1-mediated glucose uptake is limiting for B-cell acute lymphoblastic leukemia anabolic metabolism and resistance to apoptosis. *Cell Death Dis.* **2014**, *5*, e1470. [CrossRef] [PubMed]
77. Slack, C. Ras signaling in aging and metabolic regulation. *Nutr. Healthy Aging* **2017**, *4*, 195–205. [CrossRef] [PubMed]
78. Chan, L.N.; Chen, Z.; Braas, D.; Lee, J.W.; Xiao, G.; Geng, H.; Cosgun, K.N.; Hurtz, C.; Shojaee, S.; Cazzaniga, V.; et al. Metabolic gatekeeper function of B-lymphoid transcription factors. *Nature* **2017**, *542*, 479–483. [CrossRef] [PubMed]
79. Man, L.M.; Morris, A.L.; Keng, M. New Therapeutic Strategies in Acute Lymphocytic Leukemia. *Curr. Hematol. Malig. Rep.* **2017**, *12*, 197–206. [CrossRef]
80. Steelman, L.S.; Martelli, A.M.; Cocco, L.; Libra, M.; Nicoletti, F.; Abrams, S.L.; McCubrey, J.A. The therapeutic potential of mTOR inhibitors in breast cancer. *Br. J. Clin. Pharmacol.* **2016**, *82*, 1189–1212. [CrossRef] [PubMed]

81. Teachey, D.T.; Obzut, D.A.; Cooperman, J.; Fang, J.; Carroll, M.; Choi, J.K.; Houghton, P.J.; Brown, V.I.; Grupp, S.A. The mTOR inhibitor CCI-779 induces apoptosis and inhibits growth in preclinical models of primary adult human ALL. *Blood* **2006**, *107*, 1149–1155. [CrossRef] [PubMed]
82. Zhelev, Z.; Ivanova, D.; Bakalova, R.; Aoki, I.; Higashi, T. Synergistic Cytotoxicity of Melatonin and New-generation Anticancer Drugs Against Leukemia Lymphocytes But Not Normal Lymphocytes. *Anticancer Res.* **2017**, *37*, 149–159. [CrossRef] [PubMed]
83. Gazi, M.; Moharram, S.A.; Marhall, A.; Kazi, J.U. The dual specificity PI3K/mTOR inhibitor PKI-587 displays efficacy against T-cell acute lymphoblastic leukemia (T-ALL). *Cancer Lett.* **2017**, *392*, 9–16. [CrossRef] [PubMed]
84. Tasian, S.K.; Teachey, D.T.; Li, Y.; Shen, F.; Harvey, R.C.; Chen, I.M.; Ryan, T.; Vincent, T.L.; Willman, C.L.; Perl, A.E.; et al. Potent efficacy of combined PI3K/mTOR and JAK or ABL inhibition in murine xenograft models of Ph-like acute lymphoblastic leukemia. *Blood* **2017**, *129*, 177–187. [CrossRef] [PubMed]
85. Carroll, W.L.; Aifantis, I.; Raetz, E. Beating the Clock in T-cell Acute Lymphoblastic Leukemia. *Clin. Cancer Res.* **2017**, *23*, 873–875. [CrossRef] [PubMed]
86. Ballou, L.M.; Lin, R.Z. Rapamycin and mTOR kinase inhibitors. *J. Chem. Biol.* **2008**, *1*, 27–36. [CrossRef] [PubMed]
87. Chen, J.; Zheng, X.F.; Brown, E.J.; Schreiber, S.L. Identification of an 11-kDa FKBP12-rapamycin-binding domain within the 289-kDa FKBP12-rapamycin-associated protein and characterization of a critical serine residue. *Proc. Natl. Acad. Sci. USA* **1995**, *92*, 4947–4951. [CrossRef]
88. Oshiro, N.; Yoshino, K.; Hidayat, S.; Tokunaga, C.; Hara, K.; Eguchi, S.; Avruch, J.; Yonezawa, K. Dissociation of raptor from mTOR is a mechanism of rapamycin-induced inhibition of mTOR function. *Genes Cells* **2004**, *9*, 359–366. [CrossRef]
89. Pikman, Y.; Alexe, G.; Roti, G.; Conway, A.S.; Furman, A.; Lee, E.S.; Place, A.E.; Kim, S.; Saran, C.; Modiste, R.; et al. Synergistic Drug Combinations with a CDK4/6 Inhibitor in T-cell Acute Lymphoblastic Leukemia. *Clin. Cancer Res.* **2017**, *23*, 1012–1024. [CrossRef]
90. Li, H.; Kong, X.; Cui, G.; Ren, C.; Fan, S.; Sun, L.; Zhang, Y.; Cao, R.; Li, Y.; Zhou, J. Rapamycin restores p14, p15 and p57 expression and inhibits the mTOR/p70S6K pathway in acute lymphoblastic leukemia cells. *Int. J. Hematol.* **2015**, *102*, 558–568. [CrossRef]
91. Saunders, P.O.; Weiss, J.; Welschinger, R.; Baraz, R.; Bradstock, K.F.; Bendall, L.J. RAD001 (everolimus) induces dose-dependent changes to cell cycle regulation and modifies the cell cycle response to vincristine. *Oncogene* **2013**, *32*, 4789–4797. [CrossRef] [PubMed]
92. Kuwatsuka, Y.; Minami, M.; Minami, Y.; Sugimoto, K.; Hayakawa, F.; Miyata, Y.; Abe, A.; Goff, D.J.; Kiyoi, H.; Naoe, T. The mTOR inhibitor, everolimus (RAD001), overcomes resistance to imatinib in quiescent Ph-positive acute lymphoblastic leukemia cells. *Blood Cancer J.* **2011**, *1*, e17. [CrossRef] [PubMed]
93. Alameen, A.A.; Simioni, C.; Martelli, A.M.; Zauli, G.; Ultimo, S.; McCubrey, J.A.; Gonelli, A.; Marisi, G.; Ulivi, P.; Capitani, S.; et al. Healthy CD4+ T lymphocytes are not affected by targeted therapies against the PI3K/Akt/mTOR pathway in T-cell acute lymphoblastic leukemia. *Oncotarget* **2016**, *7*, 55690–55703. [CrossRef] [PubMed]
94. Badura, S.; Tesanovic, T.; Pfeifer, H.; Wystub, S.; Nijmeijer, B.A.; Liebermann, M.; Falkenburg, J.H.; Ruthardt, M.; Ottmann, O.G. Differential effects of selective inhibitors targeting the PI3K/AKT/mTOR pathway in acute lymphoblastic leukemia. *PLoS ONE* **2013**, *8*, e80070. [CrossRef] [PubMed]
95. Wong, J.; Welschinger, R.; Hewson, J.; Bradstock, K.F.; Bendall, L.J. Efficacy of dual PI-3K and mTOR inhibitors in vitro and in vivo in acute lymphoblastic leukemia. *Oncotarget* **2014**, *5*, 10460–10472. [CrossRef] [PubMed]
96. Daver, N.; Boumber, Y.; Kantarjian, H.; Ravandi, F.; Cortes, J.; Rytting, M.E.; Kawedia, J.D.; Basnett, J.; Culotta, K.S.; Zeng, Z.; et al. A Phase I/II Study of the mTOR Inhibitor Everolimus in Combination with HyperCVAD Chemotherapy in Patients with Relapsed/Refractory Acute Lymphoblastic Leukemia. *Clin. Cancer Res.* **2015**, *21*, 2704–2714. [CrossRef]
97. Beagle, B.R.; Nguyen, D.M.; Mallya, S.; Tang, S.S.; Lu, M.; Zeng, Z.; Konopleva, M.; Vo, T.T.; Fruman, D.A. mTOR kinase inhibitors synergize with histone deacetylase inhibitors to kill B-cell acute lymphoblastic leukemia cells. *Oncotarget* **2015**, *6*, 2088–2100. [CrossRef]

98. Iacovelli, S.; Ricciardi, M.R.; Allegretti, M.; Mirabilii, S.; Licchetta, R.; Bergamo, P.; Rinaldo, C.; Zeuner, A.; Foa, R.; Milella, M.; et al. Co-targeting of Bcl-2 and mTOR pathway triggers synergistic apoptosis in BH3 mimetics resistant acute lymphoblastic leukemia. *Oncotarget* **2015**, *6*, 32089–32103. [CrossRef]
99. Simioni, C.; Neri, L.M.; Tabellini, G.; Ricci, F.; Bressanin, D.; Chiarini, F.; Evangelisti, C.; Cani, A.; Tazzari, P.L.; Melchionda, F.; et al. Cytotoxic activity of the novel Akt inhibitor, MK-2206, in T-cell acute lymphoblastic leukemia. *Leukemia* **2012**, *26*, 2336–2342. [CrossRef]
100. Girardi, T.; Vereecke, S.; Sulima, S.O.; Khan, Y.; Fancello, L.; Briggs, J.W.; Schwab, C.; de Beeck, J.O.; Verbeeck, J.; Royaert, J.; et al. The T-cell leukemia-associated ribosomal RPL10 R98S mutation enhances JAK-STAT signaling. *Leukemia* **2018**, *32*, 809–819. [CrossRef]
101. Martelli, A.M.; Lonetti, A.; Buontempo, F.; Ricci, F.; Tazzari, P.L.; Evangelisti, C.; Bressanin, D.; Cappellini, A.; Orsini, E.; Chiarini, F. Targeting signaling pathways in T-cell acute lymphoblastic leukemia initiating cells. *Adv. Biol. Regul.* **2014**, *56*, 6–21. [CrossRef] [PubMed]
102. Khanna, A.; Bhushan, B.; Chauhan, P.S.; Saxena, S.; Gupta, D.K.; Siraj, F. High mTOR expression independently prognosticates poor clinical outcome to induction chemotherapy in acute lymphoblastic leukemia. *Clin. Exp. Med.* **2018**, *18*, 221–227. [CrossRef] [PubMed]
103. Cani, A.; Simioni, C.; Martelli, A.M.; Zauli, G.; Tabellini, G.; Ultimo, S.; McCubrey, J.A.; Capitani, S.; Neri, L.M. Triple Akt inhibition as a new therapeutic strategy in T-cell acute lymphoblastic leukemia. *Oncotarget* **2015**, *6*, 6597–6610. [CrossRef] [PubMed]
104. Imbert, V.; Nebout, M.; Mary, D.; Endou, H.; Wempe, M.F.; Supuran, C.T.; Winum, J.Y.; Peyron, J.F. Co-targeting intracellular pH and essential amino acid uptake cooperates to induce cell death of T-ALL/LL cells. *Leuk. Lymphoma* **2018**, *59*, 460–468. [CrossRef] [PubMed]
105. Batista, A.; Barata, J.T.; Raderschall, E.; Sallan, S.E.; Carlesso, N.; Nadler, L.M.; Cardoso, A.A. Targeting of active mTOR inhibits primary leukemia T cells and synergizes with cytotoxic drugs and signaling inhibitors. *Exp. Hematol.* **2011**, *39*, 457–472. [CrossRef] [PubMed]
106. Zhang, Y.; Hua, C.; Cheng, H.; Wang, W.; Hao, S.; Xu, J.; Wang, X.; Gao, Y.; Zhu, X.; Cheng, T.; et al. Distinct sensitivity of CD8+ CD4- and CD8+ CD4+ leukemic cell subpopulations to cyclophosphamide and rapamycin in Notch1-induced T-ALL mouse model. *Leuk. Res.* **2013**, *37*, 1592–1601. [CrossRef] [PubMed]
107. Aliper, A.; Jellen, L.; Cortese, F.; Artemov, A.; Karpinsky-Semper, D.; Moskalev, A.; Swick, A.G.; Zhavoronkov, A. Towards natural mimetics of metformin and rapamycin. *Aging* **2017**, *9*, 2245–2268. [CrossRef]
108. Paganin, M.; Ferrando, A. Molecular pathogenesis and targeted therapies for NOTCH1-induced T-cell acute lymphoblastic leukemia. *Blood Rev.* **2011**, *25*, 83–90. [CrossRef]
109. Hu, Y.; Su, H.; Liu, C.; Wang, Z.; Huang, L.; Wang, Q.; Liu, S.; Chen, S.; Zhou, J.; Li, P.; et al. DEPTOR is a direct NOTCH1 target that promotes cell proliferation and survival in T-cell leukemia. *Oncogene* **2017**, *36*, 1038–1047. [CrossRef]
110. Shepherd, C.; Banerjee, L.; Cheung, C.W.; Mansour, M.R.; Jenkinson, S.; Gale, R.E.; Khwaja, A. PI3K/mTOR inhibition upregulates NOTCH-MYC signalling leading to an impaired cytotoxic response. *Leukemia* **2013**, *27*, 650–660. [CrossRef]
111. Schult, C.; Dahlhaus, M.; Glass, A.; Fischer, K.; Lange, S.; Freund, M.; Junghanss, C. The dual kinase inhibitor NVP-BEZ235 in combination with cytotoxic drugs exerts anti-proliferative activity towards acute lymphoblastic leukemia cells. *Anticancer Res.* **2012**, *32*, 463–474. [PubMed]
112. Hall, C.P.; Reynolds, C.P.; Kang, M.H. Modulation of Glucocorticoid Resistance in Pediatric T-cell Acute Lymphoblastic Leukemia by Increasing BIM Expression with the PI3K/mTOR Inhibitor BEZ235. *Clin. Cancer Res.* **2016**, *22*, 621–632. [CrossRef] [PubMed]
113. Li, Y.; Buijs-Gladdines, J.G.; Cante-Barrett, K.; Stubbs, A.P.; Vroegindeweij, E.M.; Smits, W.K.; van Marion, R.; Dinjens, W.N.; Horstmann, M.; Kuiper, R.P.; et al. IL-7 Receptor Mutations and Steroid Resistance in Pediatric T cell Acute Lymphoblastic Leukemia: A Genome Sequencing Study. *PLoS Med.* **2016**, *13*, e1002200. [CrossRef] [PubMed]
114. Reynolds, C.; Roderick, J.E.; LaBelle, J.L.; Bird, G.; Mathieu, R.; Bodaar, K.; Colon, D.; Pyati, U.; Stevenson, K.E.; Qi, J.; et al. Repression of BIM mediates survival signaling by MYC and AKT in high-risk T-cell acute lymphoblastic leukemia. *Leukemia* **2014**, *28*, 1819–1827. [CrossRef] [PubMed]

115. Murakami, K.; Watanabe, T.; Koike, T.; Kamata, M.; Igari, T.; Kondo, S. Pharmacological properties of a novel and potent gamma-secretase modulator as a therapeutic option for the treatment of Alzheimer's disease. *Brain Res.* **2016**, *1633*, 73–86. [CrossRef] [PubMed]
116. Dastur, A.; Choi, A.; Costa, C.; Yin, X.; Williams, A.; McClanaghan, J.; Greenberg, M.; Roderick, J.; Patel, N.U.; Boisvert, J.; et al. NOTCH1 Represses MCL-1 Levels in GSI-resistant T-ALL, Making them Susceptible to ABT-263. *Clin. Cancer Res.* **2019**, *25*, 312–324. [CrossRef] [PubMed]
117. Yun, S.; Vincelette, N.D.; Knorr, K.L.; Almada, L.L.; Schneider, P.A.; Peterson, K.L.; Flatten, K.S.; Dai, H.; Pratz, K.W.; Hess, A.D.; et al. 4EBP1/c-MYC/PUMA and NF-kappaB/EGR1/BIM pathways underlie cytotoxicity of mTOR dual inhibitors in malignant lymphoid cells. *Blood* **2016**, *127*, 2711–2722. [CrossRef]
118. Martineau, Y.; Azar, R.; Muller, D.; Lasfargues, C.; El Khawand, S.; Anesia, R.; Pelletier, J.; Bousquet, C.; Pyronnet, S. Pancreatic tumours escape from translational control through 4E-BP1 loss. *Oncogene* **2014**, *33*, 1367–1374. [CrossRef]
119. Purizaca, J.; Meza, I.; Pelayo, R. Early lymphoid development and microenvironmental cues in B-cell acute lymphoblastic leukemia. *Arch. Med. Res.* **2012**, *43*, 89–101. [CrossRef]
120. Mishra, D.; Singh, S.; Narayan, G. Role of B Cell Development Marker CD10 in Cancer Progression and Prognosis. *Mol. Biol. Int.* **2016**, *2016*, 4328697. [CrossRef]
121. Lucio, P.; Gaipa, G.; van Lochem, E.G.; van Wering, E.R.; Porwit-MacDonald, A.; Faria, T.; Bjorklund, E.; Biondi, A.; van den Beemd, M.W.; Baars, E.; et al. BIOMED-I concerted action report: Flow cytometric immunophenotyping of precursor B-ALL with standardized triple-stainings. BIOMED-1 Concerted Action Investigation of Minimal Residual Disease in Acute Leukemia: International Standardization and Clinical Evaluation. *Leukemia* **2001**, *15*, 1185–1192. [PubMed]
122. Tsao, L.; Draoua, H.Y.; Osunkwo, I.; Nandula, S.V.; Murty, V.V.; Mansukhani, M.; Bhagat, G.; Alobeid, B. Mature B-cell acute lymphoblastic leukemia with t(9;11) translocation: A distinct subset of B-cell acute lymphoblastic leukemia. *Mod. Pathol.* **2004**, *17*, 832–839. [CrossRef] [PubMed]
123. Kotb, A.; El Fakih, R.; Hanbali, A.; Hawsawi, Y.; Alfraih, F.; Hashmi, S.; Aljurf, M. Philadelphia-like acute lymphoblastic leukemia: Diagnostic dilemma and management perspectives. *Exp. Hematol.* **2018**, *67*, 1–9. [CrossRef] [PubMed]
124. Dinner, S.; Platanias, L.C. Targeting the mTOR Pathway in Leukemia. *J. Cell Biochem.* **2016**, *117*, 1745–1752. [CrossRef] [PubMed]
125. Toosi, B.; Zaker, F.; Alikarami, F.; Kazemi, A.; Teremmahi Ardestanii, M. VS-5584 as a PI3K/mTOR inhibitor enhances apoptotic effects of subtoxic dose arsenic trioxide via inhibition of NF-kappaB activity in B cell precursor-acute lymphoblastic leukemia. *Biomed. Pharmacother.* **2018**, *102*, 428–437. [CrossRef] [PubMed]
126. Ultimo, S.; Simioni, C.; Martelli, A.M.; Zauli, G.; Evangelisti, C.; Celeghini, C.; McCubrey, J.A.; Marisi, G.; Ulivi, P.; Capitani, S.; et al. PI3K isoform inhibition associated with anti Bcr-Abl drugs shows in vitro increased anti-leukemic activity in Philadelphia chromosome-positive B-acute lymphoblastic leukemia cell lines. *Oncotarget* **2017**, *8*, 23213–23227. [CrossRef] [PubMed]
127. Zhang, Q.; Shi, C.; Han, L.; Jain, N.; Roberts, K.G.; Ma, H.; Cai, T.; Cavazos, A.; Tabe, Y.; Jacamo, R.O.; et al. Inhibition of mTORC1/C2 signaling improves anti-leukemia efficacy of JAK/STAT blockade in CRLF2 rearranged and/or JAK driven Philadelphia chromosome-like acute B-cell lymphoblastic leukemia. *Oncotarget* **2018**, *9*, 8027–8041. [CrossRef] [PubMed]
128. Gomes, A.M.; Soares, M.V.; Ribeiro, P.; Caldas, J.; Povoa, V.; Martins, L.R.; Melao, A.; Serra-Caetano, A.; de Sousa, A.B.; Lacerda, J.F.; et al. Adult B-cell acute lymphoblastic leukemia cells display decreased PTEN activity and constitutive hyperactivation of PI3K/Akt pathway despite high PTEN protein levels. *Haematologica* **2014**, *99*, 1062–1068. [CrossRef] [PubMed]
129. Morishita, N.; Tsukahara, H.; Chayama, K.; Ishida, T.; Washio, K.; Miyamura, T.; Yamashita, N.; Oda, M.; Morishima, T. Activation of Akt is associated with poor prognosis and chemotherapeutic resistance in pediatric B-precursor acute lymphoblastic leukemia. *Pediatr. Blood Cancer* **2012**, *59*, 83–89. [CrossRef]
130. Nemes, K.; Sebestyen, A.; Mark, A.; Hajdu, M.; Kenessey, I.; Sticz, T.; Nagy, E.; Barna, G.; Varadi, Z.; Kovacs, G.; et al. Mammalian target of rapamycin (mTOR) activity dependent phospho-protein expression in childhood acute lymphoblastic leukemia (ALL). *PLoS ONE* **2013**, *8*, e59335. [CrossRef]
131. Lazorchak, A.S.; Su, B. Perspectives on the role of mTORC2 in B lymphocyte development, immunity and tumorigenesis. *Protein Cell* **2011**, *2*, 523–530. [CrossRef] [PubMed]

132. Iwata, T.N.; Ramirez-Komo, J.A.; Park, H.; Iritani, B.M. Control of B lymphocyte development and functions by the mTOR signaling pathways. *Cytokine Growth Factor Rev.* **2017**, *35*, 47–62. [CrossRef] [PubMed]

133. Zhang, S.; Readinger, J.A.; DuBois, W.; Janka-Junttila, M.; Robinson, R.; Pruitt, M.; Bliskovsky, V.; Wu, J.Z.; Sakakibara, K.; Patel, J.; et al. Constitutive reductions in mTOR alter cell size, immune cell development, and antibody production. *Blood* **2011**, *117*, 1228–1238. [CrossRef] [PubMed]

134. Iwata, T.N.; Ramirez, J.A.; Tsang, M.; Park, H.; Margineantu, D.H.; Hockenbery, D.M.; Iritani, B.M. Conditional Disruption of Raptor Reveals an Essential Role for mTORC1 in B Cell Development, Survival, and Metabolism. *J. Immunol.* **2016**, *197*, 2250–2260. [CrossRef] [PubMed]

135. Zhang, Y.; Hu, T.; Hua, C.; Gu, J.; Zhang, L.; Hao, S.; Liang, H.; Wang, X.; Wang, W.; Xu, J.; et al. Rictor is required for early B cell development in bone marrow. *PLoS ONE* **2014**, *9*, e103970. [CrossRef] [PubMed]

136. Fuka, G.; Kantner, H.P.; Grausenburger, R.; Inthal, A.; Bauer, E.; Krapf, G.; Kaindl, U.; Kauer, M.; Dworzak, M.N.; Stoiber, D.; et al. Silencing of ETV6/RUNX1 abrogates PI3K/AKT/mTOR signaling and impairs reconstitution of leukemia in xenografts. *Leukemia* **2012**, *26*, 927–933. [CrossRef] [PubMed]

137. Jiang, K.; Zhong, B.; Ritchey, C.; Gilvary, D.L.; Hong-Geller, E.; Wei, S.; Djeu, J.Y. Regulation of Akt-dependent cell survival by Syk and Rac. *Blood* **2003**, *101*, 236–244. [CrossRef]

138. Kohrer, S.; Havranek, O.; Seyfried, F.; Hurtz, C.; Coffey, G.P.; Kim, E.; Ten Hacken, E.; Jager, U.; Vanura, K.; O'Brien, S.; et al. Pre-BCR signaling in precursor B-cell acute lymphoblastic leukemia regulates PI3K/AKT, FOXO1 and MYC, and can be targeted by SYK inhibition. *Leukemia* **2016**, *30*, 1246–1254. [CrossRef]

139. Eswaran, J.; Sinclair, P.; Heidenreich, O.; Irving, J.; Russell, L.J.; Hall, A.; Calado, D.P.; Harrison, C.J.; Vormoor, J. The pre-B-cell receptor checkpoint in acute lymphoblastic leukaemia. *Leukemia* **2015**, *29*, 1623–1631. [CrossRef]

140. Lee, J.S.; Vo, T.T.; Fruman, D.A. Targeting mTOR for the treatment of B cell malignancies. *Br. J. Clin. Pharmacol.* **2016**, *82*, 1213–1228. [CrossRef]

141. Gupta, M.; Hendrickson, A.E.; Yun, S.S.; Han, J.J.; Schneider, P.A.; Koh, B.D.; Stenson, M.J.; Wellik, L.E.; Shing, J.C.; Peterson, K.L.; et al. Dual mTORC1/mTORC2 inhibition diminishes Akt activation and induces Puma-dependent apoptosis in lymphoid malignancies. *Blood* **2012**, *119*, 476–487. [CrossRef] [PubMed]

142. Simioni, C.; Cani, A.; Martelli, A.M.; Zauli, G.; Tabellini, G.; McCubrey, J.; Capitani, S.; Neri, L.M. Activity of the novel mTOR inhibitor Torin-2 in B-precursor acute lymphoblastic leukemia and its therapeutic potential to prevent Akt reactivation. *Oncotarget* **2014**, *5*, 10034–10047. [CrossRef] [PubMed]

143. Gotesman, M.; Vo, T.T.; Herzog, L.O.; Tea, T.; Mallya, S.; Tasian, S.K.; Konopleva, M.; Fruman, D.A. mTOR inhibition enhances efficacy of dasatinib in ABL-rearranged Ph-like B-ALL. *Oncotarget* **2018**, *9*, 6562–6571. [CrossRef] [PubMed]

144. Muller, M.C.; Cervantes, F.; Hjorth-Hansen, H.; Janssen, J.; Milojkovic, D.; Rea, D.; Rosti, G. Ponatinib in chronic myeloid leukemia (CML): Consensus on patient treatment and management from a European expert panel. *Crit. Rev. Oncol. Hematol.* **2017**, *120*, 52–59. [CrossRef] [PubMed]

145. Shao, C.; Yang, J.; Kong, Y.; Cheng, C.; Lu, W.; Guan, H.; Wang, H. Overexpression of dominant-negative Ikaros 6 isoform is associated with resistance to TKIs in patients with Philadelphia chromosome positive acute lymphoblastic leukemia. *Exp. Ther. Med.* **2017**, *14*, 3874–3879. [CrossRef] [PubMed]

146. Yang, K.; Fu, L.W. Mechanisms of resistance to BCR-ABL TKIs and the therapeutic strategies: A review. *Crit. Rev. Oncol. Hematol.* **2015**, *93*, 277–292. [CrossRef]

147. Dorshkind, K.; Witte, O.N. Linking the hematopoietic microenvironment to imatinib-resistant Ph+ B-ALL. *Genes Dev.* **2007**, *21*, 2249–2252. [CrossRef]

148. Kharas, M.G.; Janes, M.R.; Scarfone, V.M.; Lilly, M.B.; Knight, Z.A.; Shokat, K.M.; Fruman, D.A. Ablation of PI3K blocks BCR-ABL leukemogenesis in mice, and a dual PI3K/mTOR inhibitor prevents expansion of human BCR-ABL+ leukemia cells. *J. Clin. Investig.* **2008**, *118*, 3038–3050. [CrossRef]

149. Oliansky, D.M.; Camitta, B.; Gaynon, P.; Nieder, M.L.; Parsons, S.K.; Pulsipher, M.A.; Dillon, H.; Ratko, T.A.; Wall, D.; McCarthy, P.L., Jr.; et al. Role of cytotoxic therapy with hematopoietic stem cell transplantation in the treatment of pediatric acute lymphoblastic leukemia: Update of the 2005 evidence-based review. *Biol. Blood Marrow Transplant.* **2012**, *18*, 505–522. [CrossRef]

150. Wu, J.H.; Shi, F.F.; Gong, Y.P.; Shi, R. The Mechanism of Combination using mTORC1/2 Inhibitor and Imatinib to Suppress Cell Proliferation of Ph (+)ALL Cell Line. *Sichuan Da Xue Xue Bao Yi Xue Ban* **2017**, *48*, 216–220.

151. Yu, G.; Chen, F.; Yin, C.; Liu, Q.; Sun, J.; Xuan, L.; Fan, Z.; Wang, Q.; Liu, X.; Jiang, Q.; et al. Upfront treatment with the first and second-generation tyrosine kinase inhibitors in Ph-positive acute lymphoblastic leukemia. *Oncotarget* **2017**, *8*, 107022–107032. [CrossRef] [PubMed]
152. Redig, A.J.; Vakana, E.; Platanias, L.C. Regulation of mammalian target of rapamycin and mitogen activated protein kinase pathways by BCR-ABL. *Leuk. Lymphoma* **2011**, *52* (Suppl. 1), 45–53. [CrossRef] [PubMed]
153. Pasic, I.; Lipton, J.H. Current approach to the treatment of chronic myeloid leukaemia. *Leuk. Res.* **2017**, *55*, 65–78. [CrossRef] [PubMed]
154. Hirase, C.; Maeda, Y.; Takai, S.; Kanamaru, A. Hypersensitivity of Ph-positive lymphoid cell lines to rapamycin: Possible clinical application of mTOR inhibitor. *Leuk. Res.* **2009**, *33*, 450–459. [CrossRef] [PubMed]
155. Short, N.J.; Kantarjian, H.; Jabbour, E.; Ravandi, F. Which tyrosine kinase inhibitor should we use to treat Philadelphia chromosome-positive acute lymphoblastic leukemia? *Best Pract. Res. Clin. Haematol.* **2017**, *30*, 193–200. [CrossRef] [PubMed]
156. Okabe, S.; Tauchi, T.; Tanaka, Y.; Ohyashiki, K. Efficacy of ponatinib against ABL tyrosine kinase inhibitor-resistant leukemia cells. *Biochem. Biophys. Res. Commun.* **2013**, *435*, 506–511. [CrossRef] [PubMed]
157. Nicolini, F.E.; Basak, G.W.; Kim, D.W.; Olavarria, E.; Pinilla-Ibarz, J.; Apperley, J.F.; Hughes, T.; Niederwieser, D.; Mauro, M.J.; Chuah, C.; et al. Overall Survival With Ponatinib Versus Allogeneic Stem Cell Transplantation in Philadelphia Chromosome-Positive Leukemias With the T315I Mutation. *Cancer* **2017**, *123*, 2875–2880. [CrossRef] [PubMed]
158. Assi, R.; Kantarjian, H.; Short, N.J.; Daver, N.; Takahashi, K.; Garcia-Manero, G.; DiNardo, C.; Burger, J.; Cortes, J.; Jain, N.; et al. Safety and Efficacy of Blinatumomab in Combination With a Tyrosine Kinase Inhibitor for the Treatment of Relapsed Philadelphia Chromosome-positive Leukemia. *Clin. Lymphoma Myeloma Leuk.* **2017**, *17*, 897–901. [CrossRef]
159. Ghelli Luserna Di Rora, A.; Beeharry, N.; Imbrogno, E.; Ferrari, A.; Robustelli, V.; Righi, S.; Sabattini, E.; Verga Falzacappa, M.V.; Ronchini, C.; Testoni, N.; et al. Targeting WEE1 to enhance conventional therapies for acute lymphoblastic leukemia. *J. Hematol. Oncol.* **2018**, *11*, 99. [CrossRef]
160. Bertamini, L.; Nanni, J.; Marconi, G.; Abbenante, M.; Robustelli, V.; Bacci, F.; Matti, A.; Paolini, S.; Sartor, C.; Monaco, S.L.; et al. Inotuzumab ozogamicin is effective in relapsed/refractory extramedullary B acute lymphoblastic leukemia. *BMC Cancer* **2018**, *18*, 1117. [CrossRef]
161. Yang, X.; He, G.; Gong, Y.; Zheng, B.; Shi, F.; Shi, R.; Yang, X. Mammalian target of rapamycin inhibitor rapamycin enhances anti-leukemia effect of imatinib on Ph+ acute lymphoblastic leukemia cells. *Eur. J. Haematol.* **2014**, *92*, 111–120. [CrossRef] [PubMed]
162. Yang, X.; Lin, J.; Gong, Y.; Ma, H.; Shuai, X.; Zhou, R.; Guo, Y.; Shan, Q.; He, G. Antileukaemia effect of rapamycin alone or in combination with daunorubicin on Ph+ acute lymphoblastic leukaemia cell line. *Hematol. Oncol.* **2012**, *30*, 123–130. [CrossRef] [PubMed]
163. Ding, J.; Romani, J.; Zaborski, M.; MacLeod, R.A.; Nagel, S.; Drexler, H.G.; Quentmeier, H. Inhibition of PI3K/mTOR overcomes nilotinib resistance in BCR-ABL1 positive leukemia cells through translational down-regulation of MDM2. *PLoS ONE* **2013**, *8*, e83510. [CrossRef] [PubMed]
164. Janes, M.R.; Vu, C.; Mallya, S.; Shieh, M.P.; Limon, J.J.; Li, L.S.; Jessen, K.A.; Martin, M.B.; Ren, P.; Lilly, M.B.; et al. Efficacy of the investigational mTOR kinase inhibitor MLN0128/INK128 in models of B-cell acute lymphoblastic leukemia. *Leukemia* **2013**, *27*, 586–594. [CrossRef] [PubMed]
165. Tran, T.H.; Loh, M.L. Ph-like acute lymphoblastic leukemia. *Hematol. Am. Soc. Hematol. Educ. Program.* **2016**, *2016*, 561–566. [CrossRef] [PubMed]
166. Maude, S.L.; Tasian, S.K.; Vincent, T.; Hall, J.W.; Sheen, C.; Roberts, K.G.; Seif, A.E.; Barrett, D.M.; Chen, I.M.; Collins, J.R.; et al. Targeting JAK1/2 and mTOR in murine xenograft models of Ph-like acute lymphoblastic leukemia. *Blood* **2012**, *120*, 3510–3518. [CrossRef]
167. Welsh, S.J.; Churchman, M.L.; Togni, M.; Mullighan, C.G.; Hagman, J. Deregulation of kinase signaling and lymphoid development in EBF1-PDGFRB ALL leukemogenesis. *Leukemia* **2018**, *32*, 38–48. [CrossRef]
168. Herzog, S.; Reth, M.; Jumaa, H. Regulation of B-cell proliferation and differentiation by pre-B-cell receptor signalling. *Nat. Rev. Immunol.* **2009**, *9*, 195–205. [CrossRef]
169. Srinivasan, L.; Sasaki, Y.; Calado, D.P.; Zhang, B.; Paik, J.H.; DePinho, R.A.; Kutok, J.L.; Kearney, J.F.; Otipoby, K.L.; Rajewsky, K. PI3 kinase signals BCR-dependent mature B cell survival. *Cell* **2009**, *139*, 573–586. [CrossRef]

170. Duy, C.; Hurtz, C.; Shojaee, S.; Cerchietti, L.; Geng, H.; Swaminathan, S.; Klemm, L.; Kweon, S.M.; Nahar, R.; Braig, M.; et al. BCL6 enables Ph+ acute lymphoblastic leukaemia cells to survive BCR-ABL1 kinase inhibition. *Nature* **2011**, *473*, 384–388. [CrossRef]
171. Ge, Z.; Zhou, X.; Gu, Y.; Han, Q.; Li, J.; Chen, B.; Ge, Q.; Dovat, E.; Payne, J.L.; Sun, T.; et al. Ikaros regulation of the BCL6/BACH2 axis and its clinical relevance in acute lymphoblastic leukemia. *Oncotarget* **2017**, *8*, 8022–8034. [CrossRef] [PubMed]
172. Kim, E.; Hurtz, C.; Koehrer, S.; Wang, Z.; Balasubramanian, S.; Chang, B.Y.; Muschen, M.; Davis, R.E.; Burger, J.A. Ibrutinib inhibits pre-BCR(+) B-cell acute lymphoblastic leukemia progression by targeting BTK and BLK. *Blood* **2017**, *129*, 1155–1165. [CrossRef] [PubMed]
173. Saba, N.S.; Angelova, M.; Lobelle-Rich, P.A.; Levy, L.S. Disruption of pre-B-cell receptor signaling jams the WNT/beta-catenin pathway and induces cell death in B-cell acute lymphoblastic leukemia cell lines. *Leuk. Res.* **2015**. [CrossRef] [PubMed]
174. Eldfors, S.; Kuusanmaki, H.; Kontro, M.; Majumder, M.M.; Parsons, A.; Edgren, H.; Pemovska, T.; Kallioniemi, O.; Wennerberg, K.; Gokbuget, N.; et al. Idelalisib sensitivity and mechanisms of disease progression in relapsed TCF3-PBX1 acute lymphoblastic leukemia. *Leukemia* **2017**, *31*, 51–57. [CrossRef] [PubMed]
175. Brown, V.I.; Fang, J.; Alcorn, K.; Barr, R.; Kim, J.M.; Wasserman, R.; Grupp, S.A. Rapamycin is active against B-precursor leukemia in vitro and in vivo, an effect that is modulated by IL-7-mediated signaling. *Proc. Natl. Acad. Sci. USA* **2003**, *100*, 15113–15118. [CrossRef] [PubMed]
176. Irving, J.; Matheson, E.; Minto, L.; Blair, H.; Case, M.; Halsey, C.; Swidenbank, I.; Ponthan, F.; Kirschner-Schwabe, R.; Groeneveld-Krentz, S.; et al. Ras pathway mutations are prevalent in relapsed childhood acute lymphoblastic leukemia and confer sensitivity to MEK inhibition. *Blood* **2014**, *124*, 3420–3430. [CrossRef] [PubMed]
177. Case, M.; Matheson, E.; Minto, L.; Hassan, R.; Harrison, C.J.; Bown, N.; Bailey, S.; Vormoor, J.; Hall, A.G.; Irving, J.A. Mutation of genes affecting the RAS pathway is common in childhood acute lymphoblastic leukemia. *Cancer Res.* **2008**, *68*, 6803–6809. [CrossRef] [PubMed]
178. Crazzolara, R.; Bradstock, K.F.; Bendall, L.J. RAD001 (Everolimus) induces autophagy in acute lymphoblastic leukemia. *Autophagy* **2009**, *5*, 727–728. [CrossRef] [PubMed]
179. Baraz, R.; Cisterne, A.; Saunders, P.O.; Hewson, J.; Thien, M.; Weiss, J.; Basnett, J.; Bradstock, K.F.; Bendall, L.J. mTOR inhibition by everolimus in childhood acute lymphoblastic leukemia induces caspase-independent cell death. *PLoS ONE* **2014**, *9*, e102494. [CrossRef] [PubMed]
180. Hirai, H.; Sootome, H.; Nakatsuru, Y.; Miyama, K.; Taguchi, S.; Tsujioka, K.; Ueno, Y.; Hatch, H.; Majumder, P.K.; Pan, B.S.; et al. MK-2206, an allosteric Akt inhibitor, enhances antitumor efficacy by standard chemotherapeutic agents or molecular targeted drugs in vitro and in vivo. *Mol. Cancer Ther.* **2010**, *9*, 1956–1967. [CrossRef] [PubMed]
181. Stefanzl, G.; Berger, D.; Cerny-Reiterer, S.; Blatt, K.; Eisenwort, G.; Sperr, W.R.; Hoermann, G.; Lind, K.; Hauswirth, A.W.; Bettelheim, P.; et al. The pan-BCL-2-blocker obatoclax (GX15-070) and the PI3-kinase/mTOR-inhibitor BEZ235 produce cooperative growth-inhibitory effects in ALL cells. *Oncotarget* **2017**, *8*, 67709–67722. [CrossRef] [PubMed]
182. Rheingold, S.R.; Tasian, S.K.; Whitlock, J.A.; Teachey, D.T.; Borowitz, M.J.; Liu, X.; Minard, C.G.; Fox, E.; Weigel, B.J.; Blaney, S.M. A phase 1 trial of temsirolimus and intensive re-induction chemotherapy for 2nd or greater relapse of acute lymphoblastic leukaemia: A Children's Oncology Group study (ADVL1114). *Br. J. Hematol.* **2017**, *177*, 467–474. [CrossRef]
183. Place, A.E.; Pikman, Y.; Stevenson, K.E.; Harris, M.H.; Pauly, M.; Sulis, M.L.; Hijiya, N.; Gore, L.; Cooper, T.M.; Loh, M.L.; et al. Phase I trial of the mTOR inhibitor everolimus in combination with multi-agent chemotherapy in relapsed childhood acute lymphoblastic leukemia. *Pediatr. Blood Cancer* **2018**, *65*, e27062. [CrossRef] [PubMed]
184. Malouf, C.; Ottersbach, K. Molecular processes involved in B cell acute lymphoblastic leukaemia. *Cell Mol. Life Sci.* **2018**, *75*, 417–446. [CrossRef] [PubMed]
185. Wu, X.; Xing, X.; Dowlut, D.; Zeng, Y.; Liu, J.; Liu, X. Integrating phosphoproteomics into kinase-targeted cancer therapies in precision medicine. *J. Proteom* **2019**, *191*, 68–79. [CrossRef]

186. Montano, A.; Forero-Castro, M.; Marchena-Mendoza, D.; Benito, R.; Hernandez-Rivas, J.M. New Challenges in Targeting Signaling Pathways in Acute Lymphoblastic Leukemia by NGS Approaches: An Update. *Cancers* **2018**, *10*, 110. [CrossRef]
187. Binnewies, M.; Roberts, E.W.; Kersten, K.; Chan, V.; Fearon, D.F.; Merad, M.; Coussens, L.M.; Gabrilovich, D.I.; Ostrand-Rosenberg, S.; Hedrick, C.C.; et al. Understanding the tumor immune microenvironment (TIME) for effective therapy. *Nat. Med.* **2018**, *24*, 541–550. [CrossRef]

Review

mTOR Regulation of Metabolism in Hematologic Malignancies

Simone Mirabilii [1], Maria Rosaria Ricciardi [1] and Agostino Tafuri [1,2,*]

1 Department of Clinical and Molecular Medicine, Sapienza University of Rome, 00185 Rome, Italy; simone.mirabilii@uniroma1.it (S.M.); mariarosaria.ricciardi@uniroma1.it (M.R.R.)
2 Hematology, "Sant' Andrea" University Hospital, Sapienza University of Rome, 00185 Rome, Italy
* Correspondence: agostino.tafuri@uniroma1.it; Tel.: +39-06-3377-5113

Received: 6 December 2019; Accepted: 7 February 2020; Published: 11 February 2020

Abstract: Neoplastic cells rewire their metabolism, acquiring a selective advantage over normal cells and a protection from therapeutic agents. The mammalian Target of Rapamycin (mTOR) is a serine/threonine kinase involved in a variety of cellular activities, including the control of metabolic processes. mTOR is hyperactivated in a large number of tumor types, and among them, in many hematologic malignancies. In this article, we summarized the evidence from the literature that describes a central role for mTOR in the acquisition of new metabolic phenotypes for different hematologic malignancies, in concert with other metabolic modulators (AMPK, HIF1α) and microenvironmental stimuli, and shows how these features can be targeted for therapeutic purposes.

Keywords: mTOR; hematologic malignancies; cell metabolism

---

1. mTOR Structure and Function

The mammalian Target of Rapamycin (mTOR) is a kinase involved in the PI3k/PTEN/Akt axis, which plays a key role in the control of many biological processes, including cell growth and survival, protein translation, ribosomal biogenesis, autophagy, and metabolism [1–3].

Originally identified in the yeast *Saccharomyces cerevisiae*, mTOR is a pleiotropic serine/threonine kinase of 289kDa, which shows a terminal COOH catalytic domain with a high sequence homology with PI3K [4].

mTOR is composed of 2549 amino acids and contains up to 20 tandem repeated HEAT motifs, a repeated structural motif composed of two tandem anti-parallel α-helices linked by a short loop, which work as a scaffold for a protein-protein interaction [5].

It operates within two multiprotein complexes, mTORC1 and mTORC2, which phosphorylate a different set of substrates coordinating different physiological cell functions. mTORC1 includes mTOR (the catalytic subunit of the complex), the regulatory-associated protein of mTOR (Raptor), the DEP domain-containing mTOR-interacting protein (Deptor), the mammalian lethal with SEC13 protein 8 (mLST8), the raptor binding protein PRAS40 and the FK506-binding protein 38 (FKBP38). mTORC2 is conversely composed of mTOR itself, the rapamycin-insensitive companion of mTOR (Rictor), mLST8, the mammalian stress-activated map kinase-interacting protein 1 (mSIN1), a protein observed with Rictor (Protor-1) and Deptor [3,6].

The two complexes display different response to rapamycin and its derivatives (rapalogs), being mTORC1 sensitive to the inhibitory effects of these immunosuppressant, while mTORC2 proved insensitive. However, in some cell types, it has been shown that prolonged treatment with rapamycin and rapalogs can indirectly inhibit the formation and activity of the TORC2 complex [7].

Various upstream events can lead to the activation of mTORC1, mostly convergent on Akt. For instance, Akt can inactivate through phosphorylation either TSC2 (tuberous sclerosis protein 2) in the

TSC1–TSC2 complex, which negatively regulates mTORC1, or PRAS40, antagonizing its activation by Rheb, respectively [8,9].

In response to nutrient availability and growth factors, activated mTORC1 regulates protein translation by phosphorylating p70S6 (p70S6K) and 4E-BP1 kinases, which in turn phosphorylate the S6 protein kinases (p70S6K1/2) and the eukaryotic initiation factor 4E (eIF4E)-binding proteins (4E-BP1/2), which are involved in the translation process [6,10]. In particular, the phosphorylated S6K enhances the translation of mRNAs that have 5′ polypyrimidine rich sequences [11,12]. Conversely, phosphorylation of 4E-BP1 causes it to release eIF4E, which binds the mRNA 5′-cap, thus allowing the translation to begin [13].

In addition, the mTORC1 complex regulates the expression of key proteins such as cyclin D1, STAT3, Bcl-2, Bcl-xL, Mcl-1, thus promoting cell proliferation and survival [14–16]. As for the metabolic function, mTORC1 is a central signaling node in coordinating the metabolic cell response (Figure 1). mTORC1 is involved in metabolic reprogramming by increasing glycolysis and macromolecules biosynthesis through transcriptional, translational, and post-translational mechanisms mediated by its substrates, p70S6K and 4E-BP [17–19]. Among these mechanisms, mTOR enhances the translation of critical metabolic mediators such as c-Myc and hypoxia-inducible factor 1 alpha (HIF1α) [20]. c-Myc upregulates many genes involved in the glycolytic process such as glucose transporters, hexokinase 2 (HK2), phosphofructokinase (PFKM), and enolase 1 (ENO1) [21]. HIF1α is an oxygen-sensing molecule that is stabilized in hypoxic condition, and translocates to the nucleus initiating the transcription of hypoxic response genes [22]. Its action on cell metabolism includes an increased glucose uptake, a higher glycolytic flux and a lower oxidative phosphorylation (OXPHOS) [23]. On the other hand, AMP-activated protein kinase (AMPK) acts as an mTOR inhibitor; it is a serine/threonine kinase that is able to respond to the fluctuating intracellular AMP levels, shutting down energy-depleting processes in favor of catabolic pathways, such as fatty acid oxidation and autophagy, when the AMP level rises [24]. Once activated, AMPK inhibits mTOR through the activation of TSC2 [24].

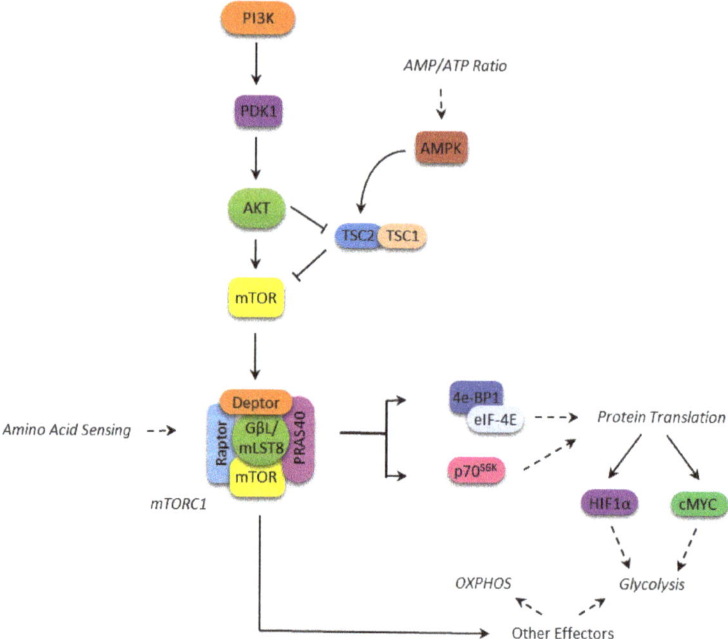

Figure 1. mammalian Target of Rapamycin (mTOR) signaling and cellular metabolism.

However, it was also reported that mTORC1 could promote anabolic metabolism independently from p70S6K and 4E-BP1 [25]. The authors demonstrated that mTOR regulates oxygen consumption and oxidative capacity independently from these effectors. Energy/nutrition depletion and stress signals seem indeed indirectly sensed by mTORC1 via the LKB1-AMPK cascade [26].

mTORC activity and, above all, its regulation mechanisms are less well known. While mTORC1 is mostly involved in sustaining cell growth, proliferation, and survival by controlling the translation machinery, autophagy or mitochondrial biogenesis, the main function attributed to mTORC2 represents the regulation of the actin cytoskeleton polarity mediated by the Rho1/Pkc1/MAPK cell cascade [27,28].

The assembly of the actin cytoskeleton in response to mitogenic signals is promoted by mTORC2 through the activation of several cytoskeletal regulators, such as TCP-1, ROM2 and Ypk [29–32]. It was observed that the deregulation of mTORC2 activity causes the alteration of cytoskeletal actin [27] and impacts cell motility in some types of tumor cells [33].

It has been widely reported that nutrient availability and cellular energy status do not seem to be necessary for the activation of mTORC2 [34–36]. In contrast, in proliferating cells, the activation of mTORC2 requires the close interaction between the protein multicomplex and the ribosome, underlining a reciprocal interaction between the two complexes [34]. Indeed, since mTORC1 regulates ribosome biogenesis, which is crucial to determine cell growth capacity, it indirectly controls mTORC2 activity [34]. In turn, mTORC2 promotes the Akt-TSC1/TSC2-mediated mTORC1 activity [37]. Connected to the PI3K/AKT signaling pathway, it is described that mTORC2 phosphorylates Akt on Ser473. However, a crosstalk between mTORC1 and mTORC2 and with other metabolic pathways has been reported [38].

2. Cancer Metabolism and Cell Signaling

We have learned from the literature that alterations in cellular metabolism pathways are manifest in cancer cells as compared to normal tissue cells [39–41].

Nuclear and mitochondrial alterations in the genome of cancer cells, pressing for an increased import of bioenergetic substrates and/or an increased generation of biosynthetic intermediates needed for cell growth and proliferation, are tightly linked to the altered cancer metabolism. On the other hand, the products of metabolism, especially ROS, damaging cells can promote oncogenic DNA mutations [42,43].

Among the altered metabolic pathways of cancer cells, the increased glycolytic flux, the exploitation of alternative carbon source like glutamine, and the increase in fatty acid metabolism are recognized [39,44]. The alteration of the metabolic processes, however, differs from tumor to tumor, because of the close interaction between the cells and the tumor microenvironment, where the concentrations of nutrients such as glucose and glutamine, or gas such as oxygen, are spatially and temporally heterogeneous [45,46].

Moreover, there are differences, in terms of energy requirements and biomass production, between the differentiated and undifferentiated cells, as well as in the metabolic needs and the related regulatory mechanisms between the proliferating and quiescent cells [40,44,47]. This concept needs to be taken into account when approaching metabolic analysis in different leukemia settings.

Recently, several authors have shown that many of the Signal Transduction Pathways (STPs) aberrantly activated in cancer cells actually converge on the deregulation of common metabolic mechanisms responsible for cell growth and survival [40,48,49]. The instructional metabolic reprogramming from signaling is critical for cellular homeostasis and cell fate. However, the relationship between altered cellular signaling and reprogrammed metabolism is not unidirectional and several feedback mechanisms, in which metabolites can control signal fluxes through specific sensor kinases that monitor the cell bioenergetic status are found to be active in cancer cells [50,51].

In this scenario, mTOR, as an orchestrator of nutrient sensor, signaling processor, and cell growth regulator, represents one of the main actors in coordinating cell growth, division, and survival with cell metabolic activity [3].

3. mTOR and the Metabolism of Hematologic Malignancies

3.1. Acute Myeloid Leukemia

Different studies ascribe an overall glycolytic metabolism to acute myeloid leukemia (AML) cells, with a conspicuous lactate production and an active tricarboxylic acid (TCA) cycle, in order to fulfil biosynthetic purposes. In a metabolomic study on over 400 patients, Chen et al. detected a higher glycolytic flux in AML samples when compared to healthy controls, along with a compensatory pyruvate generation from amino acids in order to feed the TCA cycle [52]. Increased glycolytic rates were already reported by Herst et al., although with a variable extent, dividing patients into two subgroups based on the glycolysis rate [53]. Interestingly, both groups confer to glycolysis a prognostic meaning, encouraging further studies to explore this feature. To confirm the importance of glycolysis in AML cells, a thorough work by Wang et al. demonstrated, through the use of murine models and a serial transplantation approach, that the conditional deletion of the two last enzymes of aerobic glycolysis, pyruvate kinase and lactate dehydrogenase, profoundly affect the viability of leukemic-like cells [54]. As mentioned, AML cells can utilize carbon sources other than glucose. One such sources is glutamine, as it has been long known that this amino acid is essential to AML cells [55]. Recently, fatty acid utilization has come into attention: our group, along with others, has demonstrated that the utilization of these compounds plays a major role in AML bioenergetics balance [56–58].

In this bioenergetic picture, the glycolytic phenotype is at least partially caused by mTOR hyperactivation, as it is described as a controller of the glycolytic process in several studies that were carried out on AML cells. Liu et al. observed that mTOR inhibition through rapamycin caused a decrease in glucose uptake on AML cell lines [59]. A similar experience has been reported by our group, as the inhibition of the PI3K/mTOR axis signaling caused a drop in the glycolytic fluxes, along with a reduction in OXPHOS [60]. Additionally, Poulain et al. directly correlated the extent of glycolytic flux with the activation status of mTORC1: analysis by gene expression profile of AML cell lines treated with rapamycin revealed a downregulation of genes involved in the glycolytic pathway, while from the metabolic perspective rapamycin caused a reduction of glycolytic flux with a concomitant shift on OXPHOS, as testified by the increase of TCA cycle enzymes [61]. Coherently, rapamycin protected AML cell lines from apoptosis induced by either glucose deprivation or 2-deoxy-glucose (2DG) glycolysis inhibition [61]. Much uncertainty remains on the reverse mechanism, i.e., what happens to mTOR when glycolysis is inhibited. Pradelli et al. reported the mTOR inactivation in AML cells exposed to 2DG, due to the action of AMPK [62]. To further complicate the picture, Estañ et al. observed instead that 2DG caused AMPK inhibition and a concomitant activation of the Akt/mTOR axis [63]. This opposite mechanism of action might be partially explained by the different cell models used by these studies (U937 in the first; HL60, NB4, and THP-1 in the second) that can reflect the heterogeneity in genetic and clinical features of the AML. Indeed, Pereira et al. confirmed this heterogeneity by measuring glycolysis of AML cell lines in correlation with Akt/mTOR and AMPK activation status, finding a fluctuation in the glycolytic flux across the different cell lines, correlated with a different intracellular signaling [64]. Of interest, in the KG1 cell line they found a simultaneous activation of AMPK and mTOR, subverting the canonical view that dictates an antithetic role for these two kinases [64].

Recent evidence has identified PFKFB3 (6-phosphofructo-2-kinase/fructose-2,6-biphosphatase 3) as a probable target of the mTOR signaling directly involved in the glycolytic process: Feng and Wu, in fact, observed in their work that mTOR interacts with this enzyme upregulating aerobic glycolysis [65]. Exposure to rapamycin reverts this glycolytic metabolism, downregulating PFKFB3 protein levels [65].

Previous data demonstrated that this glycolytic phenotype, mediated by mTOR, is somehow exacerbated by extracellular stimuli originating from the microenvironment. In fact, the contact with the stromal compartment appears to increase the glycolytic flux of AML cells, through a mechanism involving a chemokine, CXCL12, recognized by its receptor CXCR4, which in turn activates mTOR [66]. This observation may constitute one of the factors involved in the chemoresistance, as it is widely accepted that the microenvironment protects leukemic cells from therapeutic agents [67]. Since glycolysis is another factor associated with resistance [68], the emergence of refractoriness may be also explained by the stroma-mediated upregulation of glycolysis.

In addition to the glycolytic process, mTOR has been considered as central in the regulation of amino acid homeostasis and usually becomes inactive in case of amino acid deficiency, leading to autophagy and decreased protein biosynthesis [69]. This mechanism seems to be confirmed also in AML cells, as mTOR inactivation in case of either glutamine depletion or L-asparaginases-mediated degradation leads to the autophagic process activation in an attempt by the cells to scavenge the building blocks for survival [70].

Of interest, a peculiar metabolism seems to be associated with the resistance to PI3K/mTOR axis inhibitors: a metabolomic study carried out on 30 primary samples exposed to 4 molecular inhibitors (rapamycin, GDC-0941, human insulin, indomethacin), targeting this pathway with different selectivity, identified the proline/glutamine and the arachidonic acid metabolism as markers of AML cell resistance to these agents [71]. Table 1 summarizes the effect of mTOR and metabolic enzyme inhibitors in AML.

Table 1. Metabolic modulators in AML.

Compound	Target	Effect (Ref)
rapamycin	mTORC1	Decreased Glucose uptake (59); Shift from Glycolysis to OXPHOS (61); Decreased glycolysis through PFKB3 downregulation (65)
BKM-120	PI3K	Decreased Glycolysis and OXPHOS
2DG	Hexokinase	mTOR inactivation (62); mTOR activation (63)
L-asparaginases	Asparagine and glutamine degradation	mTOR inactivation (70)

3.2. *Chronic Myeloid Leukemia*

Earlier reports observed an upregulation of glycolysis in chronic myeloid leukemia (CML) cells, promoting the idea that the metabolism of these cells can be characterized by the Warburg effect, the glucose fermentation to lactate even in the presence of oxygen. Two independent studies detected a glycolytic phenotype in CML cells, with a reduction of the glycolytic flux following the exposure to the Bcr-Abl inhibitor Imatinib in those cells that proved sensible to its action [72,73]. This reduction has been recently confirmed by De Rosa et al., who observed the downregulation of hexokinase II and lactate dehydrogenase (two enzymes involved in the aerobic glycolysis) in CML cell models [74]. In parallel, Baldwin's group detected an upregulation of the glucose transporter GLUT-1 and a consequent increase in glucose import in hematopoietic cell lines transfected with Bcr-Abl [75]. More recently, Sontakke et al. confirmed the aerobic glycolysis and the upregulation of glucose transporters as SLC2A1/3 in normal cord blood progenitors engineered for Bcr-Abl expression [76]. Interestingly, they found a concomitant upregulation of glutaminolysis, probably to keep the TCA cycle active despite the pyruvate subtraction [76]. More generally, the increased glycolytic flux has been associated with the resistance to tyrosine kinase inhibitors [77,78]. A different picture is emerging regarding the metabolism of CML stem cells, as multiple works stated the relevance of an oxidative phenotype in this cell population. From a transcriptomic perspective, these cells seem to upregulate genes involved in the OXPHOS when compared to CD34+ cells from healthy donors. Authors found a peculiar pattern of expression, with an upregulation of mitochondrial respiratory chain complex 1 and 2, and a downregulation of complex 3, giving rise to a defective OXPHOS and a consequent production of ROS [79]. Accordingly, two recent reports focusing on CML stem cell metabolism confirmed this oxidative phenotype, fueled by fatty acid oxidation [80], and driven by a SIRT1/PGC1-α signaling axis [81]. Importantly, targeting this metabolism, either with a specific agent (tigecycline) or a SIRT1 knockout led to the impairment of CML stem cell functions, showing synergistic interaction with tyrosine kinase inhibitors [80,81].

These observations seem to suggest a reprogramming of CML cell metabolism, starting from an oxidative phenotype in CML stem cells, then shifting toward a more glycolytic one in later stages of the disease.

Relatively few reports have focused on mTOR activity on metabolism in the CML setting.

High levels of ROS, caused by a sustained mitochondrial activity, have been linked to the PI3K/Akt/mTOR axis activation by Bcr-Abl signaling. This mitochondrial activity was fueled mainly by glucose, as the exposure of CML cells to 2DG lowered the ROS levels [82]. Similarly, the inhibition of this axis with a PI3K (wortmannin) or an mTOR (rapamycin) inhibitor caused a reduction of ROS, thus reaffirming the action of this signaling module on the glycolytic metabolism [82]. The control over glucose fate by Bcr-Abl/mTOR seems to rely on the activation of S6K1 [83]. Interestingly, the inactivation of S6K1 and the consequent impairment of glycolysis did not induce apoptosis, but caused a metabolic shift to fatty acid oxidation [83]. Shinohara et al. reported that mTOR mediates the expression regulation by Bcr-Abl of the balance between the two pyruvate kinase (PKM) isoforms 1 and 2 [84]; this glycolytic kinase is crucial in the control of glucose fate, between aerobic glycolysis and OXPHOS [85]. Finally, a sustained activity of the PI3K/Akt/mTOR signaling module has been reported to participate in the glycolytic phenotype of adriamycin-resistant CML cells [86].

3.3. *Acute Lymphoblastic Leukemia*

Similar to AML and CML, acute lymphoblastic leukemia (ALL) cell metabolism seems to be driven by the aerobic glycolysis, at least in the B lineage: when compared to normal cells, ALL cells show higher expression levels of glucose transporters (GLUT-1), an increased lactate production and a vulnerability to glycolysis inhibition [87]. The presence of fusion genes, like BCR-ABL, seems to further enhance this kind of metabolism [88]. Moreover, higher glycolytic rates are involved in resistance to chemotherapeutic agents, like daunorubicin [89]. The mitochondrial energy machinery seems however to be intact, as these cells are able to shift from glycolysis to OXPHOS fueled by autophagy under stress condition like exposure to glucocorticoids [90]. T cell ALL, conversely, appear to be less glycolytic and more oxidative [91]. NOTCH-1, a transmembrane receptor commonly activated in this leukemia subtype [92], appears to be implicated in this metabolic phenotype, driving these cells towards the use of glutamine to feed the TCA cycle [93].

In B-ALL, this glycolytic phenotype emerges from the interplay between mTOR, HIF1α and the hypoxic microenvironment, as showed by Frolova et al. [94]. They demonstrated, through ALL blasts and stromal cell co-culture, that the contact with microenvironment stimulates, under hypoxic condition, a signaling through the PI3K/Akt/mTOR axis, along with MAPK activation, stabilizing HIF1α and inducing the glycolytic phenotype [94]. The inhibition of mTOR with everolimus reverted this glycolytic phenotype, downregulating hexokinase II expression and reducing lactate generation [94]. Additionally, it has been reported that mTOR reacts during metabolic stress, such as 2DG exposure, cooperating with AMPK to lower Mcl-1 expression, especially in Bcr-Abl positive ALL, a mechanism that can potentiate the effect of TKI inhibitors in this setting [95]. Beside glycolysis, mTOR seems to participate in the metabolism control of important cofactors, such as thiamine, which is required for a large number of enzymes to be functionally active. Targeting this molecule with specific thiaminases causes a series of metabolic repercussions, such as a decrease of mitochondrial respiration and an increase of glycolysis, which can be reverted by the effect of rapamycin [96]. This constitutes an indirect proof of the role of mTOR in thiamine metabolism, which prompts further investigation in the leukemic setting.

In T-ALL, Kishton et al. depicted a complex picture in which mTOR is under the strict control of AMPK: microenvironmental stimuli activates Notch, which signals through mTOR for the aerobic glycolysis [91]. This metabolism, however, cannot be sustained by those cells, causing a shortage of ATP, which activates AMPK [90]. The latter then inactivates mTOR, causing a shift to a more oxidative metabolism [91]. In this context, mTOR seems therefore to drive cells toward a more sustained metabolism, with higher fluxes of glycolysis followed by an OXPHOS upregulation. An indirect evidence has in fact been reported by Fernández-Ramos et al., who observed that 6-mercaptopurine inhibits mTOR through AMPK activation, consequently reducing glucose and glutamine consumption by T leukemia cells [97]. In Table 2, the metabolic effects of mTOR inhibition in B- and T-ALL are reported.

Table 2. Targeting mTOR and metabolism in B- and T-ALL.

Type	Compound	Target	Effect (Ref)
B-ALL	everolimus	mTORC1	Decreased glycolysis and lactate generation (94)
	rapamycin	mTORC1	Decrease of glycolysis and increase of OXPHOS, reversion of thiaminase effects (96)
T-ALL	6-mercaptopurine	mTOR through AMPK activation	Decreased glucose and glutamine consumption (97)

3.4. Chronic Lymphocytic Leukemia

The metabolic features of chronic lymphocytic leukemia (CLL) are perhaps the best characterized among the hematological malignancies, as there is a general consensus to place these cells in the oxidative metabolic phenotype, without a clear manifestation of the Warbug effect [98]. Additionally, both glutamine and fatty acids concur to fuel the TCA cycle activity [99], conferring to CLL cells a metabolic plasticity that allows them to survive in the different body districts [100]. The rate of OXPHOS has been correlated with the degree of the disease aggressiveness: Gandhi's group observed that CLL cells show variable respiration rates, and cells with higher rates were from patients characterized by unfavorable prognostic markers, such as a higher Rai score, beta 2 microglobulin (β2M), Zap70, and unmutated immunoglobulin heavy chain (IGHV) encoding genes [101]. Interestingly, they used a CRISPR/Cas approach to dissect the B cell receptor (BCR) signaling and its role in metabolism: although they did not directly focus on mTOR, interfering with PI3K signaling led to lower metabolic rates, both in term of glycolysis and OXPHOS [101]. This demonstrated how central the BCR/PI3K/Akt/mTOR axis is in regulating the bioenergetics status of these cells [101].

Despite the general agreement on their metabolism, there is however an unclear relationship between CLL cells and the stromal microenvironment: while there is evidence that the interaction with stromal cells induces a switch to glycolysis through a mechanism involving Notch and c-Myc [102], others observed an exacerbation of the respiratory rates, indicating an increase in OXPHOS after stromal contact [103].

mTOR role in the regulation of CLL cell metabolism has been studied especially in relation to the different response to therapeutic agents. Investigating the resistance to dasatinib, a second generation TKI, Marignac Martinez et al. reported a different regulation of the PI3K/Akt/mTOR pathway between sensitive and resistant primary cells: the first ones were characterized by a higher dependence on OXPHOS, a downregulation of PTEN and an upregulation of TCL1 [104]. In addition, both metformin, a respiratory chain complex I inhibitor, and rapamycin similarly synergized with dasatinib in inducing apoptosis in the sensitive subset [104]. Sharma et al. focused on the metabolic response to chemotherapeutic agent fludarabine: CLL cells, namely MEC-1 and 2 cell lines and primary samples showed an entirely similar profile to fludarabine resistant cells, a profound mTOR activation that caused an overall increase in glycolysis and OXPHOS rates, combined with an upregulation of purine biosynthesis [105]. An interesting observation was made by Siska et al.: chronic and acute B leukemia cells can induce metabolic changes in T lymphocytes by reducing their signaling through mTOR, thus slowing down their GLUT1-mediated glucose import and glycolytic rates, ultimately impairing their anti-leukemic action [106].

3.5. Multiple Myeloma

Multiple Myeloma (MM) plasma cells display a dependence on aerobic glycolysis for survival: they are characterized by an upregulation of GLUT family glucose transporter (namely GLUT4, 8 and 11) [107], a high expression of lactate dehydrogenase [108,109], and a carbon flux from glucose to lactate, with glutamine to replenish the TCA cycle [110]. The overexpression of GLUT transporter can be targeted by ritonavir, an antiretroviral protease inhibitor active against HIV [111]. The exposure to this agent can induce a downregulation of glycolysis and the concomitant dephosphorylation of mTOR, potentiated by the action of metformin [112]. Interestingly, metformin alone has been reported to deactivate mTOR signaling in MM cells, through the stimulation of AMPK [113].

Accordingly, interfering with the PI3K/Akt/mTOR pathway have a direct impact on glycolytic rates, as the exposure to BEZ235 (PI3K/mTOR dual inhibitor) impairs glycolysis, thus counteracting its upregulation by low concentration of topoisomerase inhibitors, such as doxorubicin, etoposide, and topotecan [114]. The mechanism of action of BEZ235 on glycolysis involves the downregulation of hexokinase II, the already mentioned glycolytic enzyme that is deeply involved in the acquisition of the Warburg effect metabolic rearrangement [115], which results overexpressed in myeloma cells [116]. As it occurs in AML and CLL, the microenvironment participates in the acquisition of metabolic changes, through the action of specific adhesion proteins. In MM, Reelin, a glycoprotein found on extracellular matrix, whose action is linked to cell proliferation and migration during development and in cancer [117], has been shown to stimulate mTOR signaling, which in turn increases glycolysis through the HIF1α upregulation [118]. mTOR signaling can also be inactivated in MM by targeting metabolic pathways that have been less studied, even though they are crucial in the energetic cellular balance, such as nicotinamide adenine dinucleotide (NAD) biosynthesis. NAD is in fact a cofactor that participates in a great variety of chemical reaction, acting as an electron shuttle [119]. Cea et al. reported that, inhibiting this rate-limiting enzyme involved in its formation, a reduction in the PI3K/Akt/mTOR activity can be observed, probably due to a deficit in cellular nutrient availability induced by the lack of accessible NAD [120].

3.6. Lymphomas

Lymphomas are subdivided in Hodgkin (HD) and non-Hodgkin (NHL). From the bioenergetic perspective, this categorization reflects a difference in their metabolism, as HD are associated with OXPHOS [121], while the NHL subtype are more prone to glycolysis [122], with some relevant exceptions. In Diffuse Large B Cell Lymphoma (DLBCL), belonging to the NHL class, a subset of patients' cells shows in fact a transcriptional profile enriched in mRNAs participating in mitochondrial energy production, such as OXPHOS and electron transport chain machinery [123]. This observation has been confirmed at both proteomic and metabolic levels, with glucose- and fatty acid-derived carbon converging in the TCA cycle to generate the great proportion of cellular energy [124]. In general, evidences state that mTOR directly impacts on lymphoma glycolysis, as multiple works show that its targeted inhibition reduces the glycolytic phenotype. For example, primary effusion lymphoma (PEL), follicular and Burkitt Lymphomas have all been associated to high aerobic glycolysis and fatty acid synthesis, when compared to normal B cells [125]. This phenotype appears to be driven by the PI3K/Akt/mTOR module, since PI3K inhibition with LY294002 potently reduces both glycolysis and fatty acid synthesis (FAS) [125]. Interestingly, in normal B cells this inhibition causes the decrease of glycolysis, but not that of FAS, suggesting a different FAS regulation by the PI3K/Akt/mTOR axis between normal and pathologic cells [125]. Mediani et al. extended this study, confirming the highly glycolytic phenotype of PEL cells and its inhibition using the PI3K/Akt/mTOR inhibitors [126]. Moreover, they observed a switch in cell metabolism towards an oxidative phenotype when those cells are simultaneously exposed to a glycolysis inhibitor (2DG) and a dual PI3K/mTOR inhibitor (PF-04691502) [126]. This shift is, however, not sufficient to protect cells from the apoptosis synergistically induced by the two targeted agents [126]. Glycolysis inhibition through mTOR targeting has also been observed in mantle cell lymphoma (MCL). In fact, everolimus downregulated glucose transporters, glycolytic enzymes, and lactate dehydrogenase, thus inducing a decrease in lactate production [127], while the dual mTORC1/2 inhibitor AZD-2014 caused the activation of AMPK and the downregulation of glycolysis-related proteins [128]. Despite the already mentioned metabolic difference with the other NHL, in DBLCL interfering with the glucose catabolism has been reported to generate the same results, the mTOR inactivation, that in turn causes the downregulation of the pro-survival factor Mcl-1 [129]. However, a recent work by Chiche et al. subverted this picture, showing that mTORC1 is instead involved in the acquisition of the oxidative metabolism [130]. They observed that DLBCL can be subdivided into two categories, according to the expression levels of glyceraldehyde 3-phosphate dehydrogenase (GAPDH), an enzyme directly involved in the glycolytic pathway [130]. The high-expressing GAPDH cells were accordingly characterized by a glycolytic phenotype, while those cells that expressed GAPDH at low levels showed an OXPHOS preference, coupled with mTORC1 hyperactivation, fueled by glutamine [130]. Exposure to rapamycin impaired mitochondrial respiration while increasing glycolysis-derived ATP [130]. Interestingly, GAPDH overexpression inactivated mTORC1, thus, suggesting that this enzyme is implicated in the regulation of mTOR, at least in the DLBCL setting [118]. Table 3 summarizes the metabolic changes induced by different inhibitors.

Table 3. Targeting mTOR and metabolism in Lymphomas.

Type	Compound	Target	Effect (Ref)
PEL	LY294002	PI3K	Decreased glycolysis and FAS (125)
	PF04691502	PI3K/mTOR	Reduction of lactate production (126)
	Akti 1/2	Akt	Reduction of lactate production (126)
MCL	Everolimus	mTORC1	Reduction of lactate production (127)
	AZD-2014	mTORC1/2	Downregulation of glycolytic enzymes (128)
DBLCL	2DG	hexokinase	Inactivation of Akt/mTOR and decreased expression of Mcl-1 (129)
	rapamycin	mTORC1	Reduction of OXPHOS and increase of glycolytic ATP in the oxidative DBLCL subset (130)

4. Summary and Concluding Remarks

As shown above, in hematologic malignancies the mTOR activation, working in opposition with AMPK but in concert with other oncogenes such as Bcr-Abl, or metabolic modulators like HIF1α, contributes to confer the glycolytic phenotype by directly and indirectly regulating key glycolytic enzyme activity (Figure 2). This metabolic activity has been linked to the acquisition of resistance to therapeutic agents. In our opinion, it would be important to predict and investigate the use of compounds active in reprogramming aberrant metabolic pathways, adopted in combination with standard treatments, for reducing the onset and impairing the resistance mechanisms developed by resistant cells. However, we are only beginning to understand the intricacy of the multiple regulation layers that contribute to the mTOR-mediated metabolic reprogramming. Additional studies are surely needed to shed light upon a mechanism that might constitute a major target to improve the current therapeutic arsenal of this group of hematologic malignancies.

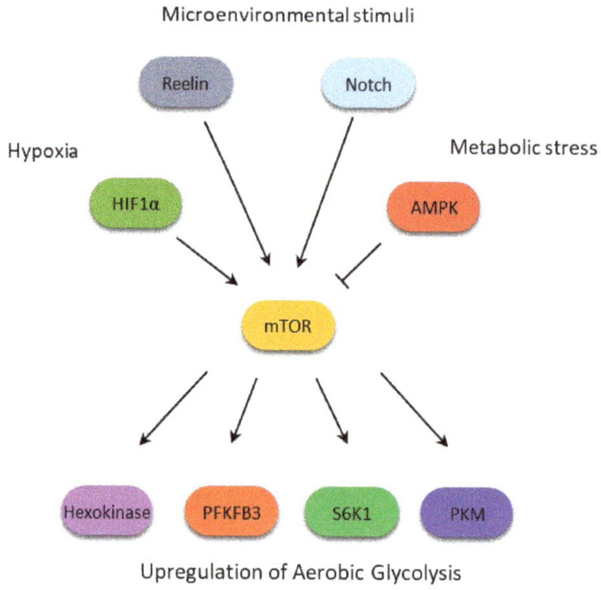

Figure 2. mTOR integrates multiple signals to confer a glycolytic phenotype on cells from hematologic malignancies.

Author Contributions: Writing—original draft preparation, S.M., M.R.R.; writing—review and editing, S.M, M.R.R, A.T. All authors have read and agreed to the published version of the manuscript.

Funding: This work was supported by the Sapienza University (RM116154EC670667, RP11715C7D08BDC2, RM11715C7D01147A, RM11816436436622) grants, A.P.E. Onlus (Associazione per i Pazienti Ematologici), and Fondazione Internazionale D'AMATO Onlus.

Acknowledgments: We are grateful to RomaAIL (Associazione Italiana contro le Leucemie-linfomi e mieloma) for providing the laboratory space.

Conflicts of Interest: The authors declare no conflict of interest.

References

1. Wullschleger, S.; Loewith, R.; Hall, M.N. TOR Signaling in Growth and Metabolism. *Cell* **2006**, *124*, 471–484. [CrossRef] [PubMed]
2. Zhao, J.; Zhai, B.; Gygi, S.P.; Goldberg, A.L. mTOR inhibition activates overall protein degradation by the ubiquitin proteasome system as well as by autophagy. *Proc. Natl. Acad. Sci. USA* **2015**, *112*, 15790–15797. [CrossRef] [PubMed]
3. Saxton, R.A.; Sabatini, D.M. mTOR Signaling in Growth, Metabolism, and Disease. *Cell* **2017**, *168*, 960–976. [CrossRef] [PubMed]
4. Martelli, A.M.; Faenza, I.; Billi, A.M.; Manzoli, L.; Evangelisti, C.; Falà, F.; Cocco, L. Intranuclear 3-phosphoinositide metabolism and Akt signaling: New mechanisms for tumori genesis and protection against apoptosis? *Cell Signal* **2006**, *18*, 1101–1107. [CrossRef]
5. Gingras, A.-C.; Raught, B.; Sonenberg, N. Regulation of translation initiation by FRAP/mTOR. *Genes Dev.* **2001**, *15*, 807–826. [CrossRef]
6. Yang, H.; Rudge, D.G.; Koos, J.D.; Vaidialingam, B.; Yang, H.J.; Pavletich, N.P. mTOR kinase structure, mechanism and regulation. *Nature* **2013**, *497*, 217–223. [CrossRef]

7. Sarbassov, S.D.; Ali, S.M.; Sengupta, S.; Sheen, J.-H.; Hsu, P.P.; Bagley, A.F.; Markhard, A.L.; Sabatini, D.M. Prolonged Rapamycin Treatment Inhibits mTORC2 Assembly and Akt/PKB. *Mol. Cell* **2006**, *22*, 159–168. [CrossRef]
8. Zhang, H.; Cicchetti, G.; Onda, H.; Koon, H.B.; Asrican, K.; Bajraszewski, N.; Vazquez, F.; Carpenter, C.L.; Kwiatkowski, D.J. Loss of Tsc1/Tsc2 activates mTOR and disrupts PI3K-Akt signaling through downregulation of PDGFR. *J. Clin. Investig.* **2003**, *112*, 1223–1233. [CrossRef]
9. Sancak, Y.; Thoreen, C.C.; Peterson, T.R.; Lindquist, R.A.; Kang, S.A.; Spooner, E.; Carr, S.A.; Sabatini, D.M. PRAS40 Is an Insulin-Regulated Inhibitor of the mTORC1 Protein Kinase. *Mol. Cell* **2007**, *25*, 903–915. [CrossRef]
10. Ma, X.M.; Blenis, J. Molecular mechanisms of mTOR-mediated translational control. *Nat. Rev. Mol. Cell Boil.* **2009**, *10*, 307–318. [CrossRef]
11. Holz, M.K.; Ballif, B.A.; Gygi, S.P.; Blenis, J. mTOR and S6K1 Mediate Assembly of the Translation Preinitiation Complex through Dynamic Protein Interchange and Ordered Phosphorylation Events. *Cell* **2005**, *123*, 569–580. [CrossRef] [PubMed]
12. Ruvinsky, I.; Meyuhas, O.; Ruvinsky, I. Ribosomal protein S6 phosphorylation: from protein synthesis to cell size. *Trends Biochem. Sci.* **2006**, *31*, 342–348. [CrossRef] [PubMed]
13. Esnault, S.; Shen, Z.-J.; Malter, J.S. Protein Translation and Signaling in Human Eosinophils. *Front. Med.* **2017**, *4*, 150. [CrossRef] [PubMed]
14. Dunlop, E.; Tee, A. Mammalian target of rapamycin complex 1: Signalling inputs, substrates and feedback mechanisms. *Cell. Signal.* **2009**, *21*, 827–835. [CrossRef]
15. Mamane, Y.; Petroulakis, E.; LeBacquer, O.; Sonenberg, N. mTOR, translation initiation and cancer. *Oncogene* **2006**, *25*, 6416–6422. [CrossRef]
16. De Benedetti, A.; Graff, J.R. eIF-4E expression and its role in malignancies and metastases. *Oncogene* **2004**, *23*, 3189–3199. [CrossRef]
17. Magnuson, B.; Ekim, B.; Fingar, D.C. Regulation and function of ribosomal protein S6 kinase (S6K) within mTOR signalling networks. *Biochem. J.* **2012**, *441*, 1–21. [CrossRef]
18. Howell, J.J.; Ricoult, S.J.; Manning, B.D.; Ben-Sahra, I. A growing role for mTOR in promoting anabolic metabolism. *Biochem. Soc. Trans.* **2013**, *41*, 906–912. [CrossRef]
19. Fruman, D.A.; Chiu, H.; Hopkins, B.D.; Bagrodia, S.; Cantley, L.C.; Abraham, R.T. The PI3K Pathway in Human Disease. *Cell* **2017**, *170*, 605–635. [CrossRef]
20. Linke, M.; Fritsch, S.D.; Sukhbaatar, N.; Hengstschläger, M.; Weichhart, T. mTORC1 and mTORC2 as regulators of cell metabolism in immunity. *FEBS Lett.* **2017**, *591*, 3089–3103. [CrossRef]
21. Miller, D.M.; Thomas, S.D.; Islam, A.; Muench, D.; Sedoris, K. c-Myc and cancer metabolism. *Clin. Cancer Res.* **2012**, *18*, 5546–5553. [CrossRef] [PubMed]
22. Thomas, L.W.; Ashcroft, M. Exploring the molecular interface between hypoxia-inducible factor signalling and mitochondria. *Cell. Mol. Life Sci.* **2019**, *76*, 1759–1777. [CrossRef] [PubMed]
23. Iurlaro, R.; León-Annicchiarico, C.L.; Munoz-Pinedo, C. Regulation of Cancer Metabolism by Oncogenes and Tumor Suppressors. *Methods Enzymol.* **2014**, *542*, 59–80. [PubMed]
24. Jacquel, A.; Luciano, F.; Robert, G.; Auberger, P. Implication and Regulation of AMPK during Physiological and Pathological Myeloid Differentiation. *Int. J. Mol. Sci.* **2018**, *19*, 2991. [CrossRef] [PubMed]
25. Schieke, S.M.; Phillips, D.; McCoy, J.P.; Aponte, A.M.; Shen, R.-F.; Balaban, R.S.; Finkel, T. The Mammalian Target of Rapamycin (mTOR) Pathway Regulates Mitochondrial Oxygen Consumption and Oxidative Capacity. *J. Boil. Chem.* **2006**, *281*, 27643–27652. [CrossRef]
26. Hardie, D.G. AMP-activated/SNF1 protein kinases: conserved guardians of cellular energy. *Nat. Rev. Mol. Cell Boil.* **2007**, *8*, 774–785. [CrossRef]
27. Jacinto, E.; Loewith, R.; Schmidt, A.; Lin, S.; Rüegg, M.A.; Hall, A.; Hall, M.N. Mammalian TOR complex 2 controls the actin cytoskeleton and is rapamycin insensitive. *Nat.* **2004**, *6*, 1122–1128. [CrossRef]
28. Sarbassov, S.D.; Ali, S.M.; Kim, D.-H.; A Guertin, D.; Latek, R.R.; Erdjument-Bromage, H.; Tempst, P.; Sabatini, D.M. Rictor, a Novel Binding Partner of mTOR, Defines a Rapamycin-Insensitive and Raptor-Independent Pathway that Regulates the Cytoskeleton. *Curr. Boil.* **2004**, *14*, 1296–1302. [CrossRef]
29. Niles, B.J.; Powers, T. TOR complex 2–Ypk1 signaling regulates actin polarization via reactive oxygen species. *Mol. Boil. Cell* **2014**, *25*, 3962–3972. [CrossRef]

30. Zoncu, R.; Sabatini, D.M.; Efeyan, A. mTOR: From growth signal integration to cancer, diabetes and ageing. *Nat. Rev. Mol. Cell Biol.* **2011**, *12*, 21–35. [CrossRef]
31. Kamada, Y.; Fujioka, Y.; Suzuki, N.N.; Inagaki, F.; Wullschleger, S.; Loewith, R.; Hall, M.N.; Ohsumi, Y. Tor2 Directly Phosphorylates the AGC Kinase Ypk2 To Regulate Actin Polarization†. *Mol. Cell. Boil.* **2005**, *25*, 7239–7248. [CrossRef] [PubMed]
32. Schmidt, A.; Bickle, M.; Beck, T.; Hall, M.N. The Yeast Phosphatidylinositol Kinase Homolog TOR2 Activates RHO1 and RHO2 via the Exchange Factor ROM2. *Cell* **1997**, *88*, 531–542. [CrossRef]
33. Dada, S.; Demartines, N.; Dormond, O. mTORC2 regulates PGE2-mediated endothelial cell survival and migration. *Biochem. Biophys. Res. Commun.* **2008**, *372*, 875–879. [CrossRef] [PubMed]
34. Zinzalla, V.; Stracka, D.; Oppliger, W.; Hall, M.N. Activation of mTORC2 by Association with the Ribosome. *Cell* **2011**, *144*, 757–768. [CrossRef]
35. Kazyken, D.; Magnuson, B.; Bodur, C.; Acosta-Jaquez, H.A.; Zhang, D.; Tong, X.; Barnes, T.M.; Steinl, G.K.; Patterson, N.E.; Altheim, C.H.; et al. AMPK directly activates mTORC2 to promote cell survival during acute energetic stress. *Sci. Signal.* **2019**, *12*, e3249. [CrossRef]
36. Moloughney, J.G.; Kim, P.K.; Vega-Cotto, N.M.; Wu, C.-C.; Zhang, S.; Adlam, M.; Lynch, T.; Chou, P.-C.; Rabinowitz, J.D.; Werlen, G.; et al. mTORC2 Responds to Glutamine Catabolite Levels to Modulate the Hexosamine Biosynthesis Enzyme GFAT1. *Mol. Cell* **2016**, *63*, 811–826. [CrossRef]
37. Willems, L.; Tamburini, J.; Chapuis, N.; Lacombe, C.; Mayeux, P.; Bouscary, D. PI3K and mTOR Signaling Pathways in Cancer: New Data on Targeted Therapies. *Curr. Oncol. Rep.* **2012**, *14*, 129–138. [CrossRef]
38. Conciatori, F.; Ciuffreda, L.; Bazzichetto, C.; Falcone, I.; Pilotto, S.; Bria, E.; Cognetti, F.; Milella, M. mTOR Cross-Talk in Cancer and Potential for Combination Therapy. *Cancers* **2018**, *10*, 23. [CrossRef]
39. Cairns, R.A.; Harris, I.S.; Mak, T.W. Regulation of cancer cell metabolism. *Nat. Rev. Cancer* **2011**, *11*, 85–95. [CrossRef]
40. Kohli, L.; Passegué, E. Surviving change: the metabolic journey of hematopoietic stem cells. *Trends Cell Boil.* **2014**, *24*, 479–487. [CrossRef]
41. Sharma, A.; Boise, L.H.; Shanmugam, M. Cancer Metabolism and the Evasion of Apoptotic Cell Death. *Cancers* **2019**, *11*, 1144. [CrossRef] [PubMed]
42. Anastasiou, D.; Poulogiannis, G.; Asara, J.M.; Boxer, M.B.; Jiang, J.-K.; Shen, M.; Bellinger, G.; Sasaki, A.T.; Locasale, J.W.; Auld, U.S.; et al. Inhibition of Pyruvate Kinase M2 by Reactive Oxygen Species Contributes to Cellular Antioxidant Responses. *Sci.* **2011**, *334*, 1278–1283. [CrossRef] [PubMed]
43. Finkel, T. Signal transduction by reactive oxygen species. *J. Cell Boil.* **2011**, *194*, 7–15. [CrossRef] [PubMed]
44. DeBerardinis, R.J. Is cancer a disease of abnormal cellular metabolism? New angles on an old idea. *Genet. Med.* **2008**, *10*, 767–777. [CrossRef] [PubMed]
45. Martinez-Outschoorn, U.E.; Lin, Z.; Trimmer, C.; Flomenberg, N.; Wang, C.; Pavlides, S.; Pestell, R.G.; Howell, A.; Sotgia, F.; Lisanti, M.P. Cancer cells metabolically 'fertilize' the tumor microenvironment with hydrogen peroxide, driving the Warburg effect: Implications for PET imaging of human tumors. *Cell Cycle* **2011**, *10*, 2504–2520. [CrossRef] [PubMed]
46. Yuneva, M.O.; Fan, T.W.M.; Allen, T.D.; Higashi, R.M.; Ferraris, D.V.; Tsukamoto, T.; Matés, J.M.; Alonso, F.J.; Wang, C.; Seo, Y.; et al. The Metabolic Profile of Tumors Depends on Both the Responsible Genetic Lesion and Tissue Type. *Cell Metab.* **2012**, *15*, 157–170. [CrossRef]
47. Heiden, M.G.V.; Cantley, L.C.; Thompson, C.B. Understanding the Warburg Effect: The Metabolic Requirements of Cell Proliferation. *Sci.* **2009**, *324*, 1029–1033. [CrossRef]
48. Li, B.; Simon, M.C. Molecular Pathways: Targeting MYC-induced metabolic reprogramming and oncogenic stress in cancer. *Clin. Cancer Res.* **2013**, *19*, 5835–5841. [CrossRef]
49. Jiang, Y.; Nakada, D. Cell intrinsic and extrinsic regulation of leukemia cell metabolism. *Int. J. Hematol.* **2016**, *103*, 607–616. [CrossRef]
50. Krejčí, A. Metabolic sensors and their interplay with cell signalling and transcription. *Biochem. Soc. Trans.* **2012**, *40*, 311–323. [CrossRef]
51. Metallo, C.M.; Heiden, M.G.V. Metabolism strikes back: metabolic flux regulates cell signaling. *Genome Res.* **2010**, *24*, 2717–2722. [CrossRef] [PubMed]
52. Chen, W.-L.; Wang, J.-H.; Zhao, A.-H.; Xu, X.; Wang, Y.-H.; Chen, T.-L.; Li, J.-M.; Mi, J.-Q.; Zhu, Y.-M.; Liu, Y.-F.; et al. A distinct glucose metabolism signature of acute myeloid leukemia with prognostic value. *Blood* **2014**, *124*, 1645–1654. [CrossRef] [PubMed]

53. Herst, P.M.; Howman, R.A.; Neeson, P.J.; Berridge, M.V.; Ritchie, D.S. The level of glycolytic metabolism in acute myeloid leukemia blasts at diagnosis is prognostic for clinical outcome. *J. Leukoc. Biol.* **2011**, *89*, 51–55. [CrossRef]
54. Wang, Y.-H.; Israelsen, W.J.; Lee, N.; Yu, V.W.C.; Jeanson, N.T.; Clish, C.B.; Cantley, L.C.; Heiden, M.G.V.; Scadden, D.T. Cell-state-specific metabolic dependency in hematopoiesis and leukemogenesis. *Cell* **2014**, *158*, 1309–1323. [CrossRef] [PubMed]
55. Petronini, P.G.; Urbani, S.; Alfieri, R.; Borghetti, A.F.; Guidotti, G.G. Cell susceptibility to apoptosis by glutamine deprivation and rescue: Survival and apoptotic death in cultured lymphoma-leukemia cell lines. *J. Cell. Physiol.* **1996**, *169*, 175–185. [CrossRef]
56. Ricciardi, M.R.; Mirabilii, S.; Allegretti, M.; Licchetta, R.; Calarco, A.; Torrisi, M.R.; Foà, R.; Nicolai, R.; Peluso, G.; Tafuri, A. Targeting the leukemia cell metabolism by the CPT1a inhibition: functional preclinical effects in leukemias. *Blood* **2015**, *126*, 1925–1929. [CrossRef] [PubMed]
57. Tabe, Y.; Saitoh, K.; Yang, H.; Sekihara, K.; Yamatani, K.; Ruvolo, V.; Taka, H.; Kaga, N.; Kikkawa, M.; Arai, H.; et al. Inhibition of FAO in AML co-cultured with BM adipocytes: mechanisms of survival and chemosensitization to cytarabine. *Sci. Rep.* **2018**, *8*, 16837. [CrossRef]
58. Samudio, I.; Harmancey, R.; Fiegl, M.; Kantarjian, H.; Konopleva, M.; Korchin, B.; Kaluarachchi, K.; Bornmann, W.; Duvvuri, S.; Taegtmeyer, H.; et al. Pharmacologic inhibition of fatty acid oxidation sensitizes human leukemia cells to apoptosis induction. *J. Clin. Investig.* **2010**, *120*, 142–156. [CrossRef]
59. Liu, L.-L.; Long, Z.-J.; Wang, L.-X.; Zheng, F.-M.; Fang, Z.-G.; Yan, M.; Xu, D.-F.; Chen, J.-J.; Wang, S.-W.; Lin, N.-J.; et al. Inhibition of mTOR Pathway Sensitizes Acute Myeloid Leukemia Cells to Aurora Inhibitors by Suppression of Glycolytic Metabolism. *Mol. Cancer Res.* **2013**, *11*, 1326–1336. [CrossRef]
60. Allegretti, M.; Ricciardi, M.R.; Licchetta, R.; Mirabilii, S.; Orecchioni, S.; Reggiani, F.; Talarico, G.; Foà, R.; Bertolini, F.; Amadori, S.; et al. The pan-class I phosphatidyl-inositol-3 kinase inhibitor NVP-BKM120 demonstrates anti-leukemic activity in acute myeloid leukemia. *Sci. Rep.* **2015**, *5*, 18137. [CrossRef]
61. Poulain, L.; Sujobert, P.; Zylbersztejn, F.; Barreau, S.; Stuani, L.; Lambert, M.; Palama, T.L.; Chesnais, V.; Birsen, R.; Vergez, F.; et al. High mTORC1 activity drives glycolysis addiction and sensitivity to G6PD inhibition in acute myeloid leukemia cells. *Leuk.* **2017**, *31*, 2326–2335. [CrossRef] [PubMed]
62. Pradelli, L.A.; Bénéteau, M.; Chauvin, C.; Jacquin, M.A.; Marchetti, S.; Muñoz-Pinedo, C.; Auberger, P.; Pende, M.; Ricci, J.E. Glycolysis inhibition sensitizes tumor cells to death receptors-induced apoptosis by AMP kinase activation leading to Mcl-1 block in translations. *Oncogene* **2010**, *29*, 1641–1652. [CrossRef] [PubMed]
63. Estañ, M.C.; Calviño, E.; De Blas, E.; Boyano-Adánez, M.D.C.; Mena, M.L.; Gómez-Gómez, M.; Rial, E.; Aller, P. 2-Deoxy-d-glucose cooperates with arsenic trioxide to induce apoptosis in leukemia cells: Involvement of IGF-1R-regulated Akt/mTOR, MEK/ERK and LKB-1/AMPK signaling pathways. *Biochem. Pharmacol.* **2012**, *84*, 1604–1616. [CrossRef] [PubMed]
64. Pereira, O.; Teixeira, A.; Sampaio-Marques, B.; Castro, I.; Girão, H.; Ludovico, P. Signalling mechanisms that regulate metabolic profile and autophagy of acute myeloid leukaemia cells. *J. Cell. Mol. Med.* **2018**, *22*, 4807–4817. [CrossRef]
65. Feng, Y.; Wu, L. mTOR up-regulation of PFKFB3 is essential for acute myeloid leukemia cell survival. *Biochem. Biophys. Res. Commun.* **2017**, *483*, 897–903. [CrossRef]
66. Braun, M.; Qorraj, M.; Büttner, M.; A Klein, F.; Saul, D.; Aigner, M.; Huber, W.; Mackensen, A.; Jitschin, R.; Mougiakakos, D. CXCL12 promotes glycolytic reprogramming in acute myeloid leukemia cells via the CXCR4/mTOR axis. *Leuk.* **2016**, *30*, 1788–1792. [CrossRef]
67. Tabe, Y.; Konopleva, M. Role of Microenvironment in Resistance to Therapy in AML. *Curr. Hematol. Malign- Rep.* **2015**, *10*, 96–103. [CrossRef]
68. Song, K.; Li, M.; Xu, X.; Xuan, L.; Huang, G.; Liu, Q. Resistance to chemotherapy is associated with altered glucose metabolism in acute myeloid leukemia. *Oncol. Lett.* **2016**, *12*, 334–342. [CrossRef]
69. Wang, X.; Proud, C.G. The mTOR Pathway in the Control of Protein Synthesis. *Physiol.* **2006**, *21*, 362–369. [CrossRef]
70. Willems, L.; Jacque, N.; Jacquel, A.; Neveux, N.; Maciel, T.T.; Lambert, M.; Schmitt, A.; Poulain, L.; Green, A.S.; Uzunov, M.; et al. Inhibiting glutamine uptake represents an attractive new strategy for treating acute myeloid leukemia. *Blood* **2013**, *122*, 3521–3532. [CrossRef]

71. Nepstad, I.; Reikvam, H.; Brenner, A.K.; Bruserud, Ø.; Hatfield, K.J. Resistance to the Antiproliferative In Vitro Effect of PI3K-Akt-mTOR Inhibition in Primary Human Acute Myeloid Leukemia Cells Is Associated with Altered Cell Metabolism. *Int. J. Mol. Sci.* **2018**, *19*, 382. [CrossRef] [PubMed]
72. Gottschalk, S. Imatinib (STI571)-Mediated Changes in Glucose Metabolism in Human Leukemia BCR-ABL-Positive Cells. *Clin. Cancer Res.* **2004**, *10*, 6661–6668. [CrossRef] [PubMed]
73. Kominsky, D.J.; Klawitter, J.; Brown, J.L.; Boros, L.G.; Melo, J.V.; Eckhardt, S.G.; Serkova, N.J. Abnormalities in Glucose Uptake and Metabolism in Imatinib-Resistant Human BCR-ABL-Positive Cells. *Clin. Cancer Res.* **2009**, *15*, 3442–3450. [CrossRef]
74. De Rosa, V.; Monti, M.; Terlizzi, C.; Fonti, R.; Del Vecchio, S.; Iommelli, F. Coordinate Modulation of Glycolytic Enzymes and OXPHOS by Imatinib in BCR-ABL Driven Chronic Myelogenous Leukemia Cells. *Int. J. Mol. Sci.* **2019**, *20*, 3134. [CrossRef] [PubMed]
75. Bentley, J.; Walker, I.; McIntosh, E.; Whetton, A.; Owen-Lynch, P.J.; Baldwin, S.A. Glucose transport regulation by p210 Bcr-Abl in a chronic myeloid leukaemia model. *Br. J. Haematol.* **2001**, *112*, 212–215. [CrossRef]
76. Sontakke, P.; Koczula, K.M.; Jaques, J.; Wierenga, A.T.; Brouwers-Vos, A.Z.; Pruis, M.; Günther, U.L.; Vellenga, E.; Schuringa, J.J. Hypoxia-Like Signatures Induced by BCR-ABL Potentially Alter the Glutamine Uptake for Maintaining Oxidative Phosphorylation. *PLoS ONE* **2016**, *11*, e0153226. [CrossRef]
77. Zhao, F.; Mancuso, A.; Bui, T.V.; Tong, X.; Gruber, J.J.; Swider, C.R.; Sanchez, P.V.; Lum, J.J.; Sayed, N.; Melo, J.V.; et al. Imatinib resistance associated with BCR-ABL upregulation is dependent on HIF-1a-induced metabolic reprograming. *Oncogene* **2010**, *29*, 2962–2972. [CrossRef]
78. Noel, B.M.; Ouellette, S.B.; Marholz, L.; Dickey, D.; Navis, C.; Yang, T.-Y.; Nguyen, V.; Parker, S.J.; Bernlohr, D.; Sachs, Z.; et al. Multiomic Profiling of Tyrosine Kinase Inhibitor-Resistant K562 Cells Suggests Metabolic Reprogramming To Promote Cell Survival. *J. Proteome Res.* **2019**, *18*, 1842–1856. [CrossRef]
79. Flis, K.; Irvine, D.; Copland, M.; Bhatia, R.; Skorski, T. Chronic myeloid leukemia stem cells display alterations in expression of genes involved in oxidative phosphorylation. *Leuk. Lymphoma* **2012**, *53*, 2474–2478. [CrossRef]
80. Kuntz, E.M.; Baquero, P.; Michie, A.M.; Dunn, K.; Tardito, S.; Holyoake, T.L.; Helgason, G.V.; Gottlieb, E. Targeting mitochondrial oxidative phosphorylation eradicates therapy-resistant chronic myeloid leukemic stem cells. *Nat. Med.* **2017**, *23*, 1234–1240. [CrossRef]
81. Abraham, A.; Qiu, S.; Chacko, B.K.; Li, H.; Paterson, A.; He, J.; Agarwal, P.; Shah, M.; Welner, R.; Darley-Usmar, V.M.; et al. SIRT1 regulates metabolism and leukemogenic potential in CML stem cells. *J. Clin. Investig.* **2019**, *129*, 2685–2701. [CrossRef] [PubMed]
82. Kim, J.H.; Chu, S.C.; Gramlich, J.L.; Pride, Y.B.; Babendreier, E.; Chauhan, D.; Salgia, R.; Podar, K.; Griffin, J.D.; Sattler, M. Activation of the PI3K/mTOR pathway by BCR-ABL contributes to increased production of reactive oxygen species. *Blood* **2005**, *105*, 1717–1723. [CrossRef] [PubMed]
83. Barger, J.F.; Gallo, C.A.; Tandon, P.; Liu, H.; Sullivan, A.; Grimes, H.L.; Plas, D.R. S6K1 determines the metabolic requirements for BCR-ABL survival. *Oncogene* **2013**, *32*, 453–461. [CrossRef] [PubMed]
84. Shinohara, H.; Taniguchi, K.; Kumazaki, M.; Yamada, N.; Ito, Y.; Otsuki, Y.; Uno, B.; Hayakawa, F.; Minami, Y.; Naoe, T.; et al. Anti-cancer fatty-acid derivative induces autophagic cell death through modulation of PKM isoform expression profile mediated by bcr-abl in chronic myeloid leukemia. *Cancer Lett.* **2015**, *360*, 28–38. [CrossRef] [PubMed]
85. Zhang, Z.; Deng, X.; Liu, Y.; Liu, Y.; Sun, L.; Chen, F. PKM2, function and expression and regulation. *Cell Biosci.* **2019**, *9*, 52. [CrossRef]
86. Zhang, X.; Chen, J.; Ai, Z.; Zhang, Z.; Lin, L.; Wei, H. Targeting glycometabolic reprogramming to restore the sensitivity of leukemia drug-resistant K562/ADM cells to adriamycin. *Life Sci.* **2018**, *215*, 1–10. [CrossRef]
87. Boag, J.M.; Beesley, A.H.; Firth, M.J.; Freitas, J.R.; Ford, J.; Hoffmann, K.; Cummings, A.J.; De Klerk, N.H.; Kees, U.R. Altered glucose metabolism in childhood pre-B acute lymphoblastic leukaemia. *Leuk.* **2006**, *20*, 1731–1737. [CrossRef]
88. Liu, T.; Kishton, R.J.; Macintyre, A.N.; Gerriets, V.A.; Xiang, H.; Liu, X.; Abel, E.D.; Rizzieri, D.; Locasale, J.W.; Rathmell, J.C. Glucose transporter 1-mediated glucose uptake is limiting for B-cell acute lymphoblastic leukemia anabolic metabolism and resistance to apoptosis. *Cell Death Dis* **2014**, *5*, e1470. [CrossRef]
89. Stäubert, C.; Bhuiyan, H.; Lindahl, A.; Broom, O.J.; Zhu, Y.; Islam, S.; Linnarsson, S.; Lehtiö, J.; Nordström, A. Rewired Metabolism in Drug-resistant Leukemia Cells: A metabolic switch hallmarked by reduced dependence on exogenous glutamine. *J. Biol. Chem.* **2015**, *290*, 8348–8359. [CrossRef]

90. Aoki, S.; Morita, M.; Hirao, T.; Yamaguchi, M.; Shiratori, R.; Kikuya, M.; Chibana, H.; Ito, K. Shift in energy metabolism caused by glucocorticoids enhances the effect of cytotoxic anti-cancer drugs against acute lymphoblastic leukemia cells. *Oncotarget* **2017**, *8*, 94271–94285. [CrossRef]
91. Kishton, R.J.; Barnes, C.E.; Nichols, A.G.; Cohen, S.; Gerriets, V.A.; Siska, P.J.; MacIntyre, A.N.; Goraksha-Hicks, P.; De Cubas, A.A.; Liu, T.; et al. AMPK Is Essential to Balance Glycolysis and Mitochondrial Metabolism to Control T-ALL Cell Stress and Survival. *Cell Metab.* **2016**, *23*, 649–662. [CrossRef] [PubMed]
92. Ferrando, A.A. The role of NOTCH1 signaling in T-ALL. *Hematol.* **2009**, *2009*, 353–361. [CrossRef] [PubMed]
93. Herranz, D.; Ambesi-Impiombato, A.; Sudderth, J.; Sanchez-Martin, M.; Belver, L.; Tosello, V.; Xu, L.; Wendorff, A.A.; Castillo, M.; Haydu, J.E.; et al. Metabolic reprogramming induces resistance to anti-NOTCH1 therapies in T cell acute lymphoblastic leukemia. *Nat. Med.* **2015**, *21*, 1182–1189. [CrossRef] [PubMed]
94. Frolova, O.; Samudio, I.; Benito, J.M.; Jacamo, R.; Kornblau, S.M.; Markovic, A.; Schober, W.; Lu, H.; Qiu, Y.H.; Buglio, D.; et al. Regulation of HIF-1α signaling and chemoresistance in acute lymphocytic leukemia under hypoxic conditions of the bone marrow microenvironment. *Cancer Boil. Ther.* **2012**, *13*, 858–870. [CrossRef] [PubMed]
95. Leclerc, G.J.; DeSalvo, J.; Du, J.; Gao, N.; Leclerc, G.M.; Lehrman, M.A.; Lampidis, T.J.; Barredo, J.C. Mcl-1 downregulation leads to the heightened sensitivity exhibited by BCR-ABL positive ALL to induction of energy and ER-stress. *Leuk. Res.* **2015**, *39*, 1246–1254. [CrossRef] [PubMed]
96. Liu, S.; Miriyala, S.; Keaton, M.A.; Jordan, C.T.; Wiedl, C.; Clair, D.K.S.; Moscow, J.A. Metabolic Effects of Acute Thiamine Depletion Are Reversed by Rapamycin in Breast and Leukemia Cells. *PLoS ONE* **2014**, *9*, e85702. [CrossRef]
97. Fernández-Ramos, A.A.; Marchetti-Laurent, C.; Poindessous, V.; Antonio, S.; Laurent-Puig, P.; Bortoli, S.; Loriot, M.-A.; Pallet, N. 6-mercaptopurine promotes energetic failure in proliferating T cells. *Oncotarget* **2017**, *8*, 43048–43060. [CrossRef]
98. Clapham, C.E.; Pettitt, A.R.; Slupsky, J.R. Targeting Cell Metabolism In Chronic Lymphocytic Leukaemia (CLL); A Viable Therapeutic Approach? *J. Hematol. Oncol. Res.* **2014**, *1*, 7–23. [CrossRef]
99. Galicia-Vázquez, G.; Aloyz, R. Metabolic rewiring beyond Warburg in chronic lymphocytic leukemia: How much do we actually know? *Crit. Rev. Oncol.* **2019**, *134*, 65–70. [CrossRef]
100. Koczula, K.M.; Ludwig, C.; Hayden, R.; Cronin, L.; Pratt, G.; Parry, H.; Tennant, D.; Drayson, M.; Bunce, C.M.; Khanim, F.L.; et al. Metabolic plasticity in CLL: Adaptation to the hypoxic niche. *Leukemia* **2016**, *30*, 65–73. [CrossRef]
101. Vangapandu, H.V.; Havranek, O.; Ayres, M.L.; Kaipparettu, B.A.; Balakrishnan, K.; Wierda, W.G.; Keating, M.J.; Davis, R.E.; Stellrecht, C.M.; Gandhi, V. B-cell Receptor Signaling Regulates Metabolism in Chronic Lymphocytic Leukemia. *Mol. Cancer Res.* **2017**, *15*, 1692–1703. [CrossRef] [PubMed]
102. Jitschin, R.; Braun, M.; Qorraj, M.; Saul, D.; Le Blanc, K.; Zenz, T.; Mougiakakos, D. Stromal cell–mediated glycolytic switch in CLL cells involves Notch-c-Myc signaling. *Blood* **2015**, *125*, 3432–3436. [CrossRef] [PubMed]
103. Vangapandu, H.V.; Ayres, M.L.; Bristow, C.A.; Wierda, W.G.; Keating, M.J.; Balakrishnan, K.; Stellrecht, C.M.; Gandhi, V. The Stromal Microenvironment Modulates Mitochondrial Oxidative Phosphorylation in Chronic Lymphocytic Leukemia Cells. *Neoplasia* **2017**, *19*, 762–771. [CrossRef]
104. Martinez-Marignac, V.; Smith, S.; Toban, N.; Bazile, M.; Aloyz, R. Resistance to Dasatinib in primary chronic lymphocytic leukemia lymphocytes involves AMPK-mediated energetic re-programming. *Oncotarget* **2013**, *4*, 2550–2566. [CrossRef] [PubMed]
105. Sharma, A.; Janocha, A.J.; Hill, B.T.; Smith, M.R.; Erzurum, S.C.; Almasan, A. Targeting mTORC1-mediated metabolic addiction overcomes fludarabine resistance in malignant B cells. *Mol. Cancer Res.* **2014**, *12*, 1205–1215. [CrossRef]
106. Siska, P.J.; Van Der Windt, G.J.W.; Kishton, R.J.; Cohen, S.; Eisner, W.; Maciver, N.J.; Kater, A.P.; Weinberg, J.B.; Rathmell, J.C. Suppression of Glut1 and Glucose Metabolism by Decreased Akt/mTORC1 Signaling Drives T Cell Impairment in B Cell Leukemia. *J. Immunol.* **2016**, *197*, 2532–2540. [CrossRef]
107. McBrayer, S.K.; Cheng, J.C.; Singhal, S.; Krett, N.L.; Rosen, S.T.; Shanmugam, M. Multiple myeloma exhibits novel dependence on GLUT4, GLUT8, and GLUT11: implications for glucose transporter-directed therapy. *Blood* **2012**, *119*, 4686–4697. [CrossRef]

108. Zhang, H.; Li, L.; Chen, Q.; Li, M.; Feng, J.; Sun, Y.; Zhao, R.; Zhu, Y.; Lv, Y.; Zhu, Z.; et al. PGC1β regulates multiple myeloma tumor growth through LDHA-mediated glycolytic metabolism. *Mol. Oncol.* **2018**, *12*, 1579–1595. [CrossRef]
109. Maiso, P.; Huynh, D.; Moschetta, M.; Sacco, A.; Aljawai, Y.; Mishima, Y.; Asara, J.M.; Roccaro, A.M.; Kimmelman, A.C.; Ghobrial, I.M. Metabolic signature identifies novel targets for drug resistance in multiple myeloma. *Cancer Res.* **2015**, *75*, 2071–2082. [CrossRef]
110. Gonsalves, W.I.; Ramakrishnan, V.; Hitosugi, T.; Ghosh, T.; Jevremovic, D.; Dutta, T.; Sakrikar, D.; Petterson, X.-M.; Wellik, L.; Kumar, S.K.; et al. Glutamine-derived 2-hydroxyglutarate is associated with disease progression in plasma cell malignancies. *JCI Insight* **2018**, *3*, 94543. [CrossRef]
111. Hresko, R.C.; Hruz, P.W. HIV Protease Inhibitors Act as Competitive Inhibitors of the Cytoplasmic Glucose Binding Site of GLUTs with Differing Affinities for GLUT1 and GLUT4. *PLoS ONE* **2011**, *6*, e25237. [CrossRef] [PubMed]
112. Dalva-Aydemir, S.; Bajpai, R.; Martinez, M.; Adekola, K.U.; Kandela, I.; Wei, C.; Shanmugam, M.; Singhal, S.; Koblinski, J.E.; Raje, N.S.; et al. Targeting the Metabolic Plasticity of Multiple Myeloma with FDA-Approved Ritonavir and Metformin. *Clin. Cancer Res.* **2015**, *21*, 1161–1171. [CrossRef] [PubMed]
113. Wang, Y.; Xu, W.; Yan, Z.; Zhao, W.; Mi, J.; Li, J.; Yan, H. Metformin induces autophagy and G0/G1 phase cell cycle arrest in myeloma by targeting the AMPK/mTORC1 and mTORC2 pathways. *J. Exp. Clin. Cancer Res.* **2018**, *37*, 63. [CrossRef] [PubMed]
114. Demel, H.-R.; Feuerecker, B.; Piontek, G.; Seidl, C.; Blechert, B.; Pickhard, A.; Essler, M. Effects of topoisomerase inhibitors that induce DNA damage response on glucose metabolism and PI3K/Akt/mTOR signaling in multiple myeloma cells. *Am. J. Cancer Res.* **2015**, *5*, 1649–1664. [PubMed]
115. Lis, P.; Dyląg, M.; Niedźwiecka, K.; Ko, Y.H.; Pedersen, P.L.; Goffeau, A.; Ułaszewski, S. The HK2 Dependent "Warburg Effect" and Mitochondrial Oxidative Phosphorylation in Cancer: Targets for Effective Therapy with 3-Bromopyruvate. *Mol.* **2016**, *21*, 1730. [CrossRef] [PubMed]
116. Nakano, A.; Miki, H.; Nakamura, S.; Harada, T.; Oda, A.; Amou, H.; Fujii, S.; Kagawa, K.; Takeuchi, K.; Ozaki, S.; et al. Up-regulation of hexokinaseII in myeloma cells: targeting myeloma cells with 3-bromopyruvate. *J. Bioenerg. Biomembr.* **2012**, *44*, 31–38. [CrossRef]
117. Khialeeva, E.; Carpenter, E.M. Nonneuronal roles for the reelin signaling pathway. *Dev. Dyn.* **2017**, *246*, 217–226. [CrossRef]
118. Qin, X.; Lin, L.; Cao, L.; Zhang, X.; Song, X.; Hao, J.; Zhang, Y.; Wei, R.; Huang, X.; Lu, J.; et al. Extracellular matrix protein Reelin promotes myeloma progression by facilitating tumor cell proliferation and glycolysis. *Sci. Rep.* **2017**, *7*, 45305. [CrossRef]
119. Zhang, D.X.; Zhang, J.P.; Hu, J.Y.; Huang, Y.S. The potential regulatory roles of NAD(+) and its metabolism in autophagy. *Metabolism* **2016**, *65*, 454–462. [CrossRef]
120. Cea, M.; Cagnetta, A.; Fulciniti, M.; Tai, Y.T.; Hideshima, T.; Chauhan, D.; Roccaro, A.; Sacco, A.; Calimeri, T.; Cottini, F.; et al. Targeting NAD+ salvage pathway induces autophagy in multiple myeloma cells via mTORC1 and extracellular signal-regulated kinase (ERK1/2) inhibition. *Blood* **2012**, *120*, 3519–3529. [CrossRef]
121. Mikkilineni, L.; Whitaker-Menezes, D.; Domingo-Vidal, M.; Sprandio, J.; Avena, P.; Cotzia, P.; Dulau-Florea, A.; Gong, J.; Uppal, G.; Zhan, T.; et al. Hodgkin lymphoma: A complex metabolic ecosystem with glycolytic reprogramming of the tumor microenvironment. *Semin. Oncol.* **2017**, *44*, 218–225. [CrossRef] [PubMed]
122. Bhatt, A.P.; Jacobs, S.R.; Freemerman, A.J.; Makowski, L.; Rathmell, J.C.; Dittmer, D.P.; Damania, B. Dysregulation of fatty acid synthesis and glycolysis in non-Hodgkin lymphoma. *Proc. Natl. Acad. Sci. USA* **2012**, *109*, 11818–11823. [CrossRef] [PubMed]
123. Monti, S.; Savage, K.J.; Kutok, J.L.; Feuerhake, F.; Kurtin, P.; Mihm, M.; Wu, B.; Pasqualucci, L.; Neuberg, N.; Aguiar, R.C.T.; et al. Molecular profiling of diffuse large B-cell lymphoma identifies robust subtypes including one characterized by host inflammatory response. *Blood* **2005**, *105*, 1851–1861. [CrossRef] [PubMed]
124. Caro, P.; Kishan, A.U.; Norberg, E.; Stanley, I.A.; Chapuy, B.; Ficarro, S.B.; Polak, K.; Tondera, D.; Gounarides, J.; Yin, H.; et al. Metabolic signatures uncover distinct targets in molecular subsets of diffuse large B cell lymphoma. *Cancer Cell* **2012**, *22*, 547–560. [CrossRef]
125. Lee, S.C.; Marzec, M.; Liu, X.; Wehrli, S.; Kantekure, K.; Ragunath, P.N.; Nelson, D.S.; Delikatny, E.J.; Glickson, J.D.; Wasik, M.A. Decreased lactate concentration and glycolytic enzyme expression reflect inhibition of mTOR signal. *NMR Biomed.* **2013**, *26*, 106–114. [CrossRef]

126. Mediani, L.; Gibellini, F.; Bertacchini, J.; Frasson, C.; Bosco, R.; Accordi, B.; Basso, G.; Bonora, M.; Calabrò, M.L.; Mattiolo, A.; et al. Reversal of the glycolytic phenotype of primary effusion lymphoma cells by combined targeting of cellular metabolism and PI3K/Akt/mTOR signaling. *Oncotarget* **2016**, *7*, 5521–5537. [CrossRef]
127. Akhenblit, P.J.; Hanke, N.T.; Gill, A.; Persky, D.O.; Howison, C.M.; Pagel, M.D.; Baker, A.F. Assessing Metabolic Changes in Response to mTOR Inhibition in a Mantle Cell Lymphoma Xenograft Model Using AcidoCEST MRI. *Mol. Imaging* **2016**, *15*, 1536012116645439. [CrossRef]
128. Sekihara, K.; Saitoh, K.; Han, L.; Ciurea, S.; Yamamoto, S.; Kikkawa, M.; Kazuno, S.; Taka, H.; Kaga, N.; Arai, H.; et al. Targeting mantle cell lymphoma metabolism and survival through simultaneous blockade of mTOR and nuclear transporter exportin-1. *Oncotarget* **2017**, *8*, 34552–34564. [CrossRef]
129. Coloff, J.L.; MacIntyre, A.N.; Nichols, A.G.; Liu, T.; Gallo, C.A.; Plas, D.R.; Rathmell, J.C. Akt-dependent glucose metabolism promotes Mcl-1 synthesis to maintain cell survival and resistance to Bcl-2 inhibition. *Cancer Res.* **2011**, *71*, 5204–5213. [CrossRef]
130. Chiche, J.; Reverso-Meinietti, J.; Mouchotte, A.; Rubio-Patiño, C.; Mhaidly, R.; Villa, E.; Bossowski, J.P.; Proics, E.; Grima-Reyes, M.; Paquet, A.; et al. GAPDH Expression Predicts the Response to R-CHOP, the Tumor Metabolic Status, and the Response of DLBCL Patients to Metabolic Inhibitor. *Cell Metab.* **2019**, *29*, 1243–1257. [CrossRef]

Review

mTOR Signalling in Head and Neck Cancer: Heads Up

Fiona H. Tan [1], Yuchen Bai [1], Pierre Saintigny [2,3] and Charbel Darido [1,4,*]

1 Division of Cancer Research, Peter MacCallum Cancer Centre, Grattan Street, Melbourne, Victoria 3000, Australia; fiona.tan@petermac.org (F.H.T.); yuchen.bai@petermac.org (Y.B.)
2 Univ Lyon, Université Claude Bernard Lyon 1, INSERM 1052, CNRS 5286, Centre Léon Bérard, Centre de recherche en cancérologie de Lyon, 69008 Lyon, France; Pierre.SAINTIGNY@lyon.unicancer.fr
3 Department of Medical Oncology, Centre Léon Bérard, 69008 Lyon, France
4 Sir Peter MacCallum Department of Oncology, The University of Melbourne, Parkville, Victoria 3052, Australia
* Correspondence: charbel.darido@petermac.org; Tel.: +613-8559-7111

Received: 9 March 2019; Accepted: 9 April 2019; Published: 9 April 2019 

Abstract: The mammalian target of rapamycin (mTOR) signalling pathway is a central regulator of metabolism in all cells. It senses intracellular and extracellular signals and nutrient levels, and coordinates the metabolic requirements for cell growth, survival, and proliferation. Genetic alterations that deregulate mTOR signalling lead to metabolic reprogramming, resulting in the development of several cancers including those of the head and neck. Gain-of-function mutations in *EGFR*, *PIK3CA*, and *HRAS*, or loss-of-function in p53 and *PTEN* are often associated with mTOR hyperactivation, whereas mutations identified from The Cancer Genome Atlas (TCGA) dataset that potentially lead to aberrant mTOR signalling are found in the *EIF4G1*, *PLD1*, *RAC1*, and *SZT2* genes. In this review, we discuss how these mutant genes could affect mTOR signalling and highlight their impact on metabolic processes, as well as suggest potential targets for therapeutic intervention, primarily in head and neck cancer.

Keywords: mTOR signalling; metabolism; head and neck cancer; mutant genes; biomarkers; targeted therapies; clinical trials

---

1. Background

Head and neck squamous cell carcinoma (HNSCC) is currently the sixth most frequently diagnosed malignancy worldwide [1]. It is the most common cancer of the head and neck, with anatomic subsites spanning the oral cavity, nasopharynx, larynx, oropharynx, and hypopharynx. HNSCC is a heterogeneous disease that harbours complex genetic defects. While the specific multiple risk factors for the development of HNSCC differ depending on the cancer site, chronic tobacco use and alcohol abuse are historically recognised as the main promoting factors associated with the overall occurrence of HNSCC [2–4]. Infection with high-risk human papillomaviruses (HPV) has also emerged as a risk factor for a subset of HNSCC (~25% of cases) but has more of a profound role in the development of oropharyngeal cancer [5–7]. Nonetheless, HPV positive HNSCCs are shown to have better prognosis compared to HPV negative patients (70–80% versus 25–40%) [8–10]. The standard of care for HNSCC patients involves surgery, radiation therapy, chemotherapy and most recently, targeted therapy and immunotherapy. However, these therapies are usually administered in the absence of accurate biomarkers of response, which often leads to treatment resistance, higher systemic toxicities and, in some cases, results in morbidity and mortality. Currently, the only Food and Drug Administration (FDA)-approved targeted therapy for recurrent or metastatic HNSCC patients is cetuximab, a monoclonal antibody that specifically binds and inhibits the epidermal growth factor receptor (EGFR). However, only ~10% of patients demonstrated a beneficial response to cetuximab therapy, while the remainder were at higher risk of relapse [11,12]. Pembrolizumab, an immune checkpoint inhibitor that targets tumour cells expressing high levels of PD-L1 has also been FDA-approved for the treatment of patients with recurrent or metastatic HNSCC. Unfortunately, the rate of pembrolizumab responders is also quite low (~20%) in the absence of patient stratification [13–15], and a significant proportion of patients may experience increased tumour growth kinetics (hyperprogressive disease) [16,17]. While significant advances in optimising therapeutic responses have been made, the five-year survival rate has remained between 25 and 60% [18]. Genetic alterations and complex signalling pathways have been shown to drive treatment resistance, allowing for continuous cancer cell survival and proliferation. These mechanisms render most HNSCC patients hard to cure, and therefore there is a need to identify biomarkers of treatment response that will serve to tailor treatment regimens in specific subsets of HNSCC patients.

2. Genomic Alterations in Head and Neck Cancer

The recent application of next generation sequencing to study patient cancer genomes has revolutionized medical oncology. In silico analyses provide great insights into the diverse genomic alterations within each cancer sample, allowing for a functional understanding of the drivers behind deregulated oncogenic pathways and biological mechanisms involved in cancer progression. Importantly, these approaches are being exploited to potentially personalise suitable treatment regimens, tailored towards targeting key oncogenic drivers based on the individual's mutational profile. On this basis, The Cancer Genome Atlas (TCGA) has been extensively interrogated for a comprehensive genomic characterization of HNSCC, whereby several reports have identified hundreds of mutations in each cancer subtype [19–22]. This has resulted in common dysregulated pathways being identified across most HNSCC patients. Multiple genetic and epigenetic alterations, including point mutations, deletions, promoter methylation, and oncogene amplification, are strongly triggered by chronic exposure to the major risk factors associated with HNSCC development. Some of the mutated genes frequently associated with HNSCC are *TP53*, *CDKN2A*, *FAT1*, *NOTCH1*, *EGFR*, *HRAS*, and *PI3KCA* [23–27], and the mutations in these genes are recognised as drivers of tumour development and progression. In HPV$^+$ subtype patients, high-risk HPV infection has been associated with the abnormal expression of proteins associated with cell cycle regulation, including p53 and p16 (CDKN2A) [21,28,29]. Functional TP53 inhibits the mammalian target of rapamycin (mTOR) pathway through AMPK in response to cellular stresses and DNA damage [21]. On the other hand, aberrant TP53 allows for persistent mTOR activity and has been associated with poor survival in HNSCC patients [30].

Furthermore, FAT1 is involved in the migration and invasion of HNSCC cells through the activation of the β-catenin pathway [31]. In addition, Notch1 has been reported to play a bimodal role as a tumour promoter and tumour suppressor [22,32,33]. Overall, genetic alterations in the above-mentioned genes and in *EGFR* and *HRAS* often lead to aberrant signalling and deregulation of important proto-oncogenic networks, such as the PI3K–mTOR pathway. For instance, *PIK3CA* mutations that directly activate the PI3K–mTOR signalling pathway have been reported in HNSCC at rates ranging from 2.6% to 19% [21]. Overall, genetic amplifications and overexpression of key proteins responsible for driving mTOR activation underlie the tumour progression that is often observed in cancers, including HNSCC. This review provides a comprehensive analysis of the driver mutations that lead to aberrant mTOR signalling in HNSCC and assesses a number of contemporary inhibitors.

3. The mTOR Complex and the Cellular Metabolism

The mammalian target of rapamycin (mTOR) is a serine-threonine kinase that senses growth factor cues, nutrient and oxygen status, and directs appropriate changes to maintain cellular and tissue homeostasis (Figure 1). The mTOR signalling pathway is recognised as a key driver and regulator of cell growth and proliferation, cell survival, metabolism, and protein synthesis. mTOR belongs to the phospho-inositide 3-kinase (PI3K)-related kinase family, and consists of two distinct complexes; mTOR complex 1 (mTORC1) and complex 2 (mTORC2). Whilst both complexes are tightly regulated in a normal context, they are often deregulated in multiple disease-associated metabolic alterations and in cancer development [34–36]. In normal conditions, activation of mTOR signalling occurs in response to the binding of specific growth factors to their cognate receptor tyrosine kinases (RTKs), including insulin-like growth factor (IGF), vascular endothelial growth factor (VEGF), platelet-derived growth factor (PDGF), and epidermal growth factor (EGF) [37]. The ligand-activated receptors recruit PI3K, which converts phosphatidylinositol bisphosphate (PIP2) to phosphatidylinositol triphosphate (PIP3), and provides binding sites for phosphoinositide-dependent protein kinase 1 (PDK1). PDK1 then phosphorylates and activates AKT, which in turn phosphorylates several downstream substrates that engage multiple pathways, including mTORC1. On the other hand, mTORC2 has been known to phosphorylate and activate members of the AGC kinase family, including AKT, serum and glucocorticoid-induced kinase (SGK1), and protein kinase C (PKC), whereby inhibition of these kinases results in tumour suppression [38–40].

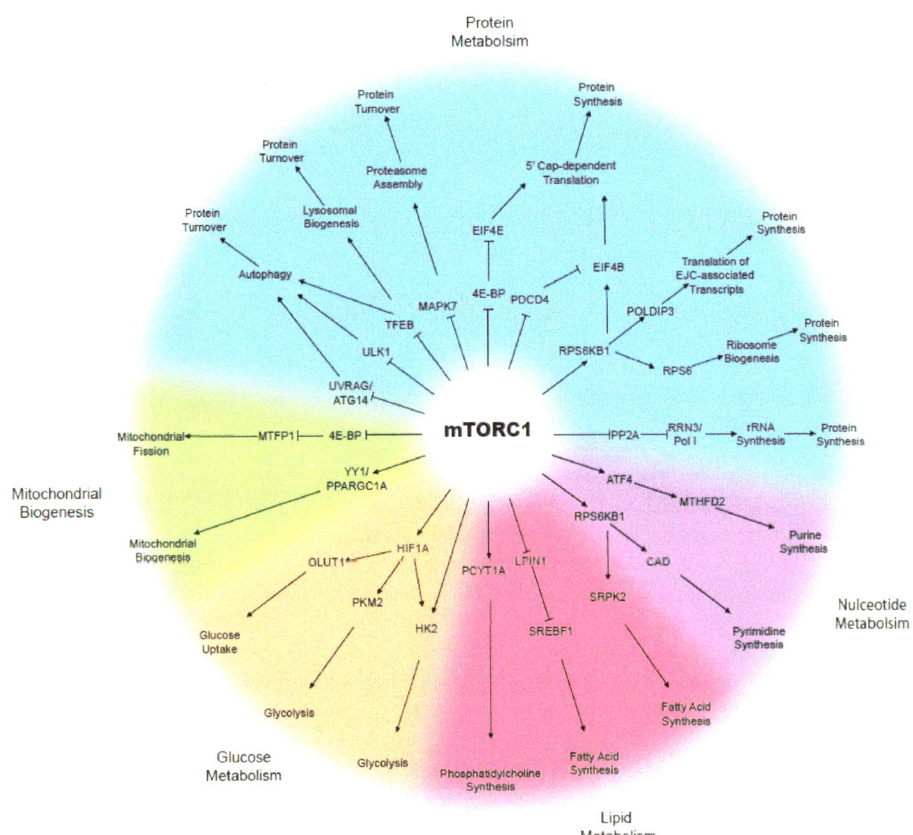

Figure 1. *Cont.*

(**B**)

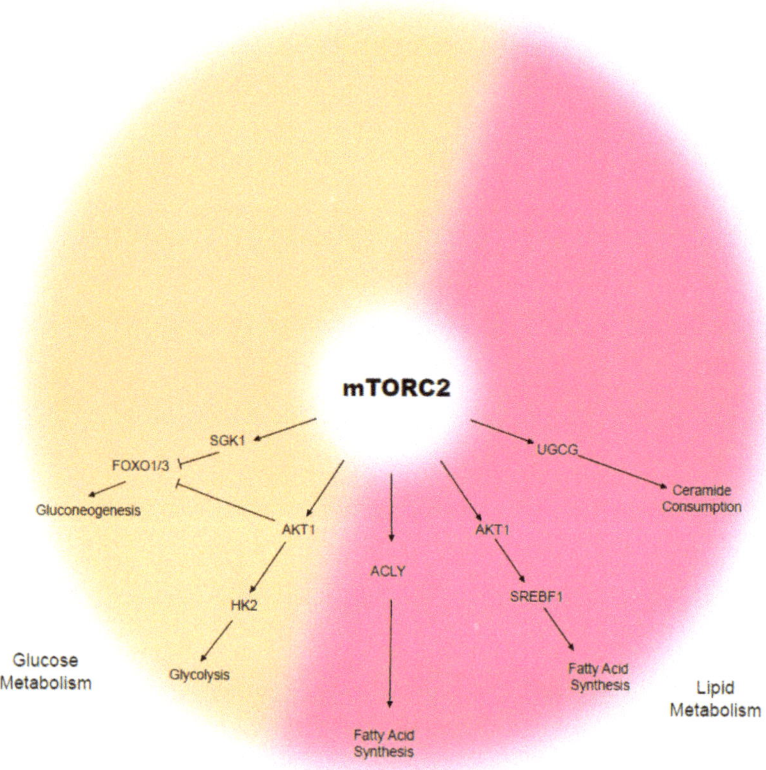

Figure 1. Biological functions of the mammalian target of rapamycin (mTOR): (**A**) mTOR complex 1 (mTORC1) and (**B**) mTOR complex 2 (mTORC2).

Constitutive mTOR activation is known to promote metabolic changes, including dysregulation in glucose, fatty acid, amino acid, and lipid metabolism. For instance, cancer cells largely rely on glucose as the major source of cellular energy to sustain proliferation and survival. In the context of aberrant mTOR signalling, glucose metabolism is dysregulated as a result of increased synthesis of glucose transporter proteins and glycolytic enzyme activation, followed by lactic acid fermentation even when oxygen is available—a phenomenon known as the "Warburg effect" [41,42]. The Warburg effect links the rewiring of metabolism to sustained cancer cell survival and growth, in which increased glucose uptake and fermentation of glucose to lactate are key processes [43]. In several cancers, including HNSCC, the expression of glucose transporter 1 (GLUT1) is often elevated, and in conjunction with enhanced mTOR signalling (mTORC1 and C2), the pair activates key oncogenic drivers, including c-MYC and HIF-1α [44–47]. GLUT1 is a protein of the GLUT family, responsible for glucose uptake into the cytoplasm [48]. GLUT1 is negatively regulated by glycogen synthase kinase-3 (GSK-3) that, in turn, exerts its inhibitory effects through a tuberous sclerosis complex (TSC)- and mTOR-dependent pathway [44]. Glycolysis is also upregulated through mTOR signalling via elevated Hexokinase 2 (HK2) expression, which further promotes the activation of c-MYC and HIF-1α [49]. Furthermore, the mTOR signalling stimulates fatty acid synthesis in cancer, via the persistent activation of sterol regulatory element-binding protein-1c (SREBP1c) [50,51]. Although not widely described in HNSCC,

Guri et al. observed that elevated lipogenesis correlated with enhanced mTOR activity in hepatocellular carcinoma patients, which in turn facilitated energy production and cancer growth [52]. Overall, deregulated or reprogrammed mTOR signalling is a key signature of cancer cellular metabolism, while the molecular manipulation of the internal and surrounding tumour environment both initiates and sustains cancer cell survival, growth, and proliferation.

4. Deregulated mTOR Signalling in HNSCC

The mTOR pathway is known to be hyperactivated in several cancers, including HNSCC, and both mTOR complexes play essential roles in HNSCC tumorigenesis. Interestingly, mTOR deregulation in HNSCC is the most commonly seen genomic alteration (~80–90% HNSCC) involved in aberrant mitogenic signalling, compared to other known pathways such as the JAK/STAT and MAPK, which harbour mutations in less than 10% of lesions [53,54].

In vivo analyses of mTOR signalling in HNSCC are commonly studied, and chemically-induced HNSCC mouse models have long been established. These include the widely used carcinogens DMBA-TPA and 4-nitroquinoline-1-oxide (4NQO), which have both been reported to result in persistent mTOR activation, leading to tumour development, and regression is observed after the administration of the mTOR inhibitor rapamycin [55–57]. Furthermore, several studies have analysed the effect of rapamycin and mTOR activity in other mouse models, including anal squamous cell carcinoma (SCC) and skin and breast cancers [58–60]. Aside from chemically-induced models, genetic mouse models have also been established for mTOR hyper activation. For instance, Sun et al. observed that conditional deletion of Tgfbr1 and Pten in an HNSCC mouse model was associated with the development of sporadic tongue tumours that were driven by mTOR activation [61]. Furthermore, tumour burden was significantly reduced following rapamycin treatment, confirming the role of mTOR in driving HNSCC [61]. Bozec et al. analysed the effect of temsirolimus, a potent mTOR inhibitor, in combination with cetuximab and conventional chemotherapeutic agents (cisplatin and 5-fluorouracil) on orthotopic CAL33 xenografts harbouring *PIK3CA* mutations [62]. The combination therapy was synergistic and resulted in almost complete tumour growth arrest, further associating a profound role between tumorigenesis and mTOR activity [62]. Furthermore, it has been reported that co-targeting mTOR and PD-L1 enhances tumour growth inhibition in a syngeneic oral cancer mouse model [63].

mTORC1 interacts with the Rag GTPases, which promote its translocation and activation at the lysosomal surface in response to amino acids [64,65]. Inhibition of mTOR reduced the lysosomal efflux of essential amino acids and converted the lysosome into a cellular depot for them [66]. This process can also be deregulated in cancers where inactivating mutations of the Rag GTPases regulators can lead to hyperactivation of mTORC1, even in the absence of amino acids [67]. mTOR not only has a major role in tumour progression but also plays a role as the central regulator of autophagy. Autophagy is an intracellular process mediated by lysosomes for the breakdown and recycling of damaged cellular components (e.g., organelles, proteins) [68]. In HNSCC, the oral cavity has been known to acquire mutations that are associated with impaired autophagy and correlate with reduced overall survival [69,70]. Whilst a number of inhibitors against the PI3K/AKT/mTOR signalling pathway have undergone extensive preclinical evaluation, the specific mechanism of the action remains elusive and successfully reversing defective autophagy has been variable [71,72].

5. HPV Status, mTOR Activation and Metabolism in HNSCC

HPV infection is known to activate mTOR signalling in HNSCC and is further sustained through deregulation of metabolic pathways. For instance, HPV-positive cells utilise mitochondrial respiration, as evidenced by increased oxygen consumption in comparison to HPV-negative HNSCC cells, which exhibit increased glucose metabolism, as evidenced by the over production of lactate [73]. HPV-negative cells express HIF-1α, which is responsible for upregulating downstream mediators involved in glucose metabolism, including hexokinase II (HKII) and carbonic anhydrase IX (CAIX), while HPV-negative cells show greater expression of cytochrome c oxidase (COX) [74,75]. Moreover, as a result of increased lactate and pyruvate production, Jung et al. found that HPV-negative HNSCC cells exhibit advantageous growth, survival and radioresistance [73]. Inhibition of pyruvate dehydrogenase kinase (PDK) sensitises HPV-negative HNSCC to irradiation, which could potentially explain why those tumours are more inclined to have an unfavourable prognosis compared to HPV-positive tumours [9,10]. Therefore, both HPV-positive and HPV-negative HNSCC cells are characterised by deregulated mTOR signalling, which impairs their metabolism and thus sustains the survival and growth of cancer cells in a vicious cycle.

mTOR inhibitors have shown promising anti-cancer effects in HPV-positive HNSCC mouse models. The mTOR inhibitors Rapamycin and RAD001 reduced tumour burden in HPV-positive HNSCC xenografts through the inhibition of mTOR activity [76]. Moreover, HPV E6/E7 mouse models develop SCC lesions with high mTOR activation, and, unsurprisingly, tumour development was abolished using the mTOR inhibitor Rapamycin [56]. Despite the link between HPV infection and mTOR signalling activation with altered metabolic processes, the potential inhibition of both mTOR and HPV-deregulated pathways in HNSCC is still not well explored.

Furthermore, the E6 and E7 HPV oncoproteins are known to correlate with *PIK3CA* mutations or amplifications in over half of HPV-positive HNSCC, leading to drug resistance. Brand et al. showed that PI3K inhibition resulted in increased expression of the HER3 receptor and, in turn, elevated the abundance of E6 and E7 oncoproteins to promote resistance to PI3K inhibition [77]. This study also assessed the targeting of HER3 with the monoclonal antibody CDX-3379, which resulted in reduced E6 and E7 expression and enhanced the treatment efficacy of PI3K-targeted inhibition. As concluded by the authors, this suggests that co-targeting HER3 and PI3K may be an effective treatment strategy for HPV+ tumours where HER3 and HPV oncoproteins promote resistance to PI3K inhibitors. In addition, Madera et al. inhibited mTOR signalling using metformin, a ubiquitous anti-diabetic drug. This resulted in reduced tumour growth that was driven by the PIK3CA and HPV oncogenes in oral SCC (OSCC) [78].

6. Validated Mutant Genes Known to Drive Activation of mTOR Signalling in HNSCC

Mutations in EGFR, *PIK3CA*, and *HRAS*, as well as others found in potential genes such as *EIF4G1*, *RAC1*, *SZT2*, and *PLD1*, can result in aberrant mTOR signalling (Figure 2). Deregulation of mTOR signalling can equally be induced by a loss of tumour suppressors such as *PTEN*, *APC*, and *NF1* [79–81]. A hyperactive mTORC1 engages downstream effectors through phosphorylation of the eukaryotic translation initiation factor 4E-binding proteins (4EBP-1), p70 and ribosomal protein S6 kinase (S6K1), promoting tumour development and progression [82–84]. In a similar manner, hyperactivation of mTORC2 drives cancer cell survival, proliferation and migration mainly through the oncogenic activation of AKT [34,85,86]. We next discuss the role of these validated and proposed genes (those that have not been well studied in HNSCC) that directly or indirectly activate mTOR signalling in HNSCC.

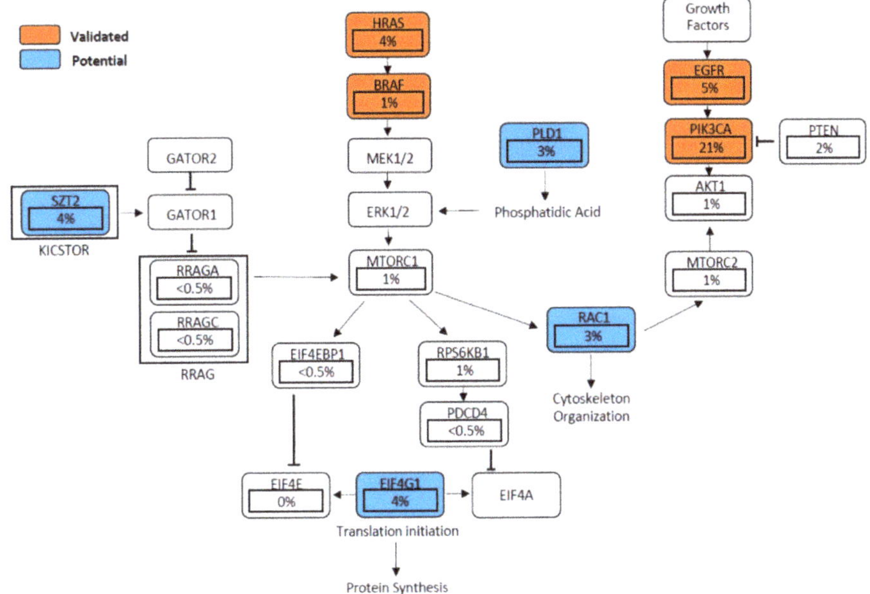

Figure 2. Validated and proposed mutant genes found to deregulate mTOR signaling in head and neck squamous cell carcinoma (HNSCC). Percentages represent frequency of mutations per gene based on The Cancer Genome Atlas (TCGA) dataset ($n = 297$).

6.1. EGFR-PI3K-AKT-mTOR Pathway

The hyperactivation of EGFR through various epigenetic and genetic mechanisms is known to activate the PI3K-AKT-mTOR pathway. This is evident in HNSCC patient samples that show high EGFR-mTOR signalling, often associated with poor clinical outcomes [87]. A study conducted by Li et al. demonstrated that a positive feedback loop involving EGFR-mTOR and the inhibitor of nuclear factor kappa-B kinase (IKK)-NF-κB signalling regulates HNSCC cell growth [88]. One regulator of the EGFR-mTOR pathway is the AXL protein, which was shown to dimerize with and phosphorylate EGFR, and to activate phospholipase Cγ (PLCγ)-protein kinase C (PKC), resulting in the hyper activation of mTOR [89]. Furthermore, genetic alterations of the EGFR gene result in a common cancer-associated variant III (EGFRvIII), in several cancers. Widely studied in gliomas, the EGFRvIII is characterized by the absence of exons 2–7, leading to disruption of the ligand-binding region and is therefore constitutively active in a ligand-independent manner [90]. However, in HNSCC cases the role of EGFRvIII has remained controversial. For instance, although the sample number analysed was low, Sok et al. reported EGFRvIII to be hyperactive in >42% of HNSCC samples (14/33) [91]. However, in a recent study conducted by Khattri et al., it was established that EGFRvIII is rarely seen in HNSCC samples (2/540, 0.37%) and the clinical significance remains unclear [92]. Moreover, EGFR amplification (chromosome 7) has been reported in 11% of HNSCC cases [93,94]. Hashmi et al. observed that EGFR copy number gain occurs in oral leukoplakia and is tightly linked with an increased risk of oral cancer development [94].

Aberrant EGFR signalling was also shown to mediate aerobic glycolysis and to upregulate GLUT1 expression. In EGFR mutated lung adenocarcinoma, Makinoshima et al. found that mTOR signalling plays a crucial role in regulating glycolysis and in upregulating GLUT1 localisation [95]. In a panel of EGFR-mutated lung cancer cell lines, mTOR inhibitors significantly suppressed glycolysis and down regulation of GLUT1 by RNAi reduced cell proliferation [95]. Conversely, Chiang et al. found that mTOR signalling contributes to metabolic reprogramming in erlotinib (EGFR inhibitor) resistant lung cancer cells and strongly correlates with poor clinical outcomes of EGFR-mutated lung cancer patients [96].

The frequent activation of the EGFR pathway led to the development of EGFR inhibitors targeting receptor function to prevent downstream signalling, including mTOR activation. To date, cetuximab is the only FDA-approved EGFR inhibitor in combination with chemotherapy or with radiation therapy. Tyrosine kinase inhibitors with reversible-binding activity, such as erlotinib and gefitinib, have been disappointing in the head and neck setting, while irreversible-binding Tyrosine Kinase Inhibitors (TKI), including afatinib, appear clinically promising [97–99]. The combination of inhibitors targeting both mTOR and EGFR has also emerged as beneficial. For instance, combinatorial treatment targeting mTOR and EGFR has been successful in other cancers, including small cell lung cancers [100,101]. Furthermore, Bozec et al. investigated combined mTOR (temsirolimus) and EGFR (cetuximab) targeting in an orthotopic xenograft model of HNSCC, which culminated in synergistic effects against tumour growth [62]. In agreement, Lattanzio et al. observed a similar result in HNSCC cell lines [100] and Wang et al. also demonstrated reduced tumour burden in both PIK3CA- and RAS-expressing HNSCC xenografts, particularly in cetuximab resistant HNSCC cell lines [53]. Overall, targeting both EGFR and mTOR related pathways could be a promising personalised targeted therapy for HNSCC patients.

6.2. PIK3CA Mutation and PTEN Loss

Mutations that activate the catalytic unit of phosphoinositide-3-kinase (PI3K) have been implicated in several cancers, including HNSCC. Gain of function mutations of PIK3CA, the most common activator of the PI3K pathway, is detected in approximately 6–20% of HNSCC cases [21,22]. Lui et al. analysed whole-exome sequencing data from 151 tumours and revealed frequent oncogenic mutations in 30.5% (46/151) of the cases affecting the PI3K-mTOR pathway, whereas only 9.3% (14/151) and 8% (46/151) of tumours harboured mutations in the JAK/STAT or the MAPK pathways, respectively [21]. Furthermore, all tumours exhibiting PI3K pathway mutations were advanced (stage IV) cancers, implying a strong role in cancer progression.

Aberrant PI3K-mTOR signalling was shown to also regulate the properties of key cancer stem cell (CSC) factors, including the sex determining region Y box 2 (SOX2) [102,103]. SOX2 is involved in cancer stem cell (CSC) maintenance and is also associated with increased levels of CSC markers, including aldehyde dehydrogenase (ALDH1) [104]. Keysar et al. characterised the CSC from patient-derived xenografts and defined the molecular features of tumours caused by tobacco smoking and HPV infection [103]. This work unraveled the consequences of deregulated PI3K signalling, such as increased SOX2 translation and expression of ALDH, resulting in enhanced spheroid and tumour formation. This study also observed reduced SOX2 levels after silencing *AKT1* (downstream of PI3K) or *EIF4E* (downstream of mTORC1), suggesting a direct link between SOX2 regulation and PI3K. Additionally, SOX2 knocks down suppressed ALDH transcripts and protein levels. Moreover, Suda et al. revealed that copy-number amplification of PIK3CA, within 3q (found in up to 30% of HNSCC) is associated with a poor prognosis of HNSCC patients [105] and partially overlaps with PIK3CA driving mutations. In addition, it has been shown that PIK3CA mutations are associated with an elevated uptake of glucose and glutamine in colorectal cancer [106], and a similar effect is observed in PIK3CA mutant breast cancer cells [107]. The elevated glucose and glutamine uptakes fuel the growth and progression of tumourigenicity.

Conversely, inactivation of phosphatase and tensin homologue (PTEN), a potent tumour suppressor and negative regulator of PI3K, also leads to hyperactivation of PI3K-driven mTOR signalling [108]. Although the penetrance of PTEN mutations in HNSCC ranges between 5 and 16%, loss of PTEN expression is observed in 29% of tongue cancers, and loss of heterozygosity of the PTEN locus occurs in 40% of HNSCC tumours [109,110]. Genetic alterations were even lower in SCC of the skin, in which loss of PTEN was mainly due to loss of gene transcription [111,112]. Deletion of the developmental transcription factor Grainyhead-like 3 (Grhl3) induces HNSCC in both humans and mice [111,113–115], and GRHL3 functions as a tumour suppressor against SCC of the skin through the direct transcriptional regulation of *Pten* [111,116,117]. Loss of Grhl3 leads to PTEN downregulation and the development of aggressive cutaneous SCC via the activation of the PI3K–mTOR signalling pathway [111]. Inhibition of PI3K/mTOR using BEZ235 was able to prevent the initiation as well as the promotion to malignancy of carcinogen-induced SCC, but was not efficient against the established cancer [118]. Interestingly, mutations in the *PTEN* gene are rare in human skin SCC and common in HNSCC, which could be a prognostic marker for patients with tongue cancer [111,114,119,120]. Moreover, suppression of PTEN in concert with other tumour suppressors, like transforming growth factor beta-receptor 1 (TGFBR1), can also contribute to deregulated PI3K-mTOR signalling. Bian et al. unraveled the relationship between TGF-β signalling and the PI3K-mTOR pathway by conditionally deleting both TGFBR1 and PTEN in HNSCC mouse models using the Cre-LoxP system. Enhanced cell proliferation and decreased apoptosis occurred, which promoted HNSCC tumour development [121].

PTEN loss also promotes cancer progression by enhancing glucose metabolism and reducing DNA repair and checkpoint pathways. Martin et al. observed PTEN loss in prostate cancer cell lines and increased pAKT expression and enhanced glucose metabolism, resulting in the survival of tumour cells [122]. Mathur et al. also observed enhanced glutamine metabolism in PTEN mutant breast cancer cells [123]. Conversely, Garcia-Cao et al. showed that transgenic overexpression of PTEN in mice decreased the levels of PFKFB3 (6-phosphofructo-2-kinase/fructose-2,6-biphosphatase 3) and glutaminase, key rate-limiting enzymes responsible for glycolysis and glutaminolysis respectively, and two important metabolic features of tumour cell growth [124]. Together, these data predict that tumours with loss of PTEN function will respond to treatment with inhibitors of glycolysis and glutaminolysis, therefore providing a potential targeted therapy for these tumours.

6.3. *HRAS*

HRAS belongs to the Ras oncogene family, and *HRAS* mutants are known to aberrantly activate mTOR signalling in HNSCC tumours [125]. The Cancer Genome Atlas analysed 279 HNSCC samples and reported that a subgroup of oral cavity tumours had favourable clinical outcomes displaying infrequent copy number alterations in conjunction with activating mutations of HRAS [20]. Nakagaki et al. utilised next-generation sequencing on a cohort of 80 Japanese OSCC patients, and identified *HRAS* mutations in 5% of samples [126]. Su et al. analysed whole-exome sequencing of 120 Taiwanese OSCC patients and identified 11.7% of the samples were positive for *HRAS* mutations [127]. Furthermore, Koumaki et al. identified *HRAS* mutations in 86 OSCC patients (8.6%) of Greek descent, and defined a role for HRAS in driving PI3K-mTOR signalling in OSCC [128]. Adding to the evaluations of several ethnicities, Murugan et al. identified mutant *HRAS* in 10 out of 56 Vietnamese OSCC patients (18%) and associated these events with an advanced tumour stage [129]. In addition to promoting mTOR activation, *HRAS* mutations have been shown to play a role in altering metabolic processes. Zheng et al. suggest that the HRAS transformed breast cancer cell line MCF10a, derived from an early stage cancer, exhibits a profound alteration in glucose metabolism and is strongly regulated by the oncogenic proteins HIF-1α and c-Myc [130].

Additional investigations have implicated the HRAS protein in driving resistance to therapies in HNSCC. In a recent study conducted by Ruicci et al., *HRAS* mutant HNSCC cell lines did not respond to PI3K inhibition (BYL719). This inhibitor induced constitutive MAPK signalling suggests feedbacks between MAPK and PI3K, resulting in persistent mTOR activity [131]. Hah et al. also outlined an association between *HRAS* mutations and resistance to the EGFR tyrosine kinase inhibitor erlotinib in a panel of HNSCC cell lines [132]. Likewise, Rampias et al. demonstrated that oncogenic HRAS leads to the activation of MAPK signalling, which results in resistance to cetuximab in HNSCC cells [133]. Furthermore, the authors evaluated a cohort of 55 HNSCC patients, and identified *HRAS* mutations in 7 out of 55 samples (12.7%) that were associated with a poorer response to Cetuximab treatment.

7. Potential Mutant Genes Activating the mTOR Signalling Pathway

7.1. EIF4G1

Eukaryotic translation initiation factor (EIF) 4 gamma 1 (EIF4G1) plays a crucial role downstream of mTOR signalling. EIF4G1 functions as a modular scaffold in the translational initiation complex, interacting with EIF3, EIF4A, EIF4E, poly (A)-binding proteins, and MNK1 [134,135]. When mTOR phosphorylates EIF4E-binding proteins (4E-BPs), it releases 4E (EIF4E) to bind to 4G1 and initiate cap-dependent translation [136]. Moreover, the downstream target of mTOR, programmed cell death 4 (PDCD4), disrupts the interaction between 4A and 4G1, leading to translational inhibition [137]. In this study, PDCD4 was shown to be downregulated by miR-21 in HNSCC, suggesting the initiation of mTOR-regulated translation by 4G1.

Aberrant over-expression of EIF4G1 is tightly linked with the prognosis of several cancers, such as lung squamous cell carcinoma, inflammatory breast cancer, cervical cancers, and nasopharyngeal carcinoma [138–141]. Although not well studied in HNSCC, the 2015 TCGA analysis of 279 HNSCC patients has provided evidence that EIF4G1 has a high alteration frequency in HNSCC, with 19% of patients harbouring EIF4G1 amplifications and 3.9% somatic mutations [20]. This is supported by studies in breast and nasopharyngeal cancers, indicating that the overexpression of 4G1 facilitates tumourigenesis, malignant transformation, and invasion, while the depletion of 4G1 remarkably leads to the inhibition of cell cycle progression, invasion, and colony formation in vitro and in vivo [139,140,142]. Although there are limited studies testing EIF4G1 inhibitors in HNSCC, small molecule inhibitors have been investigated in other cancer types. For instance, 4EGI1 has been designed to block the interaction of EIF4G1/EIF4E, resulting in decreased translation of oncogenic proteins and abrogating lung tumour growth in vivo [143–146]. Moreover, SBI-0640756 and RNA aptamers have been designed to directly target 4G1 in melanoma models [147,148]. Overall, despite limited available evidence, inhibitors of EIF4G or EIF4G1/EIF4E interactions could be emerging as novel strategies to indirectly target mTOR signalling in HNSCC.

7.2. RAC1

The Rac family of small GTPase 1 (RAC1) functions downstream of the mTOR signalling to regulate the reorganization of F-actin, lamellipodia formation, and cell motility [149]. mTORC1-mediated activation is essential to increase RAC1 expression, while mTORC2 directly facilitates RAC1 activation via the inhibition of Rho GDP dissociation inhibitor beta ARHGDIB. This results in the initiation of phosphatidylinositol-3,4,5-trisphosphate dependent rac exchange factor 1/2 (PREX1 and PREX2) and T cell lymphoma invasion and metastasis 1 (*TIAM-1*) expression, which all contribute to tumour growth [150,151]. Increasing evidence suggests that RAC1 modulates mTOR activity, whereby the binding of RAC1 to mTOR regulates the plasma membrane localization of the mTORC1/2 complex. This in turn promotes the phosphorylation of mTOR downstream substrates [152,153]. Collectively, RAC1 is therefore considered to be both an upstream and downstream effector of mTOR activity.

Overexpression of RAC1 is frequently observed in oral, breast, gastric, testicular, and prostate cancers and increased *RAC1* expression is positively associated with cancer progression [154–157]. Aberrant RAC1 activity was shown to facilitate metastasis of colorectal and lung cancer cells by multiple mechanisms, including epithelial-mesenchymal transition (EMT), migration, and invasion [158,159]. Although not well studied in HNSCC, the TCGA database reports that 3.2% of HNSCC patients harbour RAC1 somatic mutation [20]. Moreover, persistent RAC1 overexpression has been shown to drive resistance to radio/chemotherapy [20,160,161]. In recent years, limited inhibitors targeting RAC1 have been developed. These have included EHop-016 [162] and EHT 1864, which was designed to prevent RAC1-GTP interactions and the RAC1 downstream effectors in order to block RAC1-mediated metastasis [163]. Alongside monotherapy, recent reports suggest that the combined inhibition of RAC1 and mTOR could dramatically increase treatment efficacy against renal cell carcinoma by dephosphorylating the retinoblastoma transcriptional corepressor 1 (RB1) [164]. Although the exact mechanism of the action of RAC1 inhibitors has not been thoroughly explored, their use as single or combination agents seems to have a synergistic effect with mTOR inhibition that could be considered for the treatment of HNSCC.

7.3. SZT2

Deficiency of seizure threshold 2 protein homolog (SZT2) is commonly detected in patients with intellectual disability, epilepsy, and autism [165–168]. Only recently however, SZT2 was identified as a component of KICSTOR, which negatively regulates the mTOR signalling pathway [168,169]. Interestingly, the *SZT2* gene shows a relatively high somatic mutation frequency (3.6%) in HNSCC in the TCGA database. In addition, low expression of SZT2 is correlated with a low five-year survival rate of HNSCC patients [20]. Future investigation of the SZT2 function is therefore warranted to determine whether it could act as a prognostic factor for HNSCC patients and/or a possible biomarker of response to mTOR inhibitors.

7.4. PLD1

Phospholipase D1 (PLD1) is an established upstream regulator of mTOR signalling [140]. Once activated, PLD1 leads to the accumulation of phosphatidic acid (PA), resulting in mTOR activation via the ERK signaling pathway, an acquired resistance to mTORC1 inhibitors, and a feed-forward loop, resulting in constitutive PLD1 activity [170–172]. High *PLD1* expression is frequently detected in various cancers, including glioma, pancreatic ductal adenocarcinoma, colorectal cancer, hepatocellular carcinoma, breast cancer, and melanoma [173–178]. Although PLD1 has not been extensively investigated in HNSCC, data from the TCGA show that 20% of HNSCC patients harbour copy number amplification, while 2.9% of patients harbour mutant PLD1. Based on the high percentage of its genetic alteration, we anticipate that PLD1 could function as a driver or a prognostic marker for HNSCC [20].

Multiple inhibitors targeting PLD1 have been developed, such as VU-0155069 and VU-0359595, which directly bind to the N-terminus, allosterically suppressing the catalytic activity of PLD1 [179]. In addition, inhibitors such as Fifi, ML-299, VU-0155056, and VU-0285655-1 show less selectivity by targeting both PLD1 and PLD2 [180]. Kang et al. found that inhibition of PLD1 suppresses the PI3K–mTOR pathway and results in reduced cell proliferation, migration, and invasion in vitro, as well as reduced tumour growth and EMT of patient-derived xenografts in colorectal and hepatocellular carcinoma [175,178]. Since the published literature is establishing a clear relationship between PLD1 and mTOR, further investigations are required to explore the inhibition of PLD1 for mTOR-driven malignancies, as well as the inclusion of PLD1 inhibitors in HNSCC clinical trials.

8. Current Clinical Trials Targeting mTOR in HNSCC

Because multiple mutant genes are directly associated with the oncogenic activation of the mTOR pathway, it is not surprising that multiple clinical trials are currently targeting aberrant mTOR signalling in cancer (Table 1). For instance, the multicentre Phase II trial recruited platinum/cetuximab-refractory HNSCC patients for treatment with the mTOR inhibitor temsirolimus (NCT01172769). Results from this trial indicate that in a total of 40 patients, the treatment was well tolerated, and tumour shrinkage was observed in 13/40 (39.4%) patients [181]. This study indicated that mTOR inhibition alleviates tumour burden, although further molecular analysis is required to identify predictive parameters for temsirolimus guided treatment response. Patients included in this study showed no mutations of KRAS or BRAF. Following from this trial, a Phase II study of temsirolimus in combination with carboplatin and paclitaxel has been conducted on recurrent and/or metastatic HNSCC patients. This resulted in an objective response in 15/36 (41.7%) patients and stable disease progression in 19/36 (52.3%) patients (NCT01016769) [182]. This trial confirmed that a relatively high response can be observed with combination treatment and suggests that genetic alterations associated with aberrant mTOR signalling necessitate further exploration.

Table 1. Summary of clinical trials targeting mTOR in HNSCC tumours.

Inhibitor	Phase	Status	Targeted Pathway	Targeted Tumour	Reference
Palbociclib + Gedatolisib	I	Recruiting	CDK4/6 + mTOR	Advanced HNSCC	NCT03065062
BYL719	II	Recruiting	PI3K	Recurrent or Metastatic HNSCC	NCT02145312
Everolimus + Palbociclib + Trametinib	I	Recruiting	mTOR + CDK4/6 + MEK	Malignant neoplasms of Oral cavity	NCT03065387
CC-115	I	Active, not recruiting	Dual DNA-PK and TOR kinase	Advanced HNSCC	NCT01353625
PQR 309	I	Active, not recruiting	PI3K/mTOR/AKT	Advanced HNSCC	NCT02483858
Temsirolimus	II	Completed	mTOR	HNSCSC	NCT01172769
Sirolimus	I/II	Completed	mTOR	Advanced HNSCC	NCT01195922
Temsirolimus + Paclitaxel + Carboplatin	I/II	Completed	mTOR	Recurrent or Metastatic HNSCC	NCT01016769

Several clinical trials are currently recruiting HNSCC patients for the assessment of other mTOR pathway inhibitors. Following promising results obtained in vitro and in vivo with BYL719, a potent PI3Kα inhibitor, this drug has progressed to Phase II trials for the treatment of recurrent or metastatic HNSCC patients who have previously failed to respond to platinum-based therapy (NCT02145312) [183]. Moreover, in a large multi-centre clinical trial, the Phase II molecular analysis for therapy choice (MATCH) trial tailors personalised inhibitors to each patient's individual mutational status (NCT02465060). This study includes HNSCC patients with mutations that activate mTOR signalling, who received the inhibitor sapanisertib, which binds to and inhibits both mTOR complexes. Of the targeted therapies related to the mTOR pathway, patients with *PIK3CA* mutations received the PI3K inhibitor taselisib, patients harbouring *EGFR* mutations received the EGFR inhibitor afatinib, while patients with loss or mutated *PTEN* received the PI3K-beta inhibitor GSK2636771. In addition to monotherapies, combination treatments are scheduled with the PI3K/mTOR inhibitor gedatolisib and the cyclin-dependent kinase 4 and 6 (CDK4/6) inhibitor palbociclib, and HNSCC patients are currently being recruited for this Phase I trial (NCT03065062).

Despite mTOR signalling driving aberrant metabolic processes, there is currently no clinical trial investigating combinational treatment against both mTOR and dysregulated metabolism in HNSCC patients. Nonetheless, there are several studies targeting key transporters involved in metabolism and HNSCC progression. In a recent study conducted by Mehibel et al., the authors investigated the use of simvastatin (which specifically inhibits lipid and cholesterol biosynthesis) and AZD3965 (which inhibits monocarboxylate transporter 1 and results in enhanced glycolysis) in HNSCC cell lines [184]. They found that prophylactic simvastatin lead to the upregulation of xenograft tumour MCT1 expression that effectively primed these cells for MCT1 inhibition using AZD3965. The combination of these inhibitors led to a delay in tumour growth in HNSCC xenograft models and showed no signs of toxicity. Moreover, AZD3965 has been independently assessed in pre-clinical xenograft studies for other cancer types, such as small cell lung cancer, where it was shown to reduce tumour growth via the inhibition of lactate release and glycolysis [185,186].

9. Conclusions and Perspectives

The mTOR pathway integrates multiple intrinsic genetic alterations and extrinsic cues, leading to aberrant signalling and metabolic alterations. Since the validated and potential mutant genes, as identified from the TCGA dataset, directly affect mTOR activation status in cancer, they could be used as biomarkers for response and mTOR targeted inhibition in a tissue-specific manner. In fact, multiple biomarkers for predicting drug sensitivity have been proposed, including those related to *PTEN* loss, *PTEN* mutations, *NOTCH1* mutation, and EGFR expression in other cancers, but could be further established in HNSCC. Furthermore, the functional characterisation of these mutant genes and the molecular dissection of their associated oncogenic networks could provide targets for combinatorial therapies to alleviate resistance to mTOR inhibition.

Funding: This research was funded by the Australian National Health and Medical Research (NHMRC, grant numbers APP1049870, APP1106697) and The Association for International Cancer Research (AICR, 11-0060). The authors were funded by Fellowships from the Victorian Cancer Agency (Clare Oliver Memorial, COF11_04 and Mid-Career, MCRF16017) to C.D. and by the French LYriCAN grant (INCa-DGOS-Inserm_12563) to P.S.

Acknowledgments: The authors acknowledge Bryce van Denderen for his careful corrections of the article.

Conflicts of Interest: The authors declare no conflict of interest.

References

1. Siegel, R.L.; Miller, K.D.; Jemal, A. Cancer Statistics 2015. *CA Cancer J. Clin.* **2015**, *65*, 5–29. [CrossRef]
2. Wang, X.; Xu, J.; Wang, L.; Liu, C.; Wang, H. The role of cigarette smoking and alcohol consumption in the differentiation of oral squamous cell carcinoma for the males in China. *J. Cancer Res. Ther.* **2015**, *11*, 141–145. [CrossRef]
3. Lubin, J.H.; Purdue, M.; Kelsey, K.; Zhang, Z.F.; Winn, D.; Wei, Q.; Talamini, R.; Szeszenia-Dabrowska, N.; Sturgis, E.M.; Smith, E.; et al. Total exposure and exposure rate effects for alcohol and smoking and risk of head and neck cancer: A pooled analysis of case-control studies. *Am. J. Epidemiol.* **2009**, *170*, 937–947. [CrossRef]
4. Bagnardi, V.; Rota, M.; Botteri, E.; Tramacere, I.; Islami, F.; Fedirko, V.; Scotti, L.; Jenab, M.; Turati, F.; Pasquali, E.; et al. Alcohol consumption and site-specific cancer risk: A comprehensive dose-response meta-analysis. *Br. J. Cancer* **2015**, *112*, 580–593. [CrossRef]
5. Dalla Torre, D.; Burtscher, D.; Soelder, E.; Offermanns, V.; Rasse, M.; Puelacher, W. HPV prevalence in a Mid-European oral squamous cell cancer population: A cohort study. *Oral Dis.* **2018**, *24*, 948–956. [CrossRef]
6. Young, D.; Xiao, C.C.; Murphy, B.; Moore, M.; Fakhry, C.; Day, T.A. Increase in head and neck cancer in younger patients due to human papillomavirus (HPV). *Oral Oncol.* **2015**, *51*, 727–730. [CrossRef]
7. Kreimer, A.R.; Clifford, G.M.; Boyle, P.; Franceschi, S. Human papillomavirus types in head and neck squamous cell carcinomas worldwide: A systematic review. *Cancer Epidemiol. Biomark. Prev.* **2005**, *14*, 467–475. [CrossRef]
8. Chaturvedi, A.K.; Engels, E.A.; Anderson, W.F.; Gillison, M.L. Incidence trends for human papillomavirus-related and -unrelated oral squamous cell carcinomas in the United States. *J. Clin. Oncol.* **2008**, *26*, 612–619. [CrossRef]
9. Marur, S.; D'Souza, G.; Westra, W.H.; Forastiere, A.A. HPV-associated head and neck cancer: A virus-related cancer epidemic. *Lancet Oncol.* **2010**, *11*, 781–789. [CrossRef]
10. Ang, K.K.; Harris, J.; Wheeler, R.; Weber, R.; Rosenthal, D.I.; Nguyen-Tan, P.F.; Westra, W.H.; Chung, C.H.; Jordan, R.C.; Lu, C.; et al. Human papillomavirus and survival of patients with oropharyngeal cancer. *N. Engl. J. Med.* **2010**, *363*, 24–35. [CrossRef]
11. Griffin, S.; Walker, S.; Sculpher, M.; White, S.; Erhorn, S.; Brent, S.; Dyker, A.; Ferrie, L.; Gilfillan, C.; Horsley, W.; et al. Cetuximab plus radiotherapy for the treatment of locally advanced squamous cell carcinoma of the head and neck. *Health Technol. Assess.* **2009**, *13* Suppl. 1, 49–54. [CrossRef]
12. Bonner, J.A.; Harari, P.M.; Giralt, J.; Azarnia, N.; Shin, D.M.; Cohen, R.B.; Jones, C.U.; Sur, R.; Raben, D.; Jassem, J.; et al. Radiotherapy plus cetuximab for squamous-cell carcinoma of the head and neck. *N. Engl. J. Med.* **2006**, *354*, 567–578. [CrossRef]
13. Koyama, S.; Akbay, E.A.; Li, Y.Y.; Herter-Sprie, G.S.; Buczkowski, K.A.; Richards, W.G.; Gandhi, L.; Redig, A.J.; Rodig, S.J.; Asahina, H.; et al. Adaptive resistance to therapeutic PD-1 blockade is associated with upregulation of alternative immune checkpoints. *Nat. Commun.* **2016**, *7*, 10501. [CrossRef]
14. Seiwert, T.Y.; Burtness, B.; Mehra, R.; Weiss, J.; Berger, R.; Eder, J.P.; Heath, K.; McClanahan, T.; Lunceford, J.; Gause, C.; et al. Safety and clinical activity of pembrolizumab for treatment of recurrent or metastatic squamous cell carcinoma of the head and neck (KEYNOTE-012): An open-label, multicentre, phase 1b trial. *Lancet Oncol.* **2016**, *17*, 956–965. [CrossRef]
15. Chow, L.Q.M.; Haddad, R.; Gupta, S.; Mahipal, A.; Mehra, R.; Tahara, M.; Berger, R.; Eder, J.P.; Burtness, B.; Lee, S.H.; et al. Antitumor Activity of Pembrolizumab in Biomarker-Unselected Patients With Recurrent and/or Metastatic Head and Neck Squamous Cell Carcinoma: Results From the Phase Ib KEYNOTE-012 Expansion Cohort. *J. Clin. Oncol.* **2016**, *34*, 3838–3845. [CrossRef]

16. Champiat, S.; Ferrara, R.; Massard, C.; Besse, B.; Marabelle, A.; Soria, J.C.; Ferte, C. Hyperprogressive disease: Recognizing a novel pattern to improve patient management. *Nat. Rev. Clin. Oncol.* **2018**, *15*, 748–762. [CrossRef]
17. Saada-Bouzid, E.; Defaucheux, C.; Karabajakian, A.; Coloma, V.P.; Servois, V.; Paoletti, X.; Even, C.; Fayette, J.; Guigay, J.; Loirat, D.; et al. Hyperprogression during anti-PD-1/PD-L1 therapy in patients with recurrent and/or metastatic head and neck squamous cell carcinoma. *Ann. Oncol.* **2017**, *28*, 1605–1611. [CrossRef]
18. Pulte, D.; Brenner, H. Changes in survival in head and neck cancers in the late 20th and early 21st century: A period analysis. *Oncologist* **2010**, *15*, 994–1001. [CrossRef]
19. Pickering, C.R.; Zhang, J.; Yoo, S.Y.; Bengtsson, L.; Moorthy, S.; Neskey, D.M.; Zhao, M.; Ortega Alves, M.V.; Chang, K.; Drummond, J.; et al. Integrative genomic characterization of oral squamous cell carcinoma identifies frequent somatic drivers. *Cancer Discov.* **2013**, *3*, 770–781. [CrossRef]
20. Cancer Genome Atlas Network. Comprehensive genomic characterization of head and neck squamous cell carcinomas. *Nature* **2015**, *517*, 576–582.
21. Lui, V.W.; Hedberg, M.L.; Li, H.; Vangara, B.S.; Pendleton, K.; Zeng, Y.; Lu, Y.; Zhang, Q.; Du, Y.; Gilbert, B.R.; et al. Frequent mutation of the PI3K pathway in head and neck cancer defines predictive biomarkers. *Cancer Discov.* **2013**, *3*, 761–769. [CrossRef]
22. Stransky, N.; Egloff, A.M.; Tward, A.D.; Kostic, A.D.; Cibulskis, K.; Sivachenko, A.; Kryukov, G.V.; Lawrence, M.S.; Sougnez, C.; McKenna, A.; et al. The mutational landscape of head and neck squamous cell carcinoma. *Science* **2011**, *333*, 1157–1160. [CrossRef]
23. Zhang, H.; Liu, J.; Fu, X.; Yang, A. Identification of Key Genes and Pathways in Tongue Squamous Cell Carcinoma Using Bioinformatics Analysis. *Med. Sci. Monit.* **2017**, *23*, 5924–5932. [CrossRef]
24. Cuevas Gonzalez, J.C.; Gaitan Cepeda, L.A.; Borges Yanez, S.A.; Cornejo, A.D.; Mori Estevez, A.D.; Huerta, E.R. p53 and p16 in oral epithelial dysplasia and oral squamous cell carcinoma: A study of 208 cases. *Indian J. Pathol. Microbiol.* **2016**, *59*, 153–158.
25. Karpathiou, G.; Monaya, A.; Forest, F.; Froudarakis, M.; Casteillo, F.; Marc Dumollard, J.; Prades, J.M.; Peoc'h, M. p16 and p53 expression status in head and neck squamous cell carcinoma: A correlation with histological, histoprognostic and clinical parameters. *Pathology* **2016**, *48*, 341–348. [CrossRef]
26. Lin, S.C.; Lin, L.H.; Yu, S.Y.; Kao, S.Y.; Chang, K.W.; Cheng, H.W.; Liu, C.J. FAT1 somatic mutations in head and neck carcinoma are associated with tumor progression and survival. *Carcinogenesis* **2018**, *39*, 1320–1330. [CrossRef]
27. Lee, S.H.; Do, S.I.; Lee, H.J.; Kang, H.J.; Koo, B.S.; Lim, Y.C. Notch1 signaling contributes to stemness in head and neck squamous cell carcinoma. *Lab. Investig.* **2016**, *96*, 508–516. [CrossRef]
28. Plath, M.; Broglie, M.A.; Forbs, D.; Stoeckli, S.J.; Jochum, W. Prognostic significance of cell cycle-associated proteins p16, pRB, cyclin D1 and p53 in resected oropharyngeal carcinoma. *J. Otolaryngol. Head Neck Surg.* **2018**, *47*, 53. [CrossRef]
29. Belobrov, S.; Cornall, A.M.; Young, R.J.; Koo, K.; Angel, C.; Wiesenfeld, D.; Rischin, D.; Garland, S.M.; McCullough, M. The role of human papillomavirus in p16-positive oral cancers. *J. Oral Pathol. Med.: Off. Publ. Int. Assoc. Oral Pathol. Am. Acad. Oral Pathol.* **2018**, *47*, 18–24. [CrossRef]
30. Cheng, H.; Yang, X.; Si, H.; Saleh, A.D.; Xiao, W.; Coupar, J.; Gollin, S.M.; Ferris, R.L.; Issaeva, N.; Yarbrough, W.G.; et al. Genomic and Transcriptomic Characterization Links Cell Lines with Aggressive Head and Neck Cancers. *Cell Rep.* **2018**, *25*, 1332–1345.e5. [CrossRef]
31. Nishikawa, Y.; Miyazaki, T.; Nakashiro, K.; Yamagata, H.; Isokane, M.; Goda, H.; Tanaka, H.; Oka, R.; Hamakawa, H. Human FAT1 cadherin controls cell migration and invasion of oral squamous cell carcinoma through the localization of beta-catenin. *Oncol. Rep.* **2011**, *26*, 587–592.
32. Yoshida, R.; Nagata, M.; Nakayama, H.; Niimori-Kita, K.; Hassan, W.; Tanaka, T.; Shinohara, M.; Ito, T. The pathological significance of Notch1 in oral squamous cell carcinoma. *Lab. Investig.* **2013**, *93*, 1068–1081. [CrossRef]
33. Agrawal, N.; Frederick, M.J.; Pickering, C.R.; Bettegowda, C.; Chang, K.; Li, R.J.; Fakhry, C.; Xie, T.X.; Zhang, J.; Wang, J.; et al. Exome sequencing of head and neck squamous cell carcinoma reveals inactivating mutations in NOTCH1. *Science* **2011**, *333*, 1154–1157. [CrossRef]
34. Grabiner, B.C.; Nardi, V.; Birsoy, K.; Possemato, R.; Shen, K.; Sinha, S.; Jordan, A.; Beck, A.H.; Sabatini, D.M. A diverse array of cancer-associated MTOR mutations are hyperactivating and can predict rapamycin sensitivity. *Cancer Discov.* **2014**, *4*, 554–563. [CrossRef]

35. Li, S.; Wang, Z.; Huang, J.; Cheng, S.; Du, H.; Che, G.; Peng, Y. Clinicopathological and prognostic significance of mTOR and phosphorylated mTOR expression in patients with esophageal squamous cell carcinoma: A systematic review and meta-analysis. *BMC Cancer* **2016**, *16*, 877. [CrossRef]
36. Driscoll, D.R.; Karim, S.A.; Sano, M.; Gay, D.M.; Jacob, W.; Yu, J.; Mizukami, Y.; Gopinathan, A.; Jodrell, D.I.; Evans, T.R.; et al. mTORC2 Signaling Drives the Development and Progression of Pancreatic Cancer. *Cancer Res.* **2016**, *76*, 6911–6923. [CrossRef]
37. Vander Haar, E.; Lee, S.I.; Bandhakavi, S.; Griffin, T.J.; Kim, D.H. Insulin signalling to mTOR mediated by the Akt/PKB substrate PRAS40. *Nat. Cell Biol.* **2007**, *9*, 316–323. [CrossRef]
38. Liu, P.; Gan, W.; Inuzuka, H.; Lazorchak, A.S.; Gao, D.; Arojo, O.; Liu, D.; Wan, L.; Zhai, B.; Yu, Y.; et al. Sin1 phosphorylation impairs mTORC2 complex integrity and inhibits downstream Akt signalling to suppress tumorigenesis. *Nat. Cell Biol.* **2013**, *15*, 1340–1350. [CrossRef]
39. Tenkerian, C.; Krishnamoorthy, J.; Mounir, Z.; Kazimierczak, U.; Khoutorsky, A.; Staschke, K.A.; Kristof, A.S.; Wang, S.; Hatzoglou, M.; Koromilas, A.E. mTORC2 Balances AKT Activation and eIF2alpha Serine 51 Phosphorylation to Promote Survival under Stress. *Mol. Cancer Res.* **2015**, *13*, 1377–1388. [CrossRef]
40. Mossmann, D.; Park, S.; Hall, M.N. mTOR signalling and cellular metabolism are mutual determinants in cancer. *Nat. Rev. Cancer* **2018**, *18*, 744–757. [CrossRef]
41. Warburg, O. On the origin of cancer cells. *Science* **1956**, *123*, 309–314. [CrossRef]
42. Vander Heiden, M.G.; Cantley, L.C.; Thompson, C.B. Understanding the Warburg effect: The metabolic requirements of cell proliferation. *Science* **2009**, *324*, 1029–1033. [CrossRef] [PubMed]
43. Liberti, M.V.; Locasale, J.W. The Warburg Effect: How Does it Benefit Cancer Cells? *Trends Biochem. Sci.* **2016**, *41*, 211–218. [CrossRef] [PubMed]
44. Buller, C.L.; Loberg, R.D.; Fan, M.H.; Zhu, Q.; Park, J.L.; Vesely, E.; Inoki, K.; Guan, K.L.; Brosius, F.C., 3rd. A GSK-3/TSC2/mTOR pathway regulates glucose uptake and GLUT1 glucose transporter expression. *Am. J. Physiol. Cell Physiol.* **2008**, *295*, C836–C843.
45. Li, S.; Yang, X.; Wang, P.; Ran, X. The effects of GLUT1 on the survival of head and neck squamous cell carcinoma. *Cell Physiol. Biochem.* **2013**, *32*, 624–634. [CrossRef] [PubMed]
46. Koukourakis, M.I.; Giatromanolaki, A.; Sivridis, E.; Simopoulos, C.; Turley, H.; Talks, K.; Gatter, K.C.; Harris, A.L. Hypoxia-inducible factor (HIF1A and HIF2A), angiogenesis, and chemoradiotherapy outcome of squamous cell head-and-neck cancer. *Int. J. Radiat. Oncol. Biol. Phys.* **2002**, *53*, 1192–1202. [CrossRef]
47. Kleszcz, R.; Paluszczak, J.; Krajka-Kuzniak, V.; Baer-Dubowska, W. The inhibition of c-MYC transcription factor modulates the expression of glycolytic and glutaminolytic enzymes in FaDu hypopharyngeal carcinoma cells. *Adv. Clin. Exp. Med.* **2018**, *27*, 735–742. [CrossRef] [PubMed]
48. Szablewski, L. Expression of glucose transporters in cancers. *Biochim. Biophys. Acta* **2013**, *1835*, 164–169. [CrossRef]
49. Chen, J.; Zhang, S.; Li, Y.; Tang, Z.; Kong, W. Hexokinase 2 overexpression promotes the proliferation and survival of laryngeal squamous cell carcinoma. *Tumour Biol.* **2014**, *35*, 3743–3753. [CrossRef] [PubMed]
50. Li, L.; Pilo, G.M.; Li, X.; Cigliano, A.; Latte, G.; Che, L.; Joseph, C.; Mela, M.; Wang, C.; Jiang, L.; et al. Inactivation of fatty acid synthase impairs hepatocarcinogenesis driven by AKT in mice and humans. *J. Hepatol.* **2016**, *64*, 333–341. [CrossRef]
51. Ricoult, S.J.; Yecies, J.L.; Ben-Sahra, I.; Manning, B.D. Oncogenic PI3K and K-Ras stimulate de novo lipid synthesis through mTORC1 and SREBP. *Oncogene* **2016**, *35*, 1250–1260. [CrossRef]
52. Guri, Y.; Colombi, M.; Dazert, E.; Hindupur, S.K.; Roszik, J.; Moes, S.; Jenoe, P.; Heim, M.H.; Riezman, I.; Riezman, H.; et al. mTORC2 Promotes Tumorigenesis via Lipid Synthesis. *Cancer Cell* **2017**, *32*, 807–823 e12. [CrossRef]
53. Wang, Z.; Martin, D.; Molinolo, A.A.; Patel, V.; Iglesias-Bartolome, R.; Degese, M.S.; Vitale-Cross, L.; Chen, Q.; Gutkind, J.S. mTOR co-targeting in cetuximab resistance in head and neck cancers harboring PIK3CA and RAS mutations. *J. Natl. Cancer Inst.* **2014**, 106. [CrossRef]
54. Tian, T.; Li, X.; Zhang, J. mTOR Signaling in Cancer and mTOR Inhibitors in Solid Tumor Targeting Therapy. *Int. J. Mol. Sci.* **2019**, *20*, 755. [CrossRef]
55. Amornphimoltham, P.; Leelahavanichkul, K.; Molinolo, A.; Patel, V.; Gutkind, J.S. Inhibition of Mammalian target of rapamycin by rapamycin causes the regression of carcinogen-induced skin tumor lesions. *Clin. Cancer Res.* **2008**, *14*, 8094–8101. [CrossRef]

56. Callejas-Valera, J.L.; Iglesias-Bartolome, R.; Amornphimoltham, P.; Palacios-Garcia, J.; Martin, D.; Califano, J.A.; Molinolo, A.A.; Gutkind, J.S. mTOR inhibition prevents rapid-onset of carcinogen-induced malignancies in a novel inducible HPV-16 E6/E7 mouse model. *Carcinogenesis* **2016**, *37*, 1014–1025. [CrossRef]
57. Czerninski, R.; Amornphimoltham, P.; Patel, V.; Molinolo, A.A.; Gutkind, J.S. Targeting mammalian target of rapamycin by rapamycin prevents tumor progression in an oral-specific chemical carcinogenesis model. *Cancer Prev. Res.* **2009**, *2*, 27–36. [CrossRef]
58. Athar, M.; Kopelovich, L. Rapamycin and mTORC1 inhibition in the mouse: Skin cancer prevention. *Cancer Prev. Res.* **2011**, *4*, 957–961. [CrossRef]
59. Sun, Z.J.; Zhang, L.; Zhang, W.; Hall, B.; Bian, Y.; Kulkarni, A.B. Inhibition of mTOR reduces anal carcinogenesis in transgenic mouse model. *PLoS ONE* **2013**, *8*, e74888. [CrossRef]
60. Zhang, H.; Cohen, A.L.; Krishnakumar, S.; Wapnir, I.L.; Veeriah, S.; Deng, G.; Coram, M.A.; Piskun, C.M.; Longacre, T.A.; Herrler, M.; et al. Patient-derived xenografts of triple-negative breast cancer reproduce molecular features of patient tumors and respond to mTOR inhibition. *Breast Cancer Res.* **2014**, *16*, R36. [CrossRef]
61. Sun, Z.J.; Zhang, L.; Hall, B.; Bian, Y.; Gutkind, J.S.; Kulkarni, A.B. Chemopreventive and chemotherapeutic actions of mTOR inhibitor in genetically defined head and neck squamous cell carcinoma mouse model. *Clin. Cancer Res.* **2012**, *18*, 5304–5313. [CrossRef]
62. Bozec, A.; Ebran, N.; Radosevic-Robin, N.; Sudaka, A.; Monteverde, M.; Toussan, N.; Etienne-Grimaldi, M.C.; Nigro, C.L.; Merlano, M.; Penault-Llorca, F.; et al. Combination of mTOR and EGFR targeting in an orthotopic xenograft model of head and neck cancer. *Laryngoscope* **2016**, *126*, E156–E163. [CrossRef]
63. Moore, E.C.; Cash, H.A.; Caruso, A.M.; Uppaluri, R.; Hodge, J.W.; Van Waes, C.; Allen, C.T. Enhanced Tumor Control with Combination mTOR and PD-L1 Inhibition in Syngeneic Oral Cavity Cancers. *Cancer Immunol. Res.* **2016**, *4*, 611–620. [CrossRef]
64. Sancak, Y.; Bar-Peled, L.; Zoncu, R.; Markhard, A.L.; Nada, S.; Sabatini, D.M. Ragulator-Rag complex targets mTORC1 to the lysosomal surface and is necessary for its activation by amino acids. *Cell* **2010**, *141*, 290–303. [CrossRef]
65. Settembre, C.; Fraldi, A.; Medina, D.L.; Ballabio, A. Signals from the lysosome: A control centre for cellular clearance and energy metabolism. *Nat. Rev. Mol. Cell Biol.* **2013**, *14*, 283–296. [CrossRef]
66. Abu-Remaileh, M.; Wyant, G.A.; Kim, C.; Laqtom, N.N.; Abbasi, M.; Chan, S.H.; Freinkman, E.; Sabatini, D.M. Lysosomal metabolomics reveals V-ATPase- and mTOR-dependent regulation of amino acid efflux from lysosomes. *Science* **2017**, *358*, 807–813. [CrossRef]
67. Bar-Peled, L.; Chantranupong, L.; Cherniack, A.D.; Chen, W.W.; Ottina, K.A.; Grabiner, B.C.; Spear, E.D.; Carter, S.L.; Meyerson, M.; Sabatini, D.M. A Tumor suppressor complex with GAP activity for the Rag GTPases that signal amino acid sufficiency to mTORC1. *Science* **2013**, *340*, 1100–1106. [CrossRef]
68. Cosway, B.; Lovat, P. The role of autophagy in squamous cell carcinoma of the head and neck. *Oral Oncol.* **2016**, *54*, 1–6. [CrossRef]
69. Liu, J.L.; Chen, F.F.; Lung, J.; Lo, C.H.; Lee, F.H.; Lu, Y.C.; Hung, C.H. Prognostic significance of p62/SQSTM1 subcellular localization and LC3B in oral squamous cell carcinoma. *Br. J. Cancer* **2014**, *111*, 944–954. [CrossRef]
70. Tang, J.Y.; Hsi, E.; Huang, Y.C.; Hsu, N.C.; Chu, P.Y.; Chai, C.Y. High LC3 expression correlates with poor survival in patients with oral squamous cell carcinoma. *Hum. Pathol.* **2013**, *44*, 2558–2562. [CrossRef]
71. Wright, T.J.; McKee, C.; Birch-Machin, M.A.; Ellis, R.; Armstrong, J.L.; Lovat, P.E. Increasing the therapeutic efficacy of docetaxel for cutaneous squamous cell carcinoma through the combined inhibition of phosphatidylinositol 3-kinase/AKT signalling and autophagy. *Clin. Exp. Dermatol.* **2013**, *38*, 421–423. [CrossRef]
72. Zang, Y.; Thomas, S.M.; Chan, E.T.; Kirk, C.J.; Freilino, M.L.; DeLancey, H.M.; Grandis, J.R.; Li, C.; Johnson, D.E. Carfilzomib and ONX 0912 inhibit cell survival and tumor growth of head and neck cancer and their activities are enhanced by suppression of Mcl-1 or autophagy. *Clin. Cancer Res.* **2012**, *18*, 5639–5649. [CrossRef]
73. Jung, Y.S.; Najy, A.J.; Huang, W.; Sethi, S.; Snyder, M.; Sakr, W.; Dyson, G.; Huttemann, M.; Lee, I.; Ali-Fehmi, R.; et al. HPV-associated differential regulation of tumor metabolism in oropharyngeal head and neck cancer. *Oncotarget* **2017**, *8*, 51530–51541. [CrossRef]
74. Swartz, J.E.; Pothen, A.J.; van Kempen, P.M.; Stegeman, I.; Formsma, F.K.; Cann, E.M.; Willems, S.M.; Grolman, W. Poor prognosis in human papillomavirus-positive oropharyngeal squamous cell carcinomas that overexpress hypoxia inducible factor-1alpha. *Head Neck* **2016**, *38*, 1338–1346. [CrossRef]

75. Rodolico, V.; Arancio, W.; Amato, M.C.; Aragona, F.; Cappello, F.; Di Fede, O.; Pannone, G.; Campisi, G. Hypoxia inducible factor-1 alpha expression is increased in infected positive HPV16 DNA oral squamous cell carcinoma and positively associated with HPV16 E7 oncoprotein. *Infect. Agents Cancer* **2011**, *6*, 18. [CrossRef]
76. Molinolo, A.A.; Marsh, C.; El Dinali, M.; Gangane, N.; Jennison, K.; Hewitt, S.; Patel, V.; Seiwert, T.Y.; Gutkind, J.S. mTOR as a molecular target in HPV-associated oral and cervical squamous carcinomas. *Clin. Cancer Res.* **2012**, *18*, 2558–2568. [CrossRef]
77. Brand, T.M.; Hartmann, S.; Bhola, N.E.; Li, H.; Zeng, Y.; O'Keefe, R.A.; Ranall, M.V.; Bandyopadhyay, S.; Soucheray, M.; Krogan, N.J.; et al. Cross-talk Signaling between HER3 and HPV16 E6 and E7 Mediates Resistance to PI3K Inhibitors in Head and Neck Cancer. *Cancer Res.* **2018**, *78*, 2383–2395. [CrossRef]
78. Madera, D.; Vitale-Cross, L.; Martin, D.; Schneider, A.; Molinolo, A.A.; Gangane, N.; Carey, T.E.; McHugh, J.B.; Komarck, C.M.; Walline, H.M.; et al. Prevention of tumor growth driven by PIK3CA and HPV oncogenes by targeting mTOR signaling with metformin in oral squamous carcinomas expressing OCT3. *Cancer Prev. Res.* **2015**, *8*, 197–207. [CrossRef]
79. Goschzik, T.; Gessi, M.; Denkhaus, D.; Pietsch, T. PTEN mutations and activation of the PI3K/Akt/mTOR signaling pathway in papillary tumors of the pineal region. *J. Neuropathol. Exp. Neurol.* **2014**, *73*, 747–751. [CrossRef]
80. Matsumoto, C.S.; Almeida, L.O.; Guimaraes, D.M.; Martins, M.D.; Papagerakis, P.; Papagerakis, S.; Leopoldino, A.M.; Castilho, R.M.; Squarize, C.H. PI3K-PTEN dysregulation leads to mTOR-driven upregulation of the core clock gene BMAL1 in normal and malignant epithelial cells. *Oncotarget* **2016**, *7*, 42393–42407. [CrossRef]
81. Therkildsen, C.; Bergmann, T.K.; Henrichsen-Schnack, T.; Ladelund, S.; Nilbert, M. The predictive value of KRAS, NRAS, BRAF, PIK3CA and PTEN for anti-EGFR treatment in metastatic colorectal cancer: A systematic review and meta-analysis. *Acta Oncol.* **2014**, *53*, 852–864. [CrossRef]
82. Nojima, H.; Tokunaga, C.; Eguchi, S.; Oshiro, N.; Hidayat, S.; Yoshino, K.; Hara, K.; Tanaka, N.; Avruch, J.; Yonezawa, K. The mammalian target of rapamycin (mTOR) partner, raptor, binds the mTOR substrates p70 S6 kinase and 4E-BP1 through their TOR signaling (TOS) motif. *J. Biol. Chem.* **2003**, *278*, 15461–15464. [CrossRef]
83. Schalm, S.S.; Fingar, D.C.; Sabatini, D.M.; Blenis, J. TOS motif-mediated raptor binding regulates 4E-BP1 multisite phosphorylation and function. *Curr. Biol.* **2003**, *13*, 797–806. [CrossRef]
84. Wu, C.C.; Hou, S.; Orr, B.A.; Kuo, B.R.; Youn, Y.H.; Ong, T.; Roth, F.; Eberhart, C.G.; Robinson, G.W.; Solecki, D.J.; et al. mTORC1-Mediated Inhibition of 4EBP1 Is Essential for Hedgehog Signaling-Driven Translation and Medulloblastoma. *Dev. Cell* **2017**, *43*, 673–688 e5. [CrossRef]
85. Saxton, R.A.; Sabatini, D.M. mTOR Signaling in Growth, Metabolism, and Disease. *Cell* **2017**, *169*, 361–371.
86. Gkountakos, A.; Pilotto, S.; Mafficini, A.; Vicentini, C.; Simbolo, M.; Milella, M.; Tortora, G.; Scarpa, A.; Bria, E.; Corbo, V. Unmasking the impact of Rictor in cancer: Novel insights of mTORC2 complex. *Carcinogenesis* **2018**. [CrossRef]
87. Gupta, A.K.; McKenna, W.G.; Weber, C.N.; Feldman, M.D.; Goldsmith, J.D.; Mick, R.; Machtay, M.; Rosenthal, D.I.; Bakanauskas, V.J.; Cerniglia, G.J.; et al. Local recurrence in head and neck cancer: Relationship to radiation resistance and signal transduction. *Clin. Cancer Res.* **2002**, *8*, 885–892.
88. Li, Z.; Yang, Z.; Passaniti, A.; Lapidus, R.G.; Liu, X.; Cullen, K.J.; Dan, H.C. A positive feedback loop involving EGFR/Akt/mTORC1 and IKK/NF-kB regulates head and neck squamous cell carcinoma proliferation. *Oncotarget* **2016**, *7*, 31892–31906.
89. Elkabets, M.; Pazarentzos, E.; Juric, D.; Sheng, Q.; Pelossof, R.A.; Brook, S.; Benzaken, A.O.; Rodon, J.; Morse, N.; Yan, J.J.; et al. AXL mediates resistance to PI3Kalpha inhibition by activating the EGFR/PKC/mTOR axis in head and neck and esophageal squamous cell carcinomas. *Cancer Cell* **2015**, *27*, 533–546. [CrossRef]
90. Bigner, S.H.; Humphrey, P.A.; Wong, A.J.; Vogelstein, B.; Mark, J.; Friedman, H.S.; Bigner, D.D. Characterization of the epidermal growth factor receptor in human glioma cell lines and xenografts. *Cancer Res.* **1990**, *50*, 8017–8022.
91. Sok, J.C.; Coppelli, F.M.; Thomas, S.M.; Lango, M.N.; Xi, S.; Hunt, J.L.; Freilino, M.L.; Graner, M.W.; Wikstrand, C.J.; Bigner, D.D.; et al. Mutant epidermal growth factor receptor (EGFRvIII) contributes to head and neck cancer growth and resistance to EGFR targeting. *Clin. Cancer Res.* **2006**, *12*, 5064–5073. [CrossRef]

92. Khattri, A.; Zuo, Z.; Bragelmann, J.; Keck, M.K.; El Dinali, M.; Brown, C.D.; Stricker, T.; Munagala, A.; Cohen, E.E.; Lingen, M.W.; et al. Rare occurrence of EGFRvIII deletion in head and neck squamous cell carcinoma. *Oral Oncol.* **2015**, *51*, 53–58. [CrossRef]
93. Tomczak, K.; Czerwinska, P.; Wiznerowicz, M. The Cancer Genome Atlas (TCGA): An immeasurable source of knowledge. *Contemp. Oncol. (Pozn)* **2015**, *19*, A68–A77. [CrossRef]
94. Hashmi, A.A.; Hussain, Z.F.; Aijaz, S.; Irfan, M.; Khan, E.Y.; Naz, S.; Faridi, N.; Khan, A.; Edhi, M.M. Immunohistochemical expression of epidermal growth factor receptor (EGFR) in South Asian head and neck squamous cell carcinoma: Association with various risk factors and clinico-pathologic and prognostic parameters. *World J. Surg. Oncol.* **2018**, *16*, 118. [CrossRef]
95. Makinoshima, H.; Takita, M.; Saruwatari, K.; Umemura, S.; Obata, Y.; Ishii, G.; Matsumoto, S.; Sugiyama, E.; Ochiai, A.; Abe, R.; et al. Signaling through the Phosphatidylinositol 3-Kinase (PI3K)/Mammalian Target of Rapamycin (mTOR) Axis Is Responsible for Aerobic Glycolysis mediated by Glucose Transporter in Epidermal Growth Factor Receptor (EGFR)-mutated Lung Adenocarcinoma. *J. Biol. Chem.* **2015**, *290*, 17495–17504. [CrossRef]
96. Chiang, C.T.; Demetriou, A.N.; Ung, N.; Choudhury, N.; Ghaffarian, K.; Ruderman, D.L.; Mumenthaler, S.M. mTORC2 contributes to the metabolic reprogramming in EGFR tyrosine-kinase inhibitor resistant cells in non-small cell lung cancer. *Cancer Lett.* **2018**, *434*, 152–159. [CrossRef]
97. Cohen, E.E.; Davis, D.W.; Karrison, T.G.; Seiwert, T.Y.; Wong, S.J.; Nattam, S.; Kozloff, M.F.; Clark, J.I.; Yan, D.H.; Liu, W.; et al. Erlotinib and bevacizumab in patients with recurrent or metastatic squamous-cell carcinoma of the head and neck: A phase I/II study. *Lancet Oncol.* **2009**, *10*, 247–257. [CrossRef]
98. Sharp, H.; Morris, J.C.; Van Waes, C.; Gius, D.; Cooley-Zgela, T.; Singh, A.K. High incidence of oral dysesthesias on a trial of gefitinib, Paclitaxel, and concurrent external beam radiation for locally advanced head and neck cancers. *Am. J. Clin. Oncol.* **2008**, *31*, 557–560. [CrossRef]
99. Xu, M.J.; Johnson, D.E.; Grandis, J.R. EGFR-targeted therapies in the post-genomic era. *Cancer Metastasis Rev.* **2017**, *36*, 463–473. [CrossRef]
100. Porcelli, L.; Quatrale, A.E.; Mantuano, P.; Silvestris, N.; Rolland, J.F.; Biancolillo, L.; Paradiso, A.; Azzariti, A. Synergistic antiproliferative and antiangiogenic effects of EGFR and mTOR inhibitors. *Curr. Pharm. Des.* **2013**, *19*, 918–926. [CrossRef]
101. Schmid, K.; Bago-Horvath, Z.; Berger, W.; Haitel, A.; Cejka, D.; Werzowa, J.; Filipits, M.; Herberger, B.; Hayden, H.; Sieghart, W. Dual inhibition of EGFR and mTOR pathways in small cell lung cancer. *Br. J. Cancer* **2010**, *103*, 622–628. [CrossRef]
102. Boumahdi, S.; Driessens, G.; Lapouge, G.; Rorive, S.; Nassar, D.; Le Mercier, M.; Delatte, B.; Caauwe, A.; Lenglez, S.; Nkusi, E.; et al. SOX2 controls tumour initiation and cancer stem-cell functions in squamous-cell carcinoma. *Nature* **2014**, *511*, 246–250. [CrossRef]
103. Keysar, S.B.; Le, P.N.; Miller, B.; Jackson, B.C.; Eagles, J.R.; Nieto, C.; Kim, J.; Tang, B.; Glogowska, M.J.; Morton, J.J.; et al. Regulation of Head and Neck Squamous Cancer Stem Cells by PI3K and SOX2. *J. Natl. Cancer Inst.* **2017**, 109. [CrossRef]
104. Huang, C.F.; Xu, X.R.; Wu, T.F.; Sun, Z.J.; Zhang, W.F. Correlation of ALDH1, CD44, OCT4 and SOX2 in tongue squamous cell carcinoma and their association with disease progression and prognosis. *J. Oral Pathol. Med.: Off. Publ. Int. Assoc. Oral Pathol. Am. Acad. Oral Pathol.* **2014**, *43*, 492–498. [CrossRef]
105. Suda, T.; Hama, T.; Kondo, S.; Yuza, Y.; Yoshikawa, M.; Urashima, M.; Kato, T.; Moriyama, H. Copy number amplification of the PIK3CA gene is associated with poor prognosis in non-lymph node metastatic head and neck squamous cell carcinoma. *BMC Cancer* **2012**, *12*, 416. [CrossRef]
106. Hao, Y.; Samuels, Y.; Li, Q.; Krokowski, D.; Guan, B.J.; Wang, C.; Jin, Z.; Dong, B.; Cao, B.; Feng, X.; et al. Oncogenic PIK3CA mutations reprogram glutamine metabolism in colorectal cancer. *Nat. Commun.* **2016**, *7*, 11971. [CrossRef]
107. Lau, C.E.; Tredwell, G.D.; Ellis, J.K.; Lam, E.W.; Keun, H.C. Metabolomic characterisation of the effects of oncogenic PIK3CA transformation in a breast epithelial cell line. *Sci. Rep.* **2017**, *7*, 46079. [CrossRef]
108. Pedrero, J.M.; Carracedo, D.G.; Pinto, C.M.; Zapatero, A.H.; Rodrigo, J.P.; Nieto, C.S.; Gonzalez, M.V. Frequent genetic and biochemical alterations of the PI 3-K/AKT/PTEN pathway in head and neck squamous cell carcinoma. *Int. J. Cancer* **2005**, *114*, 242–248. [CrossRef]
109. Henderson, Y.C.; Wang, E.; Clayman, G.L. Genotypic analysis of tumor suppressor genes PTEN/MMAC1 and p53 in head and neck squamous cell carcinomas. *Laryngoscope* **1998**, *108*, 1553–1556. [CrossRef]

110. Shao, X.; Tandon, R.; Samara, G.; Kanki, H.; Yano, H.; Close, L.G.; Parsons, R.; Sato, T. Mutational analysis of the PTEN gene in head and neck squamous cell carcinoma. *Int. J. Cancer* **1998**, *77*, 684–688. [CrossRef]
111. Darido, C.; Georgy, S.R.; Wilanowski, T.; Dworkin, S.; Auden, A.; Zhao, Q.; Rank, G.; Srivastava, S.; Finlay, M.J.; Papenfuss, A.T.; et al. Targeting of the tumor suppressor GRHL3 by a miR-21-dependent proto-oncogenic network results in PTEN loss and tumorigenesis. *Cancer Cell* **2011**, *20*, 635–648. [CrossRef]
112. Cangkrama, M.; Ting, S.B.; Darido, C. Stem cells behind the barrier. *Int. J. Mol. Sci.* **2013**, *14*, 13670–13686. [CrossRef]
113. Cangkrama, M.; Darido, C.; Georgy, S.R.; Partridge, D.; Auden, A.; Srivastava, S.; Wilanowski, T.; Jane, S.M. Two Ancient Gene Families Are Critical for Maintenance of the Mammalian Skin Barrier in Postnatal Life. *J. Investig. Dermatol.* **2016**, *136*, 1438–1448. [CrossRef]
114. Georgy, S.R.; Cangkrama, M.; Srivastava, S.; Partridge, D.; Auden, A.; Dworkin, S.; McLean, C.A.; Jane, S.M.; Darido, C. Identification of a Novel Proto-oncogenic Network in Head and Neck Squamous Cell Carcinoma. *J. Natl. Cancer Inst.* **2015**, 107. [CrossRef]
115. Miles, L.B.; Dworkin, S.; Darido, C. Review article: Alternative splicing and start sites: Lessons from the Grainyhead-like family. *Dev. Biol.* **2017**, *429*, 12–19. [CrossRef]
116. Darido, C.; Georgy, S.R.; Jane, S.M. The role of barrier genes in epidermal malignancy. *Oncogene* **2016**, *35*, 5705–5712. [CrossRef]
117. Youssef, M.; Cuddihy, A.; Darido, C. Long-Lived Epidermal Cancer-Initiating Cells. *Int. J. Mol. Sci.* **2017**, *18*, 1369. [CrossRef]
118. Darido, C.; Georgy, S.R.; Cullinane, C.; Partridge, D.D.; Walker, R.; Srivastava, S.; Roslan, S.; Carpinelli, M.R.; Dworkin, S.; Pearson, R.B.; et al. Stage-dependent therapeutic efficacy in PI3K/mTOR-driven squamous cell carcinoma of the skin. *Cell Death Differ.* **2018**, *25*, 1146–1159. [CrossRef]
119. Lee, J.I.; Soria, J.C.; Hassan, K.A.; El-Naggar, A.K.; Tang, X.; Liu, D.D.; Hong, W.K.; Mao, L. Loss of PTEN expression as a prognostic marker for tongue cancer. *Arch. Otolaryngol. Head Neck Surg.* **2001**, *127*, 1441–1445. [CrossRef]
120. Squarize, C.H.; Castilho, R.M.; Santos Pinto, D., Jr. Immunohistochemical evidence of PTEN in oral squamous cell carcinoma and its correlation with the histological malignancy grading system. *J. Oral Pathol. Med.: Off. Publ. Int. Assoc. Oral Pathol. Am. Acad. Oral Pathol.* **2002**, *31*, 379–384. [CrossRef]
121. Bian, Y.; Hall, B.; Sun, Z.J.; Molinolo, A.; Chen, W.; Gutkind, J.S.; Waes, C.V.; Kulkarni, A.B. Loss of TGF-beta signaling and PTEN promotes head and neck squamous cell carcinoma through cellular senescence evasion and cancer-related inflammation. *Oncogene* **2012**, *31*, 3322–3332. [CrossRef]
122. Martin, P.L.; Yin, J.J.; Seng, V.; Casey, O.; Corey, E.; Morrissey, C.; Simpson, R.M.; Kelly, K. Androgen deprivation leads to increased carbohydrate metabolism and hexokinase 2-mediated survival in Pten/Tp53-deficient prostate cancer. *Oncogene* **2017**, *36*, 525–533. [CrossRef]
123. Mathur, D.; Stratikopoulos, E.; Ozturk, S.; Steinbach, N.; Pegno, S.; Schoenfeld, S.; Yong, R.; Murty, V.V.; Asara, J.M.; Cantley, L.C.; et al. PTEN Regulates Glutamine Flux to Pyrimidine Synthesis and Sensitivity to Dihydroorotate Dehydrogenase Inhibition. *Cancer Discov.* **2017**, *7*, 380–390. [CrossRef]
124. Garcia-Cao, I.; Song, M.S.; Hobbs, R.M.; Laurent, G.; Giorgi, C.; de Boer, V.C.; Anastasiou, D.; Ito, K.; Sasaki, A.T.; Rameh, L.; et al. Systemic elevation of PTEN induces a tumor-suppressive metabolic state. *Cell* **2012**, *149*, 49–62. [CrossRef]
125. Kiaris, H.; Spandidos, D.A.; Jones, A.S.; Vaughan, E.D.; Field, J.K. Mutations, expression and genomic instability of the H-ras proto-oncogene in squamous cell carcinomas of the head and neck. *Br. J. Cance* **1995**, *72*, 123–128. [CrossRef]
126. Nakagaki, T.; Tamura, M.; Kobashi, K.; Omori, A.; Koyama, R.; Idogawa, M.; Ogi, K.; Hiratsuka, H.; Tokino, T.; Sasaki, Y. Targeted next-generation sequencing of 50 cancer-related genes in Japanese patients with oral squamous cell carcinoma. *Tumour Biol.* **2018**, *40*, 1010428318800180. [CrossRef]
127. Su, S.C.; Lin, C.W.; Liu, Y.F.; Fan, W.L.; Chen, M.K.; Yu, C.P.; Yang, W.E.; Su, C.W.; Chuang, C.Y.; Li, W.H.; et al. Exome Sequencing of Oral Squamous Cell Carcinoma Reveals Molecular Subgroups and Novel Therapeutic Opportunities. *Theranostics* **2017**, *7*, 1088–1099. [CrossRef]
128. Koumaki, D.; Kostakis, G.; Koumaki, V.; Papadogeorgakis, N.; Makris, M.; Katoulis, A.; Kamakari, S.; Koutsodontis, G.; Perisanidis, C.; Lambadiari, V.; et al. Novel mutations of the HRAS gene and absence of hotspot mutations of the BRAF genes in oral squamous cell carcinoma in a Greek population. *Oncol. Rep.* **2012**, *27*, 1555–1560. [CrossRef]

129. Murugan, A.K.; Hong, N.T.; Cuc, T.T.; Hung, N.C.; Munirajan, A.K.; Ikeda, M.A.; Tsuchida, N. Detection of two novel mutations and relatively high incidence of H-RAS mutations in Vietnamese oral cancer. *Oral Oncol.* **2009**, *45*, e161–e166. [CrossRef]
130. Zheng, W.; Tayyari, F.; Gowda, G.A.; Raftery, D.; McLamore, E.S.; Porterfield, D.M.; Donkin, S.S.; Bequette, B.; Teegarden, D. Altered glucose metabolism in Harvey-ras transformed MCF10A cells. *Mol. Carcinog.* **2015**, *54*, 111–120. [CrossRef]
131. Ruicci, K.M.; Pinto, N.; Khan, M.I.; Yoo, J.; Fung, K.; MacNeil, D.; Mymryk, J.S.; Barrett, J.W.; Nichols, A.C. ERK-TSC2 signalling in constitutively-active HRAS mutant HNSCC cells promotes resistance to PI3K inhibition. *Oral Oncol.* **2018**, *84*, 95–103. [CrossRef]
132. Hah, J.H.; Zhao, M.; Pickering, C.R.; Frederick, M.J.; Andrews, G.A.; Jasser, S.A.; Fooshee, D.R.; Milas, Z.L.; Galer, C.; Sano, D.; et al. HRAS mutations and resistance to the epidermal growth factor receptor tyrosine kinase inhibitor erlotinib in head and neck squamous cell carcinoma cells. *Head & Neck* **2014**, *36*, 1547–1554.
133. Rampias, T.; Giagini, A.; Siolos, S.; Matsuzaki, H.; Sasaki, C.; Scorilas, A.; Psyrri, A. RAS/PI3K crosstalk and cetuximab resistance in head and neck squamous cell carcinoma. *Clin. Cancer Res.* **2014**, *20*, 2933–2946. [CrossRef] [PubMed]
134. Pyronnet, S.; Imataka, H.; Gingras, A.C.; Fukunaga, R.; Hunter, T.; Sonenberg, N. Human eukaryotic translation initiation factor 4G (eIF4G) recruits mnk1 to phosphorylate eIF4E. *EMBO J.* **1999**, *18*, 270–279. [CrossRef]
135. Raught, B.; Gingras, A.C.; Gygi, S.P.; Imataka, H.; Morino, S.; Gradi, A.; Aebersold, R.; Sonenberg, N. Serum-stimulated, rapamycin-sensitive phosphorylation sites in the eukaryotic translation initiation factor 4GI. *EMBO J.* **2000**, *19*, 434–444. [CrossRef]
136. Marcotrigiano, J.; Gingras, A.C.; Sonenberg, N.; Burley, S.K. Cap-dependent translation initiation in eukaryotes is regulated by a molecular mimic of eIF4G. *Mol. Cell* **1999**, *3*, 707–716. [CrossRef]
137. Yang, H.S.; Jansen, A.P.; Komar, A.A.; Zheng, X.; Merrick, W.C.; Costes, S.; Lockett, S.J.; Sonenberg, N.; Colburn, N.H. The transformation suppressor Pdcd4 is a novel eukaryotic translation initiation factor 4A binding protein that inhibits translation. *Mol. Cell Biol.* **2003**, *23*, 26–37. [CrossRef] [PubMed]
138. Comtesse, N.; Keller, A.; Diesinger, I.; Bauer, C.; Kayser, K.; Huwer, H.; Lenhof, H.P.; Meese, E. Frequent overexpression of the genes FXR1, CLAPM1 and EIF4G located on amplicon 3q26-27 in squamous cell carcinoma of the lung. *Int. J. Cancer* **2007**, *120*, 2538–2544. [CrossRef] [PubMed]
139. Silvera, D.; Arju, R.; Darvishian, F.; Levine, P.H.; Zolfaghari, L.; Goldberg, J.; Hochman, T.; Formenti, S.C.; Schneider, R.J. Essential role for eIF4GI overexpression in the pathogenesis of inflammatory breast cancer. *Nat. Cell Biol.* **2009**, *11*, 903–908. [CrossRef] [PubMed]
140. Liang, S.; Zhou, Y.; Chen, Y.; Ke, G.; Wen, H.; Wu, X. Decreased expression of EIF4A1 after preoperative brachytherapy predicts better tumor-specific survival in cervical cancer. *Int. J. Gynecol. Cancer* **2014**, *24*, 908–915. [CrossRef] [PubMed]
141. Tu, L.; Liu, Z.; He, X.; He, Y.; Yang, H.; Jiang, Q.; Xie, S.; Xiao, G.; Li, X.; Yao, K.; et al. Over-expression of eukaryotic translation initiation factor 4 gamma 1 correlates with tumor progression and poor prognosis in nasopharyngeal carcinoma. *Mol. Cancer* **2010**, *9*, 78. [CrossRef]
142. Ramirez-Valle, F.; Braunstein, S.; Zavadil, J.; Formenti, S.C.; Schneider, R.J. eIF4GI links nutrient sensing by mTOR to cell proliferation and inhibition of autophagy. *J. Cell Biol.* **2008**, *181*, 293–307. [CrossRef]
143. Moerke, N.J.; Aktas, H.; Chen, H.; Cantel, S.; Reibarkh, M.Y.; Fahmy, A.; Gross, J.D.; Degterev, A.; Yuan, J.; Chorev, M.; et al. Small-molecule inhibition of the interaction between the translation initiation factors eIF4E and eIF4G. *Cell* **2007**, *128*, 257–267. [CrossRef]
144. Chen, L.; Aktas, B.H.; Wang, Y.; He, X.; Sahoo, R.; Zhang, N.; Denoyelle, S.; Kabha, E.; Yang, H.; Freedman, R.Y.; et al. Tumor suppression by small molecule inhibitors of translation initiation. *Oncotarget* **2012**, *3*, 869–881. [CrossRef]
145. Papadopoulos, E.; Jenni, S.; Kabha, E.; Takrouri, K.J.; Yi, T.; Salvi, N.; Luna, R.E.; Gavathiotis, E.; Mahalingam, P.; Arthanari, H.; et al. Structure of the eukaryotic translation initiation factor eIF4E in complex with 4EGI-1 reveals an allosteric mechanism for dissociating eIF4G. *Proc. Natl. Acad. Sci. USA* **2014**, *111*, E3187–E3195. [CrossRef]
146. Cencic, R.; Hall, D.R.; Robert, F.; Du, Y.; Min, J.; Li, L.; Qui, M.; Lewis, I.; Kurtkaya, S.; Dingledine, R.; et al. Reversing chemoresistance by small molecule inhibition of the translation initiation complex eIF4F. *Proc. Natl. Acad. Sci. USA* **2011**, *108*, 1046–1051. [CrossRef]

147. Feng, Y.; Pinkerton, A.B.; Hulea, L.; Zhang, T.; Davies, M.A.; Grotegut, S.; Cheli, Y.; Yin, H.; Lau, E.; Kim, H.; et al. SBI-0640756 Attenuates the Growth of Clinically Unresponsive Melanomas by Disrupting the eIF4F Translation Initiation Complex. *Cancer Res.* **2015**, *75*, 5211–5218. [CrossRef]
148. Miyakawa, S.; Oguro, A.; Ohtsu, T.; Imataka, H.; Sonenberg, N.; Nakamura, Y. RNA aptamers to mammalian initiation factor 4G inhibit cap-dependent translation by blocking the formation of initiation factor complexes. *RNA* **2006**, *12*, 1825–1834. [CrossRef]
149. Ridley, A.J.; Paterson, H.F.; Johnston, C.L.; Diekmann, D.; Hall, A. The small GTP-binding protein rac regulates growth factor-induced membrane ruffling. *Cell* **1992**, *70*, 401–410. [CrossRef]
150. Agarwal, N.K.; Chen, C.H.; Cho, H.; Boulbes, D.R.; Spooner, E.; Sarbassov, D.D. Rictor regulates cell migration by suppressing RhoGDI2. *Oncogene* **2013**, *32*, 2521–2526. [CrossRef]
151. Morrison Joly, M.; Williams, M.M.; Hicks, D.J.; Jones, B.; Sanchez, V.; Young, C.D.; Sarbassov, D.D.; Muller, W.J.; Brantley-Sieders, D.; Cook, R.S. Two distinct mTORC2-dependent pathways converge on Rac1 to drive breast cancer metastasis. *Breast Cancer Res.* **2017**, *19*, 74. [CrossRef]
152. Saci, A.; Cantley, L.C.; Carpenter, C.L. Rac1 regulates the activity of mTORC1 and mTORC2 and controls cellular size. *Mol. Cell* **2011**, *42*, 50–61. [CrossRef]
153. Hampsch, R.A.; Shee, K.; Bates, D.; Lewis, L.D.; Desire, L.; Leblond, B.; Demidenko, E.; Stefan, K.; Huang, Y.H.; Miller, T.W. Therapeutic sensitivity to Rac GTPase inhibition requires consequential suppression of mTORC1, AKT, and MEK signaling in breast cancer. *Oncotarget* **2017**, *8*, 21806–21817. [CrossRef]
154. Schnelzer, A.; Prechtel, D.; Knaus, U.; Dehne, K.; Gerhard, M.; Graeff, H.; Harbeck, N.; Schmitt, M.; Lengyel, E. Rac1 in human breast cancer: Overexpression, mutation analysis, and characterization of a new isoform, Rac1b. *Oncogene* **2000**, *19*, 3013–3020. [CrossRef]
155. Liu, S.Y.; Yen, C.Y.; Yang, S.C.; Chiang, W.F.; Chang, K.W. Overexpression of Rac-1 small GTPase binding protein in oral squamous cell carcinoma. *J. Oral Maxillofac. Surg.* **2004**, *62*, 702–707. [CrossRef]
156. Kamai, T.; Yamanishi, T.; Shirataki, H.; Takagi, K.; Asami, H.; Ito, Y.; Yoshida, K. Overexpression of RhoA, Rac1, and Cdc42 GTPases is associated with progression in testicular cancer. *Clin. Cancer Res.* **2004**, *10*, 4799–4805. [CrossRef]
157. Engers, R.; Ziegler, S.; Mueller, M.; Walter, A.; Willers, R.; Gabbert, H.E. Prognostic relevance of increased Rac GTPase expression in prostate carcinomas. *Endocr. Relat. Cancer* **2007**, *14*, 245–256. [CrossRef]
158. Gulhati, P.; Bowen, K.A.; Liu, J.; Stevens, P.D.; Rychahou, P.G.; Chen, M.; Lee, E.Y.; Weiss, H.L.; O'Connor, K.L.; Gao, T.; et al. mTORC1 and mTORC2 regulate EMT, motility, and metastasis of colorectal cancer via RhoA and Rac1 signaling pathways. *Cancer Res.* **2011**, *71*, 3246–3256. [CrossRef]
159. Choi, U.J.; Jee, B.K.; Lim, Y.; Lee, K.H. KAI1/CD82 decreases Rac1 expression and cell proliferation through PI3K/Akt/mTOR pathway in H1299 lung carcinoma cells. *Cell Biochem. Funct.* **2009**, *27*, 40–47. [CrossRef]
160. Skvortsov, S.; Jimenez, C.R.; Knol, J.C.; Eichberger, P.; Schiestl, B.; Debbage, P.; Skvortsova, I.; Lukas, P. Radioresistant head and neck squamous cell carcinoma cells: Intracellular signaling, putative biomarkers for tumor recurrences and possible therapeutic targets. *Radiother. Oncol.* **2011**, *101*, 177–182. [CrossRef]
161. Patel, V.; Rosenfeldt, H.M.; Lyons, R.; Servitja, J.M.; Bustelo, X.R.; Siroff, M.; Gutkind, J.S. Persistent activation of Rac1 in squamous carcinomas of the head and neck: Evidence for an EGFR/Vav2 signaling axis involved in cell invasion. *Carcinogenesis* **2007**, *28*, 1145–1152. [CrossRef]
162. Montalvo-Ortiz, B.L.; Castillo-Pichardo, L.; Hernandez, E.; Humphries-Bickley, T.; De la Mota-Peynado, A.; Cubano, L.A.; Vlaar, C.P.; Dharmawardhane, S. Characterization of EHop-016, novel small molecule inhibitor of Rac GTPase. *J. Biol. Chem.* **2012**, *287*, 13228–13238. [CrossRef]
163. Shutes, A.; Onesto, C.; Picard, V.; Leblond, B.; Schweighoffer, F.; Der, C.J. Specificity and mechanism of action of EHT 1864, a novel small molecule inhibitor of Rac family small GTPases. *J. Biol. Chem.* **2007**, *282*, 35666–35678. [CrossRef]
164. Hagiwara, N.; Watanabe, M.; Iizuka-Ohashi, M.; Yokota, I.; Toriyama, S.; Sukeno, M.; Tomosugi, M.; Sowa, Y.; Hongo, F.; Mikami, K.; et al. Mevalonate pathway blockage enhances the efficacy of mTOR inhibitors with the activation of retinoblastoma protein in renal cell carcinoma. *Cancer Lett.* **2018**, *431*, 182–189. [CrossRef]
165. Frankel, W.N.; Yang, Y.; Mahaffey, C.L.; Beyer, B.J.; O'Brien, T.P. Szt2, a novel gene for seizure threshold in mice. *Genes Brain Behav.* **2009**, *8*, 568–576. [CrossRef]
166. Venkatesan, C.; Angle, B.; Millichap, J.J. Early-life epileptic encephalopathy secondary to SZT2 pathogenic recessive variants. *Epileptic Disord.* **2016**, *18*, 195–200.

167. Pizzino, A.; Whitehead, M.; Sabet Rasekh, P.; Murphy, J.; Helman, G.; Bloom, M.; Evans, S.H.; Murnick, J.G.; Conry, J.; Taft, R.J.; et al. Mutations in SZT2 result in early-onset epileptic encephalopathy and leukoencephalopathy. *Am. J. Med. Genet. A* **2018**, *176*, 1443–1448. [CrossRef]
168. Wolfson, R.L.; Chantranupong, L.; Wyant, G.A.; Gu, X.; Orozco, J.M.; Shen, K.; Condon, K.J.; Petri, S.; Kedir, J.; Scaria, S.M.; et al. KICSTOR recruits GATOR1 to the lysosome and is necessary for nutrients to regulate mTORC1. *Nature* **2017**, *543*, 438–442. [CrossRef]
169. Peng, M.; Yin, N.; Li, M.O. SZT2 dictates GATOR control of mTORC1 signalling. *Nature* **2017**, *543*, 433–437. [CrossRef]
170. Fang, Y.; Vilella-Bach, M.; Bachmann, R.; Flanigan, A.; Chen, J. Phosphatidic acid-mediated mitogenic activation of mTOR signaling. *Science* **2001**, *294*, 1942–1945. [CrossRef]
171. Yoon, M.S.; Sun, Y.; Arauz, E.; Jiang, Y.; Chen, J. Phosphatidic acid activates mammalian target of rapamycin complex 1 (mTORC1) kinase by displacing FK506 binding protein 38 (FKBP38) and exerting an allosteric effect. *J. Biol. Chem.* **2011**, *286*, 29568–29574. [CrossRef]
172. Winter, J.N.; Fox, T.E.; Kester, M.; Jefferson, L.S.; Kimball, S.R. Phosphatidic acid mediates activation of mTORC1 through the ERK signaling pathway. *Am. J. Physiol. Cell Physiol.* **2010**, *299*, C335–C344. [CrossRef]
173. Gozgit, J.M.; Pentecost, B.T.; Marconi, S.A.; Ricketts-Loriaux, R.S.; Otis, C.N.; Arcaro, K.F. PLD1 is overexpressed in an ER-negative MCF-7 cell line variant and a subset of phospho-Akt-negative breast carcinomas. *Br. J. Cancer* **2007**, *97*, 809–817. [CrossRef]
174. Hu, J.; Hu, H.; Hang, J.J.; Yang, H.Y.; Wang, Z.Y.; Wang, L.; Chen, D.H.; Wang, L.W. Simultaneous high expression of PLD1 and Sp1 predicts a poor prognosis for pancreatic ductal adenocarcinoma patients. *Oncotarget* **2016**, *7*, 78557–78565. [CrossRef]
175. Kang, D.W.; Lee, B.H.; Suh, Y.A.; Choi, Y.S.; Jang, S.J.; Kim, Y.M.; Choi, K.Y.; Min, D.S. Phospholipase D1 Inhibition Linked to Upregulation of ICAT Blocks Colorectal Cancer Growth Hyperactivated by Wnt/beta-Catenin and PI3K/Akt Signaling. *Clin. Cancer Res.* **2017**, *23*, 7340–7350. [CrossRef]
176. Ohguchi, K.; Banno, Y.; Nakagawa, Y.; Akao, Y.; Nozawa, Y. Negative regulation of melanogenesis by phospholipase D1 through mTOR/p70 S6 kinase 1 signaling in mouse B16 melanoma cells. *J. Cell Physiol.* **2005**, *205*, 444–451. [CrossRef]
177. Tang, W.; Liang, R.; Duan, Y.; Shi, Q.; Liu, X.; Liao, Y. PLD1 overexpression promotes invasion and migration and function as a risk factor for Chinese glioma patients. *Oncotarget* **2017**, *8*, 57039–57046. [CrossRef]
178. Xiao, J.; Sun, Q.; Bei, Y.; Zhang, L.; Dimitrova-Shumkovska, J.; Lv, D.; Yang, Y.; Cao, Y.; Zhao, Y.; Song, M.; et al. Therapeutic inhibition of phospholipase D1 suppresses hepatocellular carcinoma. *Clin. Sci. (Lond.)* **2016**, *130*, 1125–1136. [CrossRef]
179. Lewis, J.A.; Scott, S.A.; Lavieri, R.; Buck, J.R.; Selvy, P.E.; Stoops, S.L.; Armstrong, M.D.; Brown, H.A.; Lindsley, C.W. Design and synthesis of isoform-selective phospholipase D (PLD) inhibitors. Part I: Impact of alternative halogenated privileged structures for PLD1 specificity. *Bioorg. Med. Chem. Lett.* **2009**, *19*, 1916–1920.
180. Monovich, L.; Mugrage, B.; Quadros, E.; Toscano, K.; Tommasi, R.; LaVoie, S.; Liu, E.; Du, Z.; LaSala, D.; Boyar, W.; et al. Optimization of halopemide for phospholipase D2 inhibition. *Bioorg. Med. Chem. Lett.* **2007**, *17*, 2310–2311. [CrossRef]
181. Grunwald, V.; Keilholz, U.; Boehm, A.; Guntinas-Lichius, O.; Hennemann, B.; Schmoll, H.J.; Ivanyi, P.; Abbas, M.; Lehmann, U.; Koch, A.; et al. TEMHEAD: A single-arm multicentre phase II study of temsirolimus in platin- and cetuximab refractory recurrent and/or metastatic squamous cell carcinoma of the head and neck (SCCHN) of the German SCCHN Group (AIO). *Ann. Oncol.* **2015**, *26*, 561–567. [CrossRef]
182. Dunn, L.A.; Fury, M.G.; Xiao, H.; Baxi, S.S.; Sherman, E.J.; Korte, S.; Pfister, C.; Haque, S.; Katabi, N.; Ho, A.L.; et al. A phase II study of temsirolimus added to low-dose weekly carboplatin and paclitaxel for patients with recurrent and/or metastatic (R/M) head and neck squamous cell carcinoma (HNSCC). *Ann. Oncol.* **2017**, *28*, 2533–2538. [CrossRef]
183. Keam, B.; Kim, S.; Ahn, Y.O.; Kim, T.M.; Lee, S.H.; Kim, D.W.; Heo, D.S. In vitro anticancer activity of PI3K alpha selective inhibitor BYL719 in head and neck cancer. *AntiCancer Res.* **2015**, *35*, 175–182.
184. Mehibel, M.; Ortiz-Martinez, F.; Voelxen, N.; Boyers, A.; Chadwick, A.; Telfer, B.A.; Mueller-Klieser, W.; West, C.M.; Critchlow, S.E.; Williams, K.J.; et al. Statin-induced metabolic reprogramming in head and neck cancer: A biomarker for targeting monocarboxylate transporters. *Sci. Rep.* **2018**, *8*, 16804. [CrossRef]

185. Polanski, R.; Hodgkinson, C.L.; Fusi, A.; Nonaka, D.; Priest, L.; Kelly, P.; Trapani, F.; Bishop, P.W.; White, A.; Critchlow, S.E.; et al. Activity of the monocarboxylate transporter 1 inhibitor AZD3965 in small cell lung cancer. *Clin. Cancer Res.* **2014**, *20*, 926–937. [CrossRef]
186. Bola, B.M.; Chadwick, A.L.; Michopoulos, F.; Blount, K.G.; Telfer, B.A.; Williams, K.J.; Smith, P.D.; Critchlow, S.E.; Stratford, I.J. Inhibition of monocarboxylate transporter-1 (MCT1) by AZD3965 enhances radiosensitivity by reducing lactate transport. *Mol. Cancer Ther.* **2014**, *13*, 2805–2816. [CrossRef]

Review

mTOR Signaling Pathway in Cancer Targets Photodynamic Therapy In Vitro

Sandra M. Ayuk * and Heidi Abrahamse

Laser Research Centre, Faculty of Health Sciences, University of Johannesburg, P.O. Box 17011 Doornfontein, South Africa; habrahamse@uj.ac.za

* Correspondence: matabs63@yahoo.com (S.M.A.); Tel.: +27-11-559-6550 (H.A.)

Received: 31 March 2019; Accepted: 30 April 2019; Published: 9 May 2019

Abstract: The Mechanistic or Mammalian Target of Rapamycin (mTOR) is a major signaling pathway in eukaryotic cells belonging to the P13K-related kinase family of the serine/threonine protein kinase. It has been established that mTOR plays a central role in cellular processes and implicated in various cancers, diabetes, and in the aging process with very poor prognosis. Inhibition of the mTOR pathway in the cells may improve the therapeutic index in cancer treatment. Photodynamic therapy (PDT) has been established to selectively eradicate neoplasia at clearly delineated malignant lesions. This review highlights recent advances in understanding the role or regulation of mTOR in cancer therapy. It also discusses how mTOR currently contributes to cancer as well as future perspectives on targeting mTOR therapeutically in cancer in vitro.

Keywords: cancers; mTOR; inhibitors; photodynamic therapy

1. Introduction

The Mechanistic or Mammalian Target of Rapamycin (mTOR) pathway incorporates both intra and extracellular signals, and functions as a key regulator of physiological processes including in the growth, metabolism, proliferation, metastasis and malignant transformation of various human tumors [1]. Based on statistics from the Cancer Genome Atlas Pan-Cancer effort, the mTOR signaling pathway was found to be one of the highest mutated genes in 12 cancers analyzed from 3281 tumors. Examples of these cancers include breast, colon, lung, uterine corpus endometrioid, head and neck as well as ovarian [2,3]. mTOR receives signals from its effectors to control the cell function and homeostasis in normal cells. However, in cancer cells, this function is lost. Somatic mutation and gene amplification encode key components leading to the activation of the pathway that enhances cell proliferation and tumor growth [4–8]. mTOR serves as the major growth and survival pathway for cancer pathogenesis and has been an attractive target development of anticancer therapies. mTOR functions in controlling the downstream processes of ribosomes, mRNA, protein synthesis as well as translation. To achieve these functions, they interfere with various signaling pathways including nuclear factor kappa-light-chain-enhancer of activated B cells (NF-kB), phosphatidylinositol-3-kinase (PI3K)/AKT, reticular activating system (RAS), and tuberous sclerosis complex (TSC). When deregulated, they may induce uncontrolled cell growth and proliferation [9]. Furthermore, growth factors such as tyrosine kinase receptors play an important role in the downstream processes within the pathway to enhance biological processes such as angiogenesis, proliferation, metabolism, survival and differentiation [4]. The pathway may therefore be very useful in cancer pathogenesis and disease progression if it is altered and further lead to the development of molecularly targeted treatments that could advance into successful clinical trials [10].

Various inhibitors and signaling components for downstream processes have shown promising results in clinical trials. Clinically, relevant inhibitors target different pathways that present high sensitivity and needs to be studied [11–13]. Second-generation mTOR inhibitors have shown improved antitumor activity both in animal models and in vitro. Some previously studied 1st generation inhibitors have shown very little sensitivity including 1st generation rapamycin derivatives (Rapalogs) which have not proven to be very efficient due to their pharmacodynamics. There is still ongoing preclinical and clinical trials to evaluate various targets [14]. Several cancers become resistant to conventional therapies leading to poor prognostics [2,3] and in the effort to enhance therapy and curb resistance, several combination therapies are been investigated [6,15,16].

Photodynamic therapy (PDT) was originally developed about a hundred years ago for the treatment of various tumors and other non-malignant diseases [17]. The treatment mechanism involves the injection of a non-toxic photosensitizer (PS) locally, systemically or topically to a specific lesion accompanied by the absorption of visible light of a particular wavelength in the presence of oxygen from the singlet state to the triplet state as a means of generating cytotoxic reactions [18]. These reactions form reactive oxygen species (ROS) which result in tissue destruction, pathogenic microbes and cell death [19,20] (Figure 1). Photo activation may destroy cancer cells through apoptosis, necrosis or autophagy based on the organelle which the PS has accumulated [21]. PDT specifically targets malignant tumors and destroys the cell with minimal side effects [7]. Photoreactions release oxidant species which may alter the cell, its microenvironment, or even the whole organism. The process involves two types of reaction pathways namely type I (radicals and ROS) and type II (Singlet oxygen) [18] (Figure 1). More oxygen molecules are produced in the singlet state which makes type II more predominant [18]. The action of an ideal PS is based on various factors including PS concentration and localization, amount of energy released, the genetic profile, the dosage administered and wavelength [20]. mTOR has also been demonstrated as a target for PDT in vivo using the lysosomal-based phthalocyanine derivative. This was proven effective in treating 4-Nitroquinoline-1-Oxide (4-NQO) induced murine oral cancer. Velloso, et al. [22] found that the PI3K/Akt/mTOR pathway was inhibited in Human Oral Squamous Cell Carcinoma (OSCC) cells using Aluminum Phthalocyanine (AlPc)-based PDT. Furthermore, Fateye, et al. [23,24] found PI3K pathway inhibitors to significantly enhance the response of PDT [23,24]. Interactions between the mTOR signaling pathway and PDT is under research. This review focuses on targeting mTOR inhibitors in PDT of cancer cells.

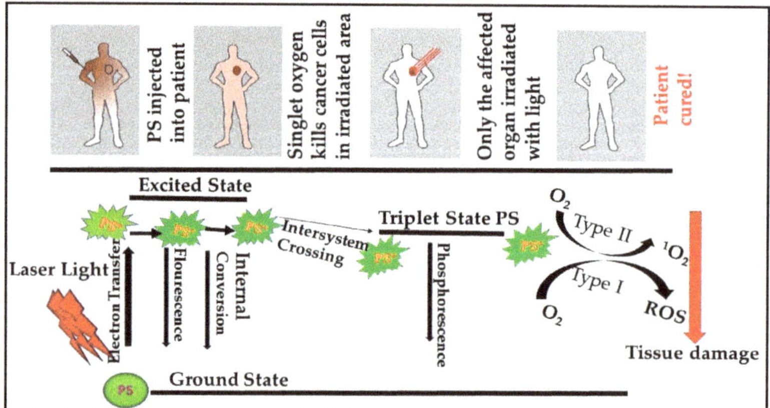

Figure 1. Schematic model of the Mechanism of Photodynamic Therapy (PDT), excitation and relaxation of a photosensitizer, and type I and type II photoreactions. Photosensitizers (PS) after an application as cream or injected become activated by light at specific wavelengths in the presence of oxygen (O_2). When activated they become excited and move from the singlet state to the triple state generating cytotoxic reactions. Some of the phytophysical reactions include electron transfer, fluorescence, internal conversion, intersystem crossing, and phosphorescence. These reactions directly generate singlet oxygen ($^1O_2$) or indirectly, reactive oxygen species (ROS) resulting in tissue damage and cell death [18].

2. The mTOR Pathway

The mTOR pathway comprises a 289 kDa serine/threonine kinase situated downstream of the PI3K-AKT signaling pathway [25]. mTOR has been revealed to be a major regulator of cell growth, proliferation, migration, differentiation, and survival [25]. Studies have also shown that mTOR is deregulated in most human cancers both upstream via the PI3K-AKT pathway and downstream via the 4E-binding protein 1 (4E-BP1) and Ribosomal protein S6 kinase beta-1 (S6 kinase) pathway, all of which make it a target for tumor suppression [26]. Being the most distorted pathway in human cancers, thePI3K signaling pathway plays a very important role in tumor cell survival and progression. AKT and mTOR are further activated downstream mechanism through the conversion of phosphatidylinositol-4, 5-biphosphate (PIP2) to phosphatidylinositol-3, 4, 5-triphosphate (PIP3) in the cell membrane to induce a cascade of protein phosphorylation (Figure 2). Abnormal activation can enhance tumorigenesis making the pathway a highly attractive target for cancer therapy [27]. mTOR consists of various domains involved in the physiological process, namely the binding or HEAT domain composed of two N-terminals and involved in protein-protein interactions, the FRB (FKPB12-rapamycin binding domain) domain of mTOR which is the binding domain for rapamycin, the FAT and c-terminal FATC (FAT Carboxyterminal) domain present in P13K-related kinases as well as the catalytic kinase domain [28].

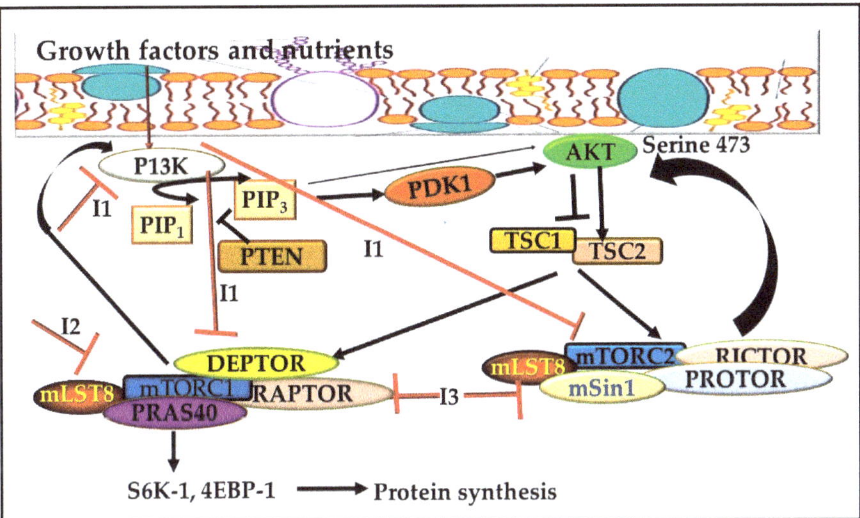

Figure 2. mTOR Signaling pathway. Activation of P13K phosphorylates phosphatidylinositol 4,5-biphosphate (PIP2) to form phosphatidylinositol-3,4,5-triphosphate (PIP3). Phosphatase and tensin homolog deleted on chromosome 10 (PTEN) regulates the function of PIP3. PIP3 prompts the activation of downstream processes such AKT, which transmits signals to effectors including mTOR complexes to enhance cellular processes. The mTORC1 is stimulated during cell activation whereby the T-cell receptor (TCR) stimulates the activation of P13K. mTORC1 comprises of three mTOR catalytic subunits, namely the regulatory associated protein of mTOR (RAPTOR), mammalian lethal with SEC13 protein 8 (MLST8), as well as the noncore components PRAS40 and DEP domain-containing mTOR-interacting protein (DEPTOR). mTORC2 comprises also of three proteins – the namely rapamycin-insensitive companion of mTOR (RICTOR), MLST8, and the mammalian stress-activated protein kinase interacting protein 1 (SIN1). Activation of mTORC2 occurs through the phosphorylation of AKT at serine-473 while that of mTORC1 when activated, phosphorylates the effectors which are major regulators of protein translation including translation-regulating factors ribosomal S6 kinase-1 (S6K-1) and eukaryote translation initiation factor 4E binding protein-1 (4EBP-1) to enhance protein synthesis.

Through interactions with nutrients, growth factors and energy stores, mTOR can directly affect cell proliferation and differentiation [29]. Furthermore, mTOR comprises a catalytic subunit of two unique protein complexes, namely mTOR complex 1 (mTORC1) and 2 (mTORC2) [30]. These complexes are unique in their function. mTORC1 is stimulated during cell activation whereby T-cell receptor (TCR) stimulates activation of P13K [31]. This activation is catalyzed by the pyruvate dehydrogenase kinase 1 enzyme (PDK1) [32]. mTORC1 comprises of three mTOR catalytic subunits, namely the regulatory associated protein of mTOR (RAPTOR), mammalian lethal with SEC13 protein 8 (MLST8), and the noncore components PRAS40 and DEP domain-containing mTOR-interacting protein (DEPTOR). When mTORC1 is activated, it phosphorylates the effectors which are major regulators of protein translation including translation-regulating factors ribosomal S6 kinase-1 (S6K-1) and eukaryote translation initiation factor 4E binding protein-1 (4EBP-1) to enhance protein synthesis [31,33,34] (Figure 2).

mTORC2, on the other hand, can be directly activated by P13K [35]. It phosphorylates and activates AKT and other related kinases [36]. Furthermore, through the PI3K-AKT signaling, co-stimulatory signals from cytokines and TCR can also activate the mTOR signaling pathway to further activate the T cells and attain energy supplies [37]. It comprises three proteins, namely the rapamycin-insensitive companion of mTOR (RICTOR), MLST8 and the mammalian stress-activated protein kinase interacting protein 1 (SIN1). Activation occurs through the phosphorylation of AKT at serine-473 [36,38]. Some cells have the same sensitivity to rapamycin [39] but rapamycin selectively inhibits mTOR with more sensitivity to mTORC1 compared to mTORC2 [40]. Studies have shown that mTORC2, as opposed to mTORC1, lacks sensitivity to rapamycin inhibition. Most cancer cells are resistant to the 1st generation mTOR inhibitors (Rapalogs) which particularly target mTORC1 which makes the insensitivity of mTORC2 a possible opening for drug discovery [41].

3. The Role of mTOR Inhibitors in Cancer

mTOR inhibitors can be classified into first and second generations depending on their mechanisms and targets. The first generation uses allosteric mechanisms to block the mTOR pathway while the second generation prevents kinase activity in both mTORC1 and 2 using their target ATP binding site. Examples of the 1st generation include the rapamycin and its analogs while the second generation includes AZD8055, Torin1, PP242 and PP30 [42]. Based on some clinical trials mTOR inhibitors are implicated in tumor cells with p53 and PTEN mutations [43]. Three generations of inhibitors has been developed namely Rapalogs (Rapamycin and derivatives), ATP-competitive inhibitors and the Rapalink [44].

Rapamycin also referred to as sirolimus was discovered as an antifungal, immunosuppressive and antitumor compound isolated from Streptomyces hygroscopicus a soil bacterium [45,46]. This drug was initially approved as an anti-host rejecter in 1997 by the food and drug administration (FDA) for kidney transplants [47]. It also functions in many human cancers mainly for the inhibition of signal transduction pathways by forming complexes with peptidyl-prolyl-isomerase FKBP12. These pathways are necessary for cell growth and proliferation [9]. According to Shafer, et al. [48], its anti-angiogenic and proliferative property can be seen in phase II preclinical studies on endometrial cancer cell lines whereby it has a synergistic effect on the paclitaxel. mTOR has been revealed to be the homolog of yeast TOR/DRR genes previously identified in genetic screens their resistance to rapamycin [49]. It has also been identified as a direct target of the complex of FKBP12-rapamycin (FRB domain) [50]. The mechanism of action for rapamycin is based on the binding of mTOR and rapamycin complex FKBP-12 with phosphatidic acid to block the function of mTOR kinase. It attaches to the FRB domain of the mTOR and finally destabilizes the mTOR–raptor–4EBP1/S6K-1 scaffold complex through the binding of mTOR and the complex FKBP-12–rapamycin. These result in dephosphorylation of 4EBP1 and S6K-1 [51,52]. The FRB domain is adjacent to the kinase domain and limits access to substrates to the kinase site [53,54]. However, rapamycin lacks sensitivity in some binding sites making them less sensitive [55].

The therapeutic development of mTOR inhibitors has improved due to their importance in cancer progression and development [56]. Several inhibitors have been approved by the FDA and are already being implemented in the treatment of various human cancers such as breast cancer (everolimus), metastatic renal cell cancer carcinoma (everolimus and temsirolimus), pancreatic neuroendocrine tumors (everolimus) and mantle cell lymphoma (temsirolimus) [57]. Temsirolimus (CCI-779), everolimus (RAD001), and ridaforolimus (MK-8669/AP23573) [6,58,59] have been improved due to their poor aqueous solubility and bioavailability. Studies have shown that rapamycin and its Rapalogs inhibit mTORC2 complex in a way that is independent on time, cell type and dose and based on interaction with newly synthesized molecules of complexes of rapamycin/Rapalogs-FKBP12 and mTOR molecules. This results in further interaction with RICTOR. Studies have shown that the inhibition of components such as RICTOR, RAPTOR, or mTOR significantly reduces the proliferation of cancer cells and offsets progression in the cell cycle [60–62]. Overexposure of cancer cells to rapamycin may encourage mTOR binding and inhibit AKT mediated signaling even before the mTORC2 complex is formed [63].

Rapalogs present antiproliferative characteristics in cells that have not been transformed and can efficiently inhibit T-cell proliferation in patients who have undergone transplants [64,65]. They have also shown antitumor responses in benign tumors of TSC [66,67] including lymphangiomyomatosis, renal angiomyolipoma, cardiac rhabdomyoma, facial angiofibroma and retinal astrocytic hamartoma [66]. Reduced efficacy was seen in sporadic cancers and when treatment was stopped [68,69]. Recently they have been approved for the treatment of various tumors including renal cell carcinoma [70,71], postmenopausal hormone receptor-positive advanced breast cancer in combination with exemestane [72], advanced pancreatic neuroendocrine tumors [73], advanced non-functional neuroendocrine tumors of the gastrointestinal tract or lung [74] and relapsed or refractory mantle cell lymphoma [75].

Gulhati, et al. [60] found that the knockdown of mTORC1and 2 mediated in colorectal cancer xenografts in vivo slows down the development of rapamycin sensitive and insensitive cell lines. In addition, the knockdown of mTORC2 increased apoptosis in colorectal cancer cells resistant to rapamycin. Guertin, et al. [76] also found that prostate cancer was mTORC2 dependent when induced in the prostate epithelium by phosphatase and tensin homolog deletion. mTOR is also vital in advanced cancer development and metastatic cancers. It alters the tumor environment to promote metastasis. The hyperactivation of mTOR by RICTOR enhanced cell proliferation in gliomas [77]. Inhibition of mTOR may also improve the way chemotherapeutic agents respond in advanced diseases. Patel, et al. [78] found that the inhibition of mTOR prevented the distribution of cancer cells to lymph nodes slowing down angiogenesis in head and neck cancer.

Everolimus was approved as an oral mTOR inhibitor for advanced renal cell cancer. It is also known for its anti-proliferative and angiogenic activity in human cancers [79,80]. This includes metastatic pancreatic neuroendocrine tumors, metastatic renal cell carcinoma, advanced estrogen receptor (ER)-positive [79] and human epidermal growth factor receptor-2 (HER2)-negative breast cancer [80]. Studies have recorded an improvement in cancer when rapamycin and its Rapalogs were used in combination with either standard chemotherapy, hormonal therapy, or alone. A current study found significant progression-free survival (PFS) when patients with HER-2 advanced stage of breast cancer, pre-treated with taxane and trastuzumab were administered both everolimus together with trastuzumab and vinorelbine [81]. Another breast cancer study by Hurvitz, et al. [82] found the combination of everolimus and paclitaxel and trastuzumab promising. Temsirolimus (Torisel®), was approved by the FDA as the 1st rapamycin analog to be used for the treatment of cancer cells. It is an intravenous injection which when injected in vivo becomes converted into rapamycin. Studies have shown an increase in progesterone mRNA and the inhibition of endoplasmic reticulum mRNA expressions when administered with bevacizumab or in combination with other chemotherapeutic agents for treating endometrial cancer cell lines [48,83]. In addition, Tinker, et al. [84] found positive results after using temsirolius for a preliminary phase II study in patients with metastatic cervical cancer. The drugs were effective when administered together with paclitaxel/carboplatin for treating stage II/IV patients with clear cell adenocarcinoma on a clinical phase II trial [85].

Not all studies have however proven positive outcomes with the drug. Behbakht, et al. [86] found decreased activity of temsirolimus with the drug efficacy failing in patients with primary peritoneal cancer or persistent/recurrent epithelial ovarian cancer. These results still need to be investigated by a phase III trial. Another inhibitor, Ridaforolimus (MK-8669/AP23573), a non-rapamycin pro-drug available in both intravenous and oral formulations has been evaluated in combination or as monotherapy on various cancers including breast, prostate, endometrial, sarcomas and non-small cell lung cancer [16]. It has had a 33% response rate when administered to patients with advanced endometrial cancer [87]. A phase two II study showed a partial response rate of 7.7% in advanced or recurrent [88]. Side effects of this drug include low toxicity with dose dependent skin rashes and mucositis [89] as well as hypertriglyceridemia, hypercholesterolemia, nausea, fatigue, anemia, and neutropenia [90]. In addition, sirolimus and temsirolimus present intense pulmonary toxicity. Other side effects include the risk of secondary lymphoma, interstitial lung disease, and the reactivation of latent infections. However, these are rare [91].

Even though Rapalogs are still been used in clinics as opposed to ATP-competitive inhibitors which have not yet been approved, and Rapalink still being developed and subject to experimentation. Several shortcomings of Rapalogs [55] have made the 2nd generation inhibitors better [92,93]. ATP-competitive inhibitors are the second-generation inhibitors. They inhibit both mTORC1 and 2 by blocking the kinase domain [94,95]. As opposed to the Rapalogs, inhibition is intense with blocking of the P13K from their kinase similarity [94]. Rapalink is the third-generation inhibitors designed to curb resistance mutations in both the rapalog and ATP-competitive inhibitors. These inhibitor crosslinks with kinase in the same molecule [96].

4. The Role of mTOR Pathway in Cancer Therapy

A major development has taken place in the last few years to understand the role of mTOR in cancer development and progression. mTOR and/or its components have been implicated in various genetic mutations of human malignant diseases [97–99]. Mutations of closely related pathways have enhanced mTOR signaling in cancers [59,95,100]. Currently, human cancer genome databases are being mined to aid identification of activated mTOR mutations [101]. Transmitted extracellular signals go through various pathways but P13K/AKT/RAS/RAF/MEK/MAPK are the most common and highly characterized. Using the same mechanism to activate PI3K/AKT/mTOR pathway has presented enhanced tumor progression and poor survival response to patients with different types of tumors [60,102]. Due to its vital function in cell growth and proliferation, its components have been increasingly used as potentials for therapeutic targets. Molecular approaches have been used to establish the role of the components of the mTOR pathway in cancer development. Components of the mTOR pathway have also been activated in various neuroendocrine tumors with a tendency of releasing bioactive products [103,104].

mTORC1 induces nucleotide and protein synthesis to regulate cellular growth via ribosome biogenesis, inhibits autophagy, protein, and nucleotide synthesis. When conditions are favorable, they sense environmental signals such as nutrients and growth factors to initiate cell growth but if conditions become unfavorable in cases of acidity and hypoxia, mTOR activity is inhibited [105,106]. When these pathways are activated, they inhibit mTORC1 through phosphorylation and inhibition of the protein complex, TSC 1 and 2. Mutation of the TSC genes causes TSC disease with benign tumors found in the brain, kidneys, heart, lungs and liver [107]. Activation could lead to the loss of phosphatase and tensin homolog (PTEN). This uncouples mTORC1 activation from growth factor signaling such as mutations of liver kinase B1/serine/threonine kinase 11 (LKB1/STK11) in nutrient-deprived vascular tumors but allows activation of mTORC1. The mutation of P53 inhibits bioenergetics processes and cell cycle arrest uncoupling DNA damage [95]. In addition, it could lead to hyperactivation of S6K-1, 4EBP1 and eukaryotic translation initiation factor 4E (eIF4E), as well as cancer growth through the activation of lipid and protein biosynthesis. Upon activation of S6K-1, 4EBP1 and other substrates are phosphorylated enhancing cell proliferation and growth from an anabolic cellular response [31,35,108]. Through the stimulation of the activity and expressions in small GTPases such as Rac1, cdc42 and Rho to control the activities of the actin cytoskeleton and motility [58,109]. Furthermore, S6K-1 and 4EBP1 mediated by mTORC1 extend vital roles in focal adhesion proteins phosphorylation including paxillion, p130 Cas and focal adhesion kinase as well as reorganization of F-actin [110].

Hyperactivation of mTORC1 results from mutations of mTOR or upstream effectors. This occurs in sporadic cancers [111–113]. Furthermore, in hamartoma syndromes, they are characterized by the growth of benign tumors and mutations in tumor suppressor genes [114]. The association of phosphorylated mTOR with AKT signaling and acquired cisplatin resistance affects primary platinum resistance and sensitivity to ovarian cancer cells [115]. Furthermore, these inhibitors restore chemosensitivity to platinum derivate both in vitro and in xenograft models [16,116]. Gulhati, Bowen, Liu, Stevens, Rychahou, Chen, Lee, Weiss, O'Connor and Gao [109] found that mTORC1 was associated with motility, metastasis, and epithelial-mesenchymal transition in colorectal cancer. mTORC1 activity has also been studied in breast cancer and gliomas [77,117]. Despite all these discoveries, more research needs to be conducted to understand how these components are regulated. Gulhati, Bowen, Liu, Stevens, Rychahou, Chen, Lee, Weiss, O'Connor and Gao [109] found that using oxaliplatin in colorectal cancer cells induced apoptosis as a result of the knockdown of mTORC1 and 2. mTORC1 is found to be associated with the transport hormone and, peptide-containing vesicles. They also regulate intestinal hormones which play a vital role in the gastrointestinal tract as well as other secreted neuroendocrine tumors to regulate neurotensin [104].

mTORC2, on the other hand, is activated via growth factors [118]. It phosphorylates and activates the AGC protein kinases including SGK1 (Ser422) and AKT (Ser473). Inhibiting mTORC2 activities will enhance the antitumor effect in several preclinical trials [61,69,76,119,120]. Varied molecular adjustments occur with this pathway which may suggest strategic therapy against cancer cells if targeted. The onset of cancer is provoked by enhanced cell growth and immune escape due to a build-up of genetic and epigenetic changes. Therefore an approach to cancer therapy would be to prevent these changes [121]. Tumor heterogeneity, as well as cellular resistance, are some of the hindrances to targeted cancer therapies. Activation of bypass mechanisms as well as making secondary reforms in the target are resistance mechanisms which have been identified [122]. Nonetheless, most of the targeted treatments have not been beneficial in the long run despite all the preclinical trials.

mTOR Signaling Pathway and PDT

The PS are composed of natural occurring macrocycles including hemoglobin, vitamin B12 and chlorophyll. These compounds consist of nitrogen, oxygen, or sulfur atoms locked in a hollow ring containing metals such as iron or magnesium. Currently, PDT makes use of plant extracts to complex synthetic macrocycles. These different agents can selectively target and accumulate in the tumor. The widely investigated PS includes tetrapyrroles such as bacteriochlorins, chlorins, porphyrins, and phthalocyanines [123]. To improve efficacy, clinical considerations have been given to other compounds such as synthetic dyes and targeted therapies which use various drug delivery systems to improve the penetration of light. The fate and effectiveness of PDT on tumors are based on the oxygen concentration, wavelength, types of photosensitizer and the genotype of the cell. This can affect certain organelles and specific target tissue [20]. Dual-specificity of the PS would depend on accumulation and localization of the PS in diseased tissue. The PS if hydrophobic accumulates in the mitochondria and endoplasmic reticulum, other polar compounds may Golgi apparatus, lysosomes and plasma membrane [124].

PDT down-regulates AKT-mTOR pathway because of ROS production (Figure 3). In modern oncology, a combination of different therapeutic modalities with non-overlapping toxic effects are strategies used to improve the therapeutic index of treatment. Combination therapies target different disease pathways, which represents an alternative approach that might offer potential advantages over a single therapy.

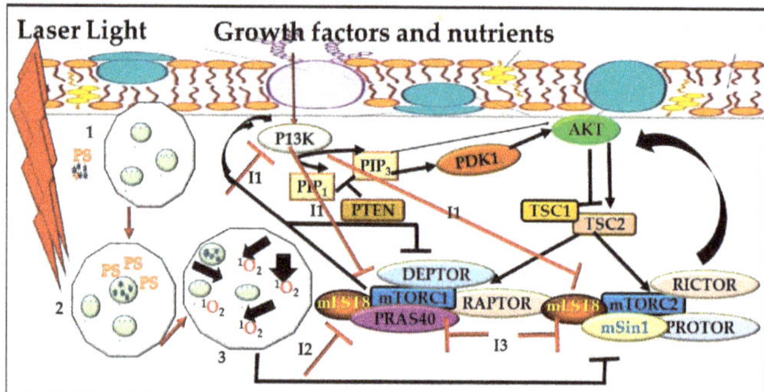

Figure 3. Proposed mechanism between mTOR Signaling Pathway, Inhibitors and Photodynamic Therapy (PDT). PDT down-regulates AKT-mTOR pathway because of reactive oxygen species (ROS) production. 1. Photosensitizer is injected into a targeted tumor. 2. Laser light is emitted at a particular wavelength. 3. Cells become activated and release reactive oxygen species, which results to tissue destruction and cell death. Interaction with inhibitors phosphoinositide 3-kinase (I1), rapamycin (I2) and mTOR kinase (I3) to enhance cell death through the P13K/AKT-mTOR pathway.

Few studies have shown these combining effects on PDT. Kraus, et al. [125] found that combining P13K/mTOR inhibitors (BYL719, BKM120, and BEZ235) with verteporfin-PDT to synergistically enhance PDT response with BEZ235 presenting the strongest. Antiapoptotic inhibition of the Bcl-2 family protein Mcl-1 and P13K pathway was critical. Fateye, et al. [24] assessed the effect of combination of P13K/mTOR inhibitor (BEZ 235 (BEZ)) on PDT efficacy using prostate tumor (PC3) and SV40-transformed mouse endothelial cell lines (SVEC-40) and found that the sub-lethal PDT was enhanced in both cell lines. Combination of PDT with pan-PI3/ mTOR kinase inhibitor LY294002 (LY) also enhanced PDT effect with PC3. However, it produced a synergistic effect in SVEC-40. In contrast, Sasore and Kennedy [126] found that there are some combinations of PI3K/AKT/mTOR pathway inhibitors, which actually interrupt developmental angiogenesis due to their additive or synergistic effect. Tuo, et al. [127] used human SZ95 sebocytes to find out the potential pharmaceutical effect of combining ALA-PDT and rapamycin through the mTOR pathway and found that cell growth was suppressed, protein levels of P-mTOR, and P-Raptor were reduced as well as lipogenesis. Their study concluded that rapamycin enhanced aminolevulinic acid hydrochloride (ALA)-PDT in SZ95 cells. mTOR inhibition can induce autophagy in various ways: direct induction, pre-condition cells, or by stressor induction. A study by Weyergang, et al. [26] using colon adenocarcinoma cell line and amphiphilic endolysosome-localizing photosensitizer Al(II) phthalocyanine chloride disulfonic acid (AlPcS(2a)) showed that targeting mTOR signaling pathway in PDT caused partial loss of both total and phosphorylated mTOR in both tumor xenografts and cultured cells in vitro and in vivo. According to Weyergang, et al. [26] combining rapamycin potentiates cytotoxicity in vitro post-PDT. The interest in the combination of PDT and other therapeutic modulates in cancer treatment is to provide a platform for potential treatment options and limited adverse effects of chemotherapy since PDT does not have the inherent dose-limiting toxicity [128]. Combination therapies are aimed at increasing responses, improving patient tolerability, decreasing drug dosages and the emergence of drug resistance [129]. Combined effects of PI3K/AKT/mTOR and PDT as a treatment regimen for cancers still needs further investigation.

5. Perspective

Despite its promising minimal and non-toxic side effects, it is still unlikely that administering conventional chemotherapies and/or inhibitors alone will completely cure cancer. There are still challenges in cancer therapy including the activation of other proliferation signaling pathways, treatment-resistant mutations as well as the intramural heterogeneity of mTOR activities. Inhibitors alone have failed to induce tumor regression but are seen as cytostatic causing disease stability rather than death [130]. Another limitation might provide negative feedback loops in the mTOR pathway which have limited the efficiency of these Rapalogs. Taking into high consideration the level of toxicity, combined therapies would be the way forward. New generation inhibitors are being produced which can prevent the catalytic activity of both mTORC1 and mTORC2 complexes and enhance therapeutic indexes.

Author Contributions: Conceptualization, S.M.A.; writing—original draft preparation, S.M.A.; writing—review and editing, S.M.A., H.A.; supervision, H.A.

Funding: This work is based on the research supported by the South African Research Chairs Initiative of the Department of Science and Technology and National Research Foundation (NRF) of South Africa (Grant No. 98337), as well as grants received from the University of Johannesburg (URC), the African Laser Centre (ALC) (student scholarship), and Council for Scientific and Industrial Research (CSIR)—National Laser Centre (NLC) Laser Rental Pool Program. All lasers were supplied and set up the NLC.

Conflicts of Interest: The authors declare no conflict of interest.

References

1. Gomez-Pinillos, A.; Ferrari, A.C. mTOR signaling pathway and mTOR inhibitors in cancer therapy. *Hematol. Oncol. N. Am.* **2012**, *26*, 483–505. [CrossRef]
2. Bartholomeusz, C.; Gonzalez-Angulo, A.M. Targeting the PI3K signaling pathway in cancer therapy. *Expert Opin. Ther. Targets* **2012**, *16*, 121–130. [CrossRef]
3. Kandoth, C.; McLellan, M.D.; Vandin, F.; Ye, K.; Niu, B.; Lu, C.; Xie, M.; Zhang, Q.; McMichael, J.F.; Wyczalkowski, M.A. Mutational landscape and significance across 12 major cancer types. *Nature* **2013**, *502*, 333. [CrossRef]
4. Yuan, T.L.; Cantley, L.C. PI3K pathway alterations in cancer: Variations on a theme. *Oncogene* **2008**, *27*, 5497. [CrossRef]
5. Willems, L.; Tamburini, J.; Chapuis, N.; Lacombe, C.; Mayeux, P.; Bouscary, D. PI3K and mTOR signaling pathways in cancer: New data on targeted therapies. *Curr. Oncol. Rep.* **2012**, *14*, 129–138. [CrossRef]
6. Zaytseva, Y.Y.; Valentino, J.D.; Gulhati, P.; Evers, B.M. mTOR inhibitors in cancer therapy. *Cancer Lett.* **2012**, *319*, 1–7. [CrossRef]
7. Cheng, H.; Walls, M.; Baxi, S.; Yin, M.-J. Targeting the mTOR pathway in tumor malignancy. *Curr. Cancer Drug Targets* **2013**, *13*, 267–277. [CrossRef]
8. Li, T.; Wang, G. Computer-aided targeting of the PI3K/Akt/mTOR pathway: Toxicity reduction and therapeutic opportunities. *Int. J. Mol. Sci.* **2014**, *15*, 18856–18891. [CrossRef]
9. Guertin, D.A.; Sabatini, D.M. Defining the role of mTOR in cancer. *Cancer Cell* **2007**, *12*, 9–22. [CrossRef]
10. Don, A.S.; Zheng, X.F. Recent clinical trials of mTOR-targeted cancer therapies. *Rev. Recent Clin. Trials* **2011**, *6*, 24–35.
11. Volanti, C.; Hendrickx, N.; Van Lint, J.; Matroule, J.-Y.; Agostinis, P.; Piette, J. Distinct transduction mechanisms of cyclooxygenase 2 gene activation in tumour cells after photodynamic therapy. *Oncogene* **2005**, *24*, 2981. [CrossRef]
12. Bozkulak, O.; Wong, S.; Luna, M.; Ferrario, A.; Rucker, N.; Gulsoy, M.; Gomer, C.J. Multiple components of photodynamic therapy can phosphorylate Akt. *Photochem. Photobiol.* **2007**, *83*, 1029–1033. [CrossRef]
13. Koon, H.K.; Chan, P.S.; Wong, R.N.S.; Wu, Z.G.; Lung, M.L.; Chang, C.K.; Mak, N.K. Targeted inhibition of the EGFR pathways enhances Zn-BC-AM PDT-induced apoptosis in well-differentiated nasopharyngeal carcinoma cells. *J. Cell. Biochem.* **2009**, *108*, 1356–1363. [CrossRef]
14. Rodon, J.; Dienstmann, R.; Serra, V.; Tabernero, J. Development of PI3K inhibitors: Lessons learned from early clinical trials. *Nat. Rev. Clin. Oncol.* **2013**, *10*, 143. [CrossRef]
15. Meric-Bernstam, F.; Gonzalez-Angulo, A.M. Targeting the mTOR signaling network for cancer therapy. *J. Clin. Oncol.* **2009**, *27*, 2278. [CrossRef]
16. Zhang, Y.-J.; Duan, Y.; Zheng, X.F.S. Targeting the mTOR kinase domain: The second generation of mTOR inhibitors. *Drug Discov. Today* **2011**, *16*, 325–331. [CrossRef]
17. van Straten, D.; Mashayekhi, V.; de Bruijn, H.; Oliveira, S.; Robinson, D. Oncologic photodynamic therapy: Basic principles, current clinical status and future directions. *Cancers* **2017**, *9*, 19. [CrossRef]

18. Agostinis, P.; Berg, K.; Cengel, K.A.; Foster, T.H.; Girotti, A.W.; Gollnick, S.O.; Hahn, S.M.; Hamblin, M.R.; Juzeniene, A.; Kessel, D. Photodynamic therapy of cancer: An update. *CA Cancer J. Clin.* **2011**, *61*, 250–281. [CrossRef]
19. Ahn, J.-C.; Biswas, R.; Chung, P.-S. Synergistic effect of radachlorin mediated photodynamic therapy on propolis induced apoptosis in AMC-HN-4 cell lines via caspase dependent pathway. *Photodiagn. Photodyn. Ther.* **2013**, *10*, 236–243. [CrossRef]
20. Allison, R.R.; Moghissi, K. Photodynamic therapy (PDT): PDT mechanisms. *Clin. Endosc.* **2013**, *46*, 24. [CrossRef]
21. Baptista, M.S.; Cadet, J.; Di Mascio, P.; Ghogare, A.A.; Greer, A.; Hamblin, M.R.; Lorente, C.; Nunez, S.C.; Ribeiro, M.S.; Thomas, A.H. Type I and type II photosensitized oxidation reactions: Guidelines and mechanistic pathways. *Photochem. Photobiol.* **2017**, *93*, 912–919. [CrossRef]
22. Velloso, N.V.; Muehlmann, L.A.; Longo, J.P.F.; Silva, J.R.; Zancanela, D.C.; Tedesco, A.C.; Azevedo, R.B.D. Aluminum-phthalocyanine chloride-based photodynamic therapy inhibits PI3K/Akt/Mtor pathway in oral squamous cell carcinoma cells in vitro. *Chemotherapy* **2012**, *1*, 5.
23. Fateye, B.; Li, W.; Wang, C.; Chen, B. Combination of Phosphatidylinositol 3-Kinases Pathway Inhibitor and Photodynamic Therapy in Endothelial and Tumor Cells. *Photochem. Photobiol.* **2012**, *88*, 1265–1272. [CrossRef]
24. Fateye, B.; Wan, A.; Yang, X.; Myers, K.; Chen, B. Comparison between endothelial and tumor cells in the response to verteporfin-photodynamic therapy and a PI3K pathway inhibitor. *Photodiagn. Photodyn. Ther.* **2015**, *12*, 19–26. [CrossRef]
25. Sun, Q.; Chen, X.; Ma, J.; Peng, H.; Wang, F.; Zha, X.; Wang, Y.; Jing, Y.; Yang, H.; Chen, R. Mammalian target of rapamycin up-regulation of pyruvate kinase isoenzyme type M2 is critical for aerobic glycolysis and tumor growth. *Proc. Natl Acad. Sci. USA* **2011**, *108*, 4129–4134. [CrossRef]
26. Weyergang, A.; Berg, K.; Kaalhus, O.; Peng, Q.; Selbo, P.K. Photodynamic therapy targets the mTOR signaling network in vitro and in vivo. *Mol. Pharm.* **2008**, *6*, 255–264. [CrossRef]
27. Thorpe, L.M.; Yuzugullu, H.; Zhao, J.J. PI3K in cancer: Divergent roles of isoforms, modes of activation and therapeutic targeting. *Nat. Rev. Cancer* **2015**, *15*, 7. [CrossRef]
28. Weichhart, T.; Hengstschläger, M.; Linke, M. Regulation of innate immune cell function by mTOR. *Nat. Rev. Immunol.* **2015**, *15*, 599–614. [CrossRef]
29. Blouet, C.; Ono, H.; Schwartz, G.J. Mediobasal hypothalamic p70 S6 kinase 1 modulates the control of energy homeostasis. *Cell Metab.* **2008**, *8*, 459–467. [CrossRef]
30. Qian, Y.; Kun-liang, G. Expanding mTOR signaling. *Cell Res.* **2007**, *17*, 666–681. [CrossRef]
31. Laplante, M.; Sabatini, D.M. mTOR Signaling in Growth Control and Disease. *Cell* **2012**, *149*, 274–293. [CrossRef]
32. Saxton, R.A.; Sabatini, D.M. mTOR Signaling in Growth, Metabolism, and Disease. *Cell* **2017**, *168*, 960–976. [CrossRef]
33. Cantley, L.C.; Engelman, J.A.; Luo, J. The evolution of phosphatidylinositol 3-kinases as regulators of growth and metabolism. *Nat. Rev. Genet.* **2006**, *7*, 606–619. [CrossRef]
34. Pópulo, H.; Lopes, J.M.; Soares, P. The mTOR signalling pathway in human cancer. *Int. J. Mol. Sci.* **2012**, *13*, 1886–1918. [CrossRef]
35. Shimobayashi, M.; Hall, M.N. Making new contacts: The mTOR network in metabolism and signalling crosstalk. *Nat. Rev. Mol. Cell Biol.* **2014**, *15*, 155. [CrossRef]
36. Russell, R.C.; Fang, C.; Guan, K.-L. An emerging role for TOR signaling in mammalian tissue and stem cell physiology. *Development* **2011**, *138*, 3343–3356. [CrossRef]
37. Chi, H. Regulation and function of mTOR signalling in T cell fate decisions. *Nat. Rev. Immunol.* **2012**, *12*, 325. [CrossRef]
38. Jacinto, E.; Loewith, R.; Schmidt, A.; Lin, S.; Rüegg, M.A.; Hall, A.; Hall, M.N. Mammalian TOR complex 2 controls the actin cytoskeleton and is rapamycin insensitive. *Nat. Cell Biol.* **2004**, *6*, 1122. [CrossRef]
39. Delgoffe, G.M.; Pollizzi, K.N.; Waickman, A.T.; Heikamp, E.; Meyers, D.J.; Horton, M.R.; Xiao, B.; Worley, P.F.; Powell, J.D. The kinase mTOR regulates the differentiation of helper T cells through the selective activation of signaling by mTORC1 and mTORC2. *Nat. Immunol.* **2011**, *12*, 295. [CrossRef]

40. Verbist, K.C.; Guy, C.S.; Milasta, S.; Liedmann, S.; Kamiński, M.M.; Wang, R.; Green, D.R. Metabolic maintenance of cell asymmetry following division in activated T lymphocytes. *Nature* **2016**, *532*, 389. [CrossRef]
41. Vadlakonda, L.; Dash, A.; Pasupuleti, M.; Kotha, A.K.; Reddanna, P. The paradox of Akt-mTOR interactions. *Front. Oncol.* **2013**, *3*, 165. [CrossRef]
42. Guertin, D.A.; Sabatini, D.M. The pharmacology of mTOR inhibition. *Sci. Signal.* **2009**, *2*, pe24. [CrossRef]
43. Feng, Z.; Zhang, H.; Levine, A.J.; Jin, S. The coordinate regulation of the p53 and mTOR pathways in cells. *Proc. Natl. Acad. Sci. USA* **2005**, *102*, 8204–8209. [CrossRef]
44. Xie, J.; Wang, X.; Proud, C.G. mTOR inhibitors in cancer therapy. *F1000Research* **2016**, *5*, F1000. [CrossRef]
45. Vezina, C.; Kudelski, A.; Sehgal, S.N. Rapamycin (AY-22, 989), a new antifungal antibiotic. *J. Antibiot.* **1975**, *28*, 721–726. [CrossRef]
46. Singh, K.; Sun, S.; Vezina, C. Rapamycin (AY-22, 989), A New Antifungal Antibiotic. *J. Antibiot.* **1979**, *32*, 630–645. [CrossRef]
47. Tsang, C.K.; Qi, H.; Liu, L.F.; Zheng, X.F.S. Targeting mammalian target of rapamycin (mTOR) for health and diseases. *Drug Discov. Today* **2007**, *12*, 112–124. [CrossRef]
48. Shafer, A.; Zhou, C.; Gehrig, P.A.; Boggess, J.F.; Bae-Jump, V.L. Rapamycin potentiates the effects of paclitaxel in endometrial cancer cells through inhibition of cell proliferation and induction of apoptosis. *Int. J. Cancer* **2010**, *126*, 1144–1154. [CrossRef]
49. Kunz, J.; Henriquez, R.; Schneider, U.; Deuter-Reinhard, M.; Movva, N.R.; Hall, M.N. Target of rapamycin in yeast, TOR2, is an essential phosphatidylinositol kinase homolog required for G1 progression. *Cell* **1993**, *73*, 585–596. [CrossRef]
50. Sabers, C.J.; Martin, M.M.; Brunn, G.J.; Williams, J.M.; Dumont, F.J.; Wiederrecht, G.; Abraham, R.T. Isolation of a protein target of the FKBP12-rapamycin complex in mammalian cells. *J. Biol. Chem.* **1995**, *270*, 815–822. [CrossRef]
51. Hara, K.; Maruki, Y.; Long, X.; Yoshino, K.-i.; Oshiro, N.; Hidayat, S.; Tokunaga, C.; Avruch, J.; Yonezawa, K. Raptor, a binding partner of target of rapamycin (TOR), mediates TOR action. *Cell* **2002**, *110*, 177–189. [CrossRef]
52. Kim, D.-H.; Sarbassov, D.D.; Ali, S.M.; King, J.E.; Latek, R.R.; Erdjument-Bromage, H.; Tempst, P.; Sabatini, D.M. mTOR interacts with raptor to form a nutrient-sensitive complex that signals to the cell growth machinery. *Cell* **2002**, *110*, 163–175. [CrossRef]
53. Aylett, C.H.S.; Sauer, E.; Imseng, S.; Boehringer, D.; Hall, M.N.; Ban, N.; Maier, T. Architecture of human mTOR complex 1. *Science* **2016**, *351*, 48–52. [CrossRef]
54. Yuan, H.-X.; Guan, K.-L. Structural insights of mTOR complex 1. *Cell Res.* **2016**, *26*, 267. [CrossRef]
55. Thoreen, C.C.; Sabatini, D.M. Rapamycin inhibits mTORC1, but not completely. *Autophagy* **2009**, *5*, 725–726. [CrossRef]
56. Calimeri, T.; Ferreri, A.J.M. m-TOR inhibitors and their potential role in haematological malignancies. *Br. J. Haematol.* **2017**, *177*, 684–702. [CrossRef]
57. Pinto-Leite, R.; Arantes-Rodrigues, R.; Sousa, N.; Oliveira, P.A.; Santos, L. mTOR inhibitors in urinary bladder cancer. *Tumour Biol.* **2016**, *37*, 11541–11551. [CrossRef]
58. Liu, L.; Luo, Y.; Chen, L.; Shen, T.; Xu, B.; Chen, W.; Zhou, H.; Han, X.; Huang, S. Rapamycin inhibits cytoskeleton reorganization and cell motility by suppressing RhoA expression and activity. *J. Biol. Chem.* **2010**, *285*, 38362–38373. [CrossRef]
59. Alvarado, Y.; Mita, M.M.; Vemulapalli, S.; Mahalingam, D.; Mita, A.C. Clinical activity of mammalian target of rapamycin inhibitors in solid tumors. *Target. Oncol.* **2011**, *6*, 69–94. [CrossRef]
60. Gulhati, P.; Cai, Q.; Li, J.; Liu, J.; Rychahou, P.G.; Qiu, S.; Lee, E.Y.; Silva, S.R.; Bowen, K.A.; Gao, T. Targeted inhibition of mammalian target of rapamycin signaling inhibits tumorigenesis of colorectal cancer. *Clin. Cancer Res.* **2009**, *15*, 7207–7216. [CrossRef]
61. Roulin, D.; Cerantola, Y.; Dormond-Meuwly, A.; Demartines, N.; Dormond, O. Targeting mTORC2 inhibits colon cancer cell proliferation in vitro and tumor formation in vivo. *Mol. Cancer* **2010**, *9*, 57. [CrossRef]
62. Wu, W.K.K.; Lee, C.W.; Cho, C.H.; Chan, F.K.L.; Yu, J.; Sung, J.J.Y. RNA interference targeting raptor inhibits proliferation of gastric cancer cells. *Exp. Cell Res.* **2011**, *317*, 1353–1358. [CrossRef]

63. Sarbassov, D.D.; Ali, S.M.; Sengupta, S.; Sheen, J.-H.; Hsu, P.P.; Bagley, A.F.; Markhard, A.L.; Sabatini, D.M. Prolonged rapamycin treatment inhibits mTORC2 assembly and Akt/PKB. *Mol. Cell* **2006**, *22*, 159–168. [CrossRef]
64. Thomson, A.W.; Turnquist, H.R.; Raimondi, G. Immunoregulatory functions of mTOR inhibition. *Nature Reviews Immunology* **2009**, *9*, 324. [CrossRef] [PubMed]
65. Waldner, M.; Fantus, D.; Solari, M.; Thomson, A.W. New perspectives on mTOR inhibitors (rapamycin, rapalogs and TORKinibs) in transplantation. *Br. J. Clin. Pharmacol.* **2016**, *82*, 1158–1170. [CrossRef]
66. Habib, S.L.; Al-Obaidi, N.Y.; Nowacki, M.; Pietkun, K.; Zegarska, B.; Kloskowski, T.; Zegarski, W.; Drewa, T.; Medina, E.A.; Zhao, Z. Is mTOR inhibitor good enough for treatment all tumors in TSC patients? *J. Cancer* **2016**, *7*, 1621. [CrossRef]
67. Sasongko, T.H.; Ismail, N.F.D.; Zabidi-Hussin, Z. Rapamycin and rapalogs for tuberous sclerosis complex. *Cochrane Database Syst. Rev.* **2016**. [CrossRef]
68. Chiarini, F.; Evangelisti, C.; McCubrey, J.A.; Martelli, A.M. Current treatment strategies for inhibiting mTOR in cancer. *Trends Pharmacol. Sci.* **2015**, *36*, 124–135. [CrossRef]
69. Kim, L.C.; Cook, R.S.; Chen, J. mTORC1 and mTORC2 in cancer and the tumor microenvironment. *Oncogene* **2017**, *36*, 2191. [CrossRef] [PubMed]
70. Hudes, G.; Carducci, M.; Tomczak, P.; Dutcher, J.; Figlin, R.; Kapoor, A.; Staroslawska, E.; Sosman, J.; McDermott, D.; Bodrogi, I. Temsirolimus, interferon alfa, or both for advanced renal-cell carcinoma. *N. Engl. J. Med.* **2007**, *356*, 2271–2281. [CrossRef]
71. Motzer, R.J.; Escudier, B.; Oudard, S.; Hutson, T.E.; Porta, C.; Bracarda, S.; Grünwald, V.; Thompson, J.A.; Figlin, R.A.; Hollaender, N. Efficacy of everolimus in advanced renal cell carcinoma: A double-blind, randomised, placebo-controlled phase III trial. *Lancet* **2008**, *372*, 449–456. [CrossRef]
72. Baselga, J.; Campone, M.; Piccart, M.; Burris Iii, H.A.; Rugo, H.S.; Sahmoud, T.; Noguchi, S.; Gnant, M.; Pritchard, K.I.; Lebrun, F. Everolimus in postmenopausal hormone-receptor–positive advanced breast cancer. *N. Engl. J. Med.* **2012**, *366*, 520–529. [CrossRef]
73. Yao, J.C.; Shah, M.H.; Ito, T.; Bohas, C.L.; Wolin, E.M.; Van Cutsem, E.; Hobday, T.J.; Okusaka, T.; Capdevila, J.; De Vries, E.G.E. Everolimus for advanced pancreatic neuroendocrine tumors. *N. Engl. J. Med.* **2011**, *364*, 514–523. [CrossRef]
74. Yao, J.C.; Fazio, N.; Singh, S.; Buzzoni, R.; Carnaghi, C.; Wolin, E.; Tomasek, J.; Raderer, M.; Lahner, H.; Voi, M. Everolimus for the treatment of advanced, non-functional neuroendocrine tumours of the lung or gastrointestinal tract (RADIANT-4): A randomised, placebo-controlled, phase 3 study. *Lancet* **2016**, *387*, 968–977. [CrossRef]
75. Hess, G.; Herbrecht, R.; Romaguera, J.; Verhoef, G.; Crump, M.; Gisselbrecht, C.; Laurell, A.; Offner, F.; Strahs, A.; Berkenblit, A. Phase III study to evaluate temsirolimus compared with investigator's choice therapy for the treatment of relapsed or refractory mantle cell lymphoma. *J. Clin. Oncol.* **2009**, *27*, 3822–3829. [CrossRef] [PubMed]
76. Guertin, D.A.; Stevens, D.M.; Saitoh, M.; Kinkel, S.; Crosby, K.; Sheen, J.-H.; Mullholland, D.J.; Magnuson, M.A.; Wu, H.; Sabatini, D.M. mTOR complex 2 is required for the development of prostate cancer induced by Pten loss in mice. *Cancer Cell* **2009**, *15*, 148–159. [CrossRef]
77. Masri, J.; Bernath, A.; Martin, J.; Jo, O.D.; Vartanian, R.; Funk, A.; Gera, J. mTORC2 activity is elevated in gliomas and promotes growth and cell motility via overexpression of rictor. *Cancer Res.* **2007**, *67*, 11712–11720. [CrossRef] [PubMed]
78. Patel, V.; Marsh, C.A.; Dorsam, R.T.; Mikelis, C.M.; Masedunskas, A.; Amornphimoltham, P.; Nathan, C.A.; Singh, B.; Weigert, R.; Molinolo, A.A. Decreased lymphangiogenesis and lymph node metastasis by mTOR inhibition in head and neck cancer. *Cancer Res.* **2011**, *71*, 7103–7112. [CrossRef]
79. Pavel, M.E.; Hainsworth, J.D.; Baudin, E.; Peeters, M.; Hörsch, D.; Winkler, R.E.; Klimovsky, J.; Lebwohl, D.; Jehl, V.; Wolin, E.M. Everolimus plus octreotide long-acting repeatable for the treatment of advanced neuroendocrine tumours associated with carcinoid syndrome (RADIANT-2): A randomised, placebo-controlled, phase 3 study. *Lancet* **2011**, *378*, 2005–2012. [CrossRef]
80. Burris Iii, H.A.; Lebrun, F.; Rugo, H.S.; Beck, J.T.; Piccart, M.; Neven, P.; Baselga, J.; Petrakova, K.; Hortobagyi, G.N.; Komorowski, A. Health-related quality of life of patients with advanced breast cancer treated with everolimus plus exemestane versus placebo plus exemestane in the phase 3, randomized, controlled, BOLERO-2 trial. *Cancer* **2013**, *119*, 1908–1915. [CrossRef]

81. André, F.; O'Regan, R.; Ozguroglu, M.; Toi, M.; Xu, B.; Jerusalem, G.; Masuda, N.; Wilks, S.; Arena, F.; Isaacs, C. Everolimus for women with trastuzumab-resistant, HER2-positive, advanced breast cancer (BOLERO-3): A randomised, double-blind, placebo-controlled phase 3 trial. *The lancet oncology* **2014**, *15*, 580–591. [CrossRef]
82. Hurvitz, S.A.; Dalenc, F.; Campone, M.; O'Regan, R.M.; Tjan-Heijnen, V.C.; Gligorov, J.; Llombart, A.; Jhangiani, H.; Mirshahidi, H.R.; Tan-Chiu, E. A phase 2 study of everolimus combined with trastuzumab and paclitaxel in patients with HER2-overexpressing advanced breast cancer that progressed during prior trastuzumab and taxane therapy. *Breast Cancer Res. Treat.* **2013**, *141*, 437–446. [CrossRef]
83. Temkin, S.M.; Fleming, G. Current treatment of metastatic endometrial cancer. *Cancer Control.* **2009**, *16*, 38–45. [CrossRef]
84. Tinker, A.V.; Ellard, S.; Welch, S.; Moens, F.; Allo, G.; Tsao, M.S.; Squire, J.; Tu, D.; Eisenhauer, E.A.; MacKay, H. Phase II study of temsirolimus (CCI-779) in women with recurrent, unresectable, locally advanced or metastatic carcinoma of the cervix. A trial of the NCIC Clinical Trials Group (NCIC CTG IND 199). *Gynecol. Oncol.* **2013**, *130*, 269–274. [CrossRef] [PubMed]
85. Takatori, E.; Shoji, T.; Miura, Y.; Takada, A.; Takeuchi, S.; Sugiyama, T. Effective use of everolimus as salvage chemotherapy for ovarian clear cell carcinoma: A case report. *Onco Targets Ther.* **2014**, *7*, 165.
86. Behbakht, K.; Sill, M.W.; Darcy, K.M.; Rubin, S.C.; Mannel, R.S.; Waggoner, S.; Schilder, R.J.; Cai, K.Q.; Godwin, A.K.; Alpaugh, R.K. Phase II trial of the mTOR inhibitor, temsirolimus and evaluation of circulating tumor cells and tumor biomarkers in persistent and recurrent epithelial ovarian and primary peritoneal malignancies: A Gynecologic Oncology Group study. *Gynecol. Oncol.* **2011**, *123*, 19–26. [CrossRef] [PubMed]
87. Colombo, N.; McMeekin, S.; Schwartz, P.; Kostka, J.; Sessa, C.; Gehrig, P.; Holloway, R.; Braly, P.; Matei, D.; Einstein, M. A phase II trial of the mTOR inhibitor AP23573 as a single agent in advanced endometrial cancer. *J. Clin. Oncol.* **2007**, *25*, 5516.
88. Mackay, H.; Welch, S.; Tsao, M.S.; Biagi, J.J.; Elit, L.; Ghatage, P.; Martin, L.A.; Tonkin, K.S.; Ellard, S.; Lau, S.K. Phase II study of oral ridaforolimus in patients with metastatic and/or locally advanced recurrent endometrial cancer: NCIC CTG IND 192. *J. Clin. Oncol.* **2011**, *29*, 5013. [CrossRef]
89. Vignot, S.; Faivre, S.; Aguirre, D.; Raymond, E. mTOR-targeted therapy of cancer with rapamycin derivatives. *Ann. Oncol.* **2005**, *16*, 525–537. [CrossRef] [PubMed]
90. Duran, I.; Siu, L.L.; Oza, A.M.; Chung, T.B.; Sturgeon, J.; Townsley, C.A.; Pond, G.R.; Seymour, L.; Niroumand, M. Characterisation of the lung toxicity of the cell cycle inhibitor temsirolimus. *Eur. J. Cancer* **2006**, *42*, 1875–1880. [CrossRef]
91. Leary, A.; Auclin, E.; Pautier, P.; Lhommé, C. The PI3K/Akt/mTOR pathway in ovarian cancer: Biological rationale and therapeutic opportunities. In *Ovarian Cancer-A Clinical and Translational Update*; Díaz-Padilla, I., Ed.; IntechOpen: London, UK, 2013. [CrossRef]
92. Cho, D.C.; Cohen, M.B.; Panka, D.J.; Collins, M.; Ghebremichael, M.; Atkins, M.B.; Signoretti, S.; Mier, J.W. The efficacy of the novel dual PI3-kinase/mTOR inhibitor NVP-BEZ235 compared with rapamycin in renal cell carcinoma. *Clin. Cancer Res.* **2010**, *16*, 3628–3638. [CrossRef]
93. Blaser, B.; Waselle, L.; Dormond-Meuwly, A.; Dufour, M.; Roulin, D.; Demartines, N.; Dormond, O. Antitumor activities of ATP-competitive inhibitors of mTOR in colon cancer cells. *BMC Cancer* **2012**, *12*, 86. [CrossRef]
94. Benjamin, D.; Colombi, M.; Moroni, C.; Hall, M.N. Rapamycin passes the torch: A new generation of mTOR inhibitors. *Nat. Rev. Drug Discov.* **2011**, *10*, 868. [CrossRef] [PubMed]
95. Wander, S.A.; Hennessy, B.T.; Slingerland, J.M. Next-generation mTOR inhibitors in clinical oncology: How pathway complexity informs therapeutic strategy. *J. Clin. Investig.* **2011**, *121*, 1231–1241. [CrossRef] [PubMed]
96. Rodrik-Outmezguine, V.S.; Okaniwa, M.; Yao, Z.; Novotny, C.J.; McWhirter, C.; Banaji, A.; Won, H.; Wong, W.; Berger, M.; de Stanchina, E. Overcoming mTOR resistance mutations with a new-generation mTOR inhibitor. *Nature* **2016**, *534*, 272. [CrossRef] [PubMed]
97. Steuer-Vogt, M.K.; Bonkowsky, V.; Ambrosch, P.; Scholz, M.; Neiβ, A.; Strutz, J.; Hennig, M.; Lenarz, T.; Arnold, W. The effect of an adjuvant mistletoe treatment programme in resected head and neck cancer patients: A randomised controlled clinical trial. *Eur. J. Cancer* **2001**, *37*, 23–31. [CrossRef]
98. Huang, S.; Houghton, P.J. Inhibitors of mammalian target of rapamycin as novel antitumor agents: From bench to clinic. *Curr. Opin. Investig. Drugs* **2002**, *3*, 295–304.
99. Dancey, J.E. Clinical development of mammalian target of rapamycin inhibitors. *Hematol./Oncol. Clin.* **2002**, *16*, 1101–1114. [CrossRef]

100. Zoncu, R.; Efeyan, A.; Sabatini, D.M. mTOR: From growth signal integration to cancer, diabetes and ageing. *Nat. Rev. Mol. Cell Biol.* **2011**, *12*, 21. [CrossRef] [PubMed]
101. Hardt, M.; Chantaravisoot, N.; Tamanoi, F. Activating mutations of TOR (target of rapamycin). *Genes Cells* **2011**, *16*, 141–151. [CrossRef] [PubMed]
102. Chiang, G.G.; Abraham, R.T. Targeting the mTOR signaling network in cancer. *Trends Mol. Med.* **2007**, *13*, 433–442. [CrossRef] [PubMed]
103. Capdevila, J.; Salazar, R.; Halperín, I.; Abad, A.; Yao, J.C. Innovations therapy: Mammalian target of rapamycin (mTOR) inhibitors for the treatment of neuroendocrine tumors. *Cancer Metastasis Rev.* **2011**, *30*, 27–34. [CrossRef]
104. Li, J.; Liu, J.; Song, J.; Wang, X.; Weiss, H.L.; Townsend, C.M., Jr.; Gao, T.; Evers, B.M. mTORC1 inhibition increases neurotensin secretion and gene expression through activation of the MEK/ERK/c-Jun pathway in the human endocrine cell line BON. *Am. J. Phys.-Cell Phys.* **2011**, *301*, C213–C226.
105. Arsham, A.M.; Howell, J.J.; Simon, M.C. A novel hypoxia-inducible factor-independent hypoxic response regulating mammalian target of rapamycin and its targets. *J. Biol. Chem.* **2003**, *278*, 29655–29660. [CrossRef]
106. Balgi, A.D.; Diering, G.H.; Donohue, E.; Lam, K.K.Y.; Fonseca, B.D.; Zimmerman, C.; Numata, M.; Roberge, M. Regulation of mTORC1 signaling by pH. *PLoS ONE* **2011**, *6*, e21549. [CrossRef] [PubMed]
107. Crino, P.B.; Nathanson, K.L.; Henske, E.P. The tuberous sclerosis complex. *N. Engl. J. Med.* **2006**, *355*, 1345–1356. [CrossRef] [PubMed]
108. Kennedy, B.K.; Lamming, D.W. The mechanistic target of rapamycin: The grand conducTOR of metabolism and aging. *Cell Metab.* **2016**, *23*, 990–1003. [CrossRef] [PubMed]
109. Gulhati, P.; Bowen, K.A.; Liu, J.; Stevens, P.D.; Rychahou, P.G.; Chen, M.; Lee, E.Y.; Weiss, H.L.; O'Connor, K.L.; Gao, T. mTORC1 and mTORC2 regulate EMT, motility, and metastasis of colorectal cancer via RhoA and Rac1 signaling pathways. *Cancer Res.* **2011**, *71*, 3246–3256. [CrossRef]
110. Zhou, H.; Huang, S. mTOR signaling in cancer cell motility and tumor metastasis. *Crit. Rev. Eukaryot. Gene Expr.* **2010**, *20*, 1–16. [CrossRef]
111. McCubrey, J.A.; Steelman, L.S.; Chappell, W.H.; Abrams, S.L.; Montalto, G.; Cervello, M.; Nicoletti, F.; Fagone, P.; Malaponte, G.; Mazzarino, M.C. Mutations and deregulation of Ras/Raf/MEK/ERK and PI3K/PTEN/Akt/mTOR cascades which alter therapy response. *Oncotarget* **2012**, *3*, 954. [CrossRef]
112. Grabiner, B.C.; Nardi, V.; Birsoy, K.; Possemato, R.; Shen, K.; Sinha, S.; Jordan, A.; Beck, A.H.; Sabatini, D.M. A diverse array of cancer-associated MTOR mutations are hyperactivating and can predict rapamycin sensitivity. *Cancer Discov.* **2014**, *4*, 554–563. [CrossRef] [PubMed]
113. Wagle, N.; Grabiner, B.C.; Van Allen, E.M.; Hodis, E.; Jacobus, S.; Supko, J.G.; Stewart, M.; Choueiri, T.K.; Gandhi, L.; Cleary, J.M. Activating mTOR mutations in a patient with an extraordinary response on a phase I trial of everolimus and pazopanib. *Cancer Discov.* **2014**, *4*, 546–553. [CrossRef]
114. Inoki, K.; Corradetti, M.N.; Guan, K.-L. Dysregulation of the TSC-mTOR pathway in human disease. *Nat. Genet.* **2005**, *37*, 19. [CrossRef]
115. Mabuchi, S.; Kawase, C.; Altomare, D.A.; Morishige, K.; Sawada, K.; Hayashi, M.; Tsujimoto, M.; Yamoto, M.; Klein-Szanto, A.J.; Schilder, R.J. mTOR is a promising therapeutic target both in cisplatin-sensitive and cisplatin-resistant clear cell carcinoma of the ovary. *Clin. Cancer Res.* **2009**, *15*, 5404–5413. [CrossRef] [PubMed]
116. Peng, C.-L.; Lai, P.-S.; Lin, F.-H.; Wu, S.Y.-H.; Shieh, M.-J. Dual chemotherapy and photodynamic therapy in an HT-29 human colon cancer xenograft model using SN-38-loaded chlorin-core star block copolymer micelles. *Biomaterials* **2009**, *30*, 3614–3625. [CrossRef] [PubMed]
117. Zhang, P.; Hu, L.; Yin, Q.; Zhang, Z.; Feng, L.; Li, Y. Transferrin-conjugated polyphosphoester hybrid micelle loading paclitaxel for brain-targeting delivery: Synthesis, preparation and in vivo evaluation. *J. Control. Release* **2012**, *159*, 429–434. [CrossRef] [PubMed]
118. Oh, W.J.; Jacinto, E. mTOR complex 2 signaling and functions. *Cell Cycle* **2011**, *10*, 2305–2316. [CrossRef]
119. Tanaka, K.; Babic, I.; Nathanson, D.; Akhavan, D.; Guo, D.; Gini, B.; Dang, J.; Zhu, S.; Yang, H.; De Jesus, J. Oncogenic EGFR signaling activates an mTORC2–NF-κB pathway that promotes chemotherapy resistance. *Cancer Discov.* **2011**, *1*, 524–538. [CrossRef] [PubMed]
120. Driscoll, D.R.; Karim, S.A.; Sano, M.; Gay, D.M.; Jacob, W.; Yu, J.; Mizukami, Y.; Gopinathan, A.; Jodrell, D.I.; Evans, T.R.J. mTORC2 signaling drives the development and progression of pancreatic cancer. *Cancer Res.* **2016**, *76*, 6911–6923. [CrossRef]

121. Weinstein, I.B.; Joe, A. Oncogene addiction. *Cancer Res.* **2008**, *68*, 3077–3080. [CrossRef] [PubMed]
122. Garraway, L.A.; Jänne, P.A. Circumventing cancer drug resistance in the era of personalized medicine. *Cancer Discov.* **2012**, *2*, 214–226. [CrossRef]
123. Abrahamse, H.; Hamblin, M.R. New photosensitizers for photodynamic therapy. *Biochem. J.* **2016**, *473*, 347–364. [CrossRef]
124. Almeida, R.D.; Manadas, B.J.; Carvalho, A.P.; Duarte, C.B. Intracellular signaling mechanisms in photodynamic therapy. *Biochim. Biophys. Acta (BBA)-Rev. Cancer* **2004**, *1704*, 59–86. [CrossRef]
125. Kraus, D.; Palasuberniam, P.; Chen, B. Targeting phosphatidylinositol 3-kinase signaling pathway for therapeutic enhancement of vascular-targeted photodynamic therapy. *Mol. Cancer Ther.* **2017**, *16*, 2422–2431. [CrossRef] [PubMed]
126. Sasore, T.; Kennedy, B. Deciphering Combinations of PI3K/AKT/mTOR Pathway Drugs Augmenting Anti-Angiogenic Efficacy In Vivo. *PLoS ONE* **2014**, *9*, e105280. [CrossRef] [PubMed]
127. Tuo, J.; Wang, Q.; Zouboulis, C.C.; Liu, Y.; Ma, Y.; Ma, L.; Ying, J.; Zhang, C.; Xiang, L. ALA-PDT suppressing the cell growth and reducing the lipogenesis in human SZ95 sebocytes by mTOR signaling pathway in vitro. *Photodiagn. Photodyn. Ther.* **2017**, *18*, 295–301. [CrossRef] [PubMed]
128. Brodin, N.P.; Guha, C.; Tomé, W.A. Photodynamic therapy and its role in combined modality anticancer treatment. *Technol. Cancer Res. Treat.* **2015**, *14*, 355–368. [CrossRef] [PubMed]
129. Chou, T.-C. Drug combination studies and their synergy quantification using the Chou-Talalay method. *Cancer Res.* **2010**, *70*, 440–446. [CrossRef] [PubMed]
130. Sadowski, K.; Kotulska, K.; Jozwiak, S. Management of side effects of mTOR inhibitors in tuberous sclerosis patients. *Pharmacol. Rep.* **2016**, *68*, 536–542. [CrossRef] [PubMed]

Review

Role and Therapeutic Targeting of the PI3K/Akt/mTOR Signaling Pathway in Skin Cancer: A Review of Current Status and Future Trends on Natural and Synthetic Agents Therapy

Jean Christopher Chamcheu [1,*], Tithi Roy [1], Mohammad Burhan Uddin [1], Sergette Banang-Mbeumi [1,2,3], Roxane-Cherille N. Chamcheu [1], Anthony L. Walker [1], Yong-Yu Liu [1] and Shile Huang [4,5]

1 College of Pharmacy, University of Louisiana at Monroe, Monroe, LA 71209-0497, USA
2 Division for Research and Innovation, POHOFI Inc., P.O. Box 44067, Madison, WI 53744, USA
3 School of Nursing and Allied Health Sciences, Louisiana Delta Community College, Monroe, LA 71203, USA
4 Department of Biochemistry and Molecular Biology, Louisiana State University Health Sciences Center, 1501 Kings Highway, Shreveport, LA 71130-3932, USA
5 Feist-Weiller Cancer Center, Louisiana State University Health Sciences Center, Shreveport, LA 71130-3932, USA
* Correspondence: chamcheu@ulm.edu; Tel.: +1-318-342-6820; Fax: +1-318-342-1737

Received: 7 July 2019; Accepted: 29 July 2019; Published: 31 July 2019

Abstract: The mammalian or mechanistic target of rapamycin (mTOR) and associated phosphatidylinositiol 3-kinase (PI3K)/protein kinase B (Akt) pathways regulate cell growth, differentiation, migration, and survival, as well as angiogenesis and metabolism. Dysregulation of these pathways is frequently associated with genetic/epigenetic alterations and predicts poor treatment outcomes in a variety of human cancers including cutaneous malignancies like melanoma and non-melanoma skin cancers. Recently, the enhanced understanding of the molecular and genetic basis of skin dysfunction in patients with skin cancers has provided a strong basis for the development of novel therapeutic strategies for these obdurate groups of skin cancers. This review summarizes recent advances in the roles of PI3K/Akt/mTOR and their targets in the development and progression of a broad spectrum of cutaneous cancers and discusses the current progress in preclinical and clinical studies for the development of PI3K/Akt/mTOR targeted therapies with nutraceuticals and synthetic small molecule inhibitors.

Keywords: PI3K; Akt; mTOR; skin cancers; phytochemicals; melanoma; basal cell carcinoma; squamous cell carcinoma; Merkel cell carcinoma; targeted therapy

---

1. Introduction

1.1. Structure and Function of the Human Skin

The human skin is the outermost and largest organ constituting about 20% of the total body weight and measures a total surface area of approximately 2 m^2 in an adult human [1]. The skin is strategically positioned at the interface between the internal and external worlds. While the skin looks simple on the surface, underneath it presents a unique and complex biochemical structure with properties that confer multiple functions vis-à-vis shaping the body [2]. The skin provides a major dynamic, mechanical, and physical defensive barrier against external insults such as toxic chemical agents, infectious microorganisms, ultraviolet (UV) radiation, as well as mechanical stressors [2–5]. In addition to barrier function, the skin also regulates the inward and outward passages of body fluids, including water, electrolytes, and various substances; mediates immune- and thermo-regulatory responses; coordinates sensory perception; and serves as a metabolic sink for the storage of energy in the hypodermis [6–8]. Histologically, the skin is mainly composed of three distinct and unifying tissue components. The first one is the epidermis, an ectodermally derived outermost, non-vascularized, stratified, keratinizing squamous epithelium, with multiple layers ranging in thickness between 75 and 150 µm (interfollicular), and up to 600 µm (the palms and soles). The second one is the dermis, a mesodermally derived, dense collagen, and elastin-rich fibrous connective tissue inundated with blood and lymph vessels, nerve elements, and embedded with disseminated cells including fibroblasts, mast cells, macrophages, and lymphocytes. The third one is the hypodermis or subcutaneous tissue, which underlies the dermis, and mainly consists of fatty or adipose tissues that are regularly crisscrossed by blood vessels [7] (Figure 1).

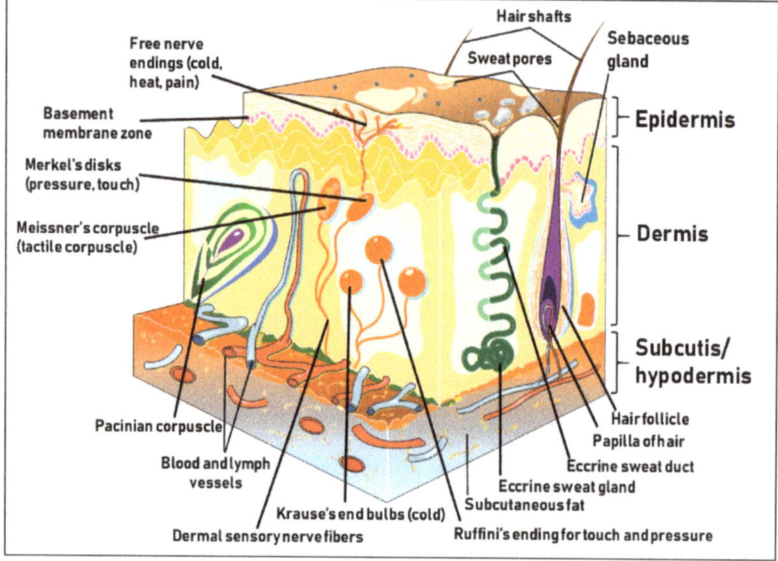

Figure 1. Schematic representation of the cross section of the skin. The skin is composed of epidermis, dermis, and hypodermis. The dermis supports and provides nourishment to the overlying epidermis through its constituent blood and lymphatic vessels. The skin also comprises sensory elements, various resident cell types like fibroblasts, macrophages, and lymphocytes. The ectodermally derived appendage such as sweat glands, sebaceous glands, and hair follicles arises as an invagination of the epidermis into the dermis. The epidermal stem cells are found in the bulge region of the hair follicles, in the basal layer of the interfollicular epidermis, and to some extent the sebaceous glands.

1.2. The Epidermis

The epidermis, an outermost skin tissue and a semi-permeable covering, functions as the major contact point of the body with the external environment. In adults, the epidermis is mainly composed of over 90% keratinocytes, epidermal cells that synthesize keratin, a diverse group of cytoskeletal scaffolding proteins that form 10–12 nm intermediate filament networks at different stages of cornification with functions that determine the condition of the epidermis and make the skin impermeable [7]. Under normal physiologic condition, there is a balance between the renewal and the death of the keratinocytes in maintaining homeostasis [9,10]. There exist other cell types housed in the viable epidermis including melanocytes, Langerhans cells, and Merkel cells, which equally contribute to ensuring skin homeostasis. Melanocytes are the skin pigment producing cells located at the dermal-epidermis junction and hair follicles, which synthesize melanin within melanosomes, and ensure skin pigmentation and photo-protection and other physiological regulatory and protective action [11–15]. Langerhans cells are antigen-presenting cells responsible for immune function of the skin and provide protection against external invading substances and microorganisms [7,16–21]. Merkel cells are oval-shaped mechanoreceptors essential for light touch sensation [18]. On the basis of the skin site, histological skin cross-section reveals that the epidermis is divided into four (hairy or interfollicular skin) or five (palmo-plantar or glabrous skin such as soles and palms) distinct cell layers or strata characterized by different stages of keratinocytes maturation (Figure 2) [9,22]. From inward-out to the skin surface, the deepest layer of the epidermis is the basal or germinative cell layer (stratum basale; SB), which harbors resident stem cells and their progenitor transient amplifying (mitotically active) keratinocytes. There exist three different epidermal keratinocyte stem cell pools located at: 1) the basal compartment, 2) the tip of the dermal papillae, and 3) the hair follicle bulge that is attached to the basement membrane zone (BMZ) via hemidesmosomal protein complexes [6,22–25]. Therefore, the skin serves as a local reservoir of various adult/multipotent stem cell populations, both within the basal epidermal and dermal tissue compartments [9,24,26]. Epidermal homeostasis is maintained by these multipotent stem cells from hair follicle to non-follicular skin, which possess the capacity to regenerate and differentiate into multiple cell lineages, for epidermal, hair/non-hair follicles, and sebaceous glands [27]. During mitosis, a sister keratinocyte stem cell maintains the stem cell pool, while the other pool of daughter keratinocytes divides asymmetrically into progenitor keratinocytes committed to terminal differentiation [9,25,27]. Upon commitment to differentiation, the progenitor basal keratinocytes proliferate, detach, and move towards the skin surface by sequentially developing into the spinous, granular, and cornified layers in a cornification process and subsequently are desquamated [9,28]. During cornification, as basal keratinocytes expressing keratin 5 (K5), keratin 14 (K14), and integrins detach and transition upward, several biochemical processes lead to gene expression of keratinocyte differentiation-related markers such as keratins 1 (K1), 10 (K10), filaggrin, involucrin, loricrin, and transglutaminase [28,29]. Overlying the basal layer is the spinous or prickle cell layer (stratum spinosum; SS) that is several layers thick and is characterized by increased numbers of desmosomes. Sitting on this layer is the granular cell layer (stratum granulosum; SG) that comprises about 3–5 cell layers encircling lamellar bodies and keratohyalin granules [29]). Here, the cells gradually flatten and collapse, which is associated with initiation of nuclear and other organelle degradation, and active lipid and protein secretion. In palmoplantar skin sites, the granular layer is overlaid by the clear layer (stratum lucidum, SL), which corresponds to a transition phase between the granular layer and the stratum corneum, but the SL is absent in interfollicular epidermis as discussed above. Finally, the cornified or horny cell layer (stratum corneum, SC) is the outermost layer of the epidermis that is comprised of 3–10 layers of flattened corneocytes ranging 10–30 µm thick, as well as intercellular lipids/protein complexes on the skin surface and provides the vital skin barrier against water loss and external insults [7,30]. The process of cornification culminates in the formation of the stratum corneum (SC), the last line of defense from the external environment, which is composed of corneocytes, dead keratinocytes containing a highly specialized protein and lipid matrix and forms a vital part of the skin surface that provides the barrier [7,30]. Corneocytes are ultimately lost through desquamation and replaced by newly differentiated cells, a process resulting in the regeneration of this tissue every 6–8 weeks in humans and 8–10 days in mice [31]. By readily diffusing through the intercellular layers, the SC may allow the transportation of some small, lipid-soluble compounds from the surface inwards into the skin, and hence, the integrity of the skin as a dynamic organ is maintained through epidermal homeostasis. The balance between epidermal keratinocyte proliferation and differentiation is tightly regulated, and deregulation of this balance is regarded as the cause of diverse skin pathologies including cutaneous cancers, and inflammatory skin disease.

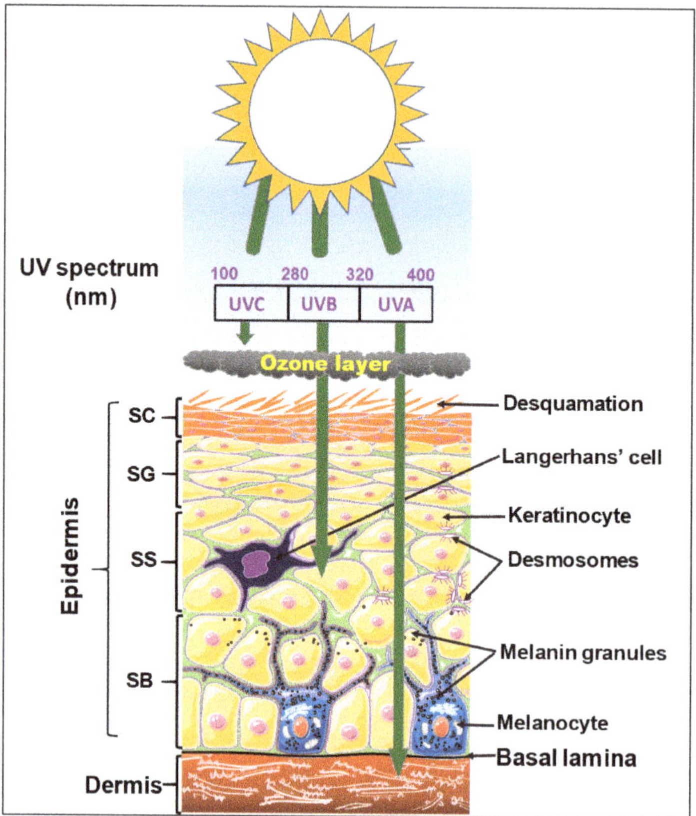

Figure 2. Schematic representation of the structure of the epidermis and papillary dermis. This stratified epithelium consists of basal, spinous, granular, lucidum, and stratum corneum layers. The basal layer contains the stem cells that sit or attach to the basal lamina via hemidesmosomes. In normal epidermis, basal keratinocytes express K5, K14, and integrins. The suprabasal and differentiating layers of keratinocytes are committed to differentiation, expressing K1, K10, and other markers, like filaggrin and loricrin. Other resident epidermal cell types include Melanocytes, Langerhans cells, and Merkel cells. The ultraviolet rays (UVR) from the sun penetrate the atmosphere and the different layers of the skin. UVC with the greatest energy photons (shorter wavelengths) is totally absorbed by the ozone layer while UVB with intermediate energy photons (wavelengths) that damage DNA in epidermal cells penetrates the upper layers of the epidermis. UVA with low energy photons (longer wavelengths) penetrates deeper into the upper dermis causing damage to collagen and elastic tissue, even skin cancers and other skin manifestations.

1.3. The Dermis

Beneath the epidermis sits the mesodermally derived dermis (intermediate skin layer), a thick layer of dense connective tissue mostly consisting of the ground substance or extra cellular matrix (ECM) particularly made up of collagen, elastin, fibrillin, and glycoproteins (non-structural) that give the skin its suppleness and mechanical strength [7,32,33]. The dermis is comprised of two major layers: a) the papillary dermis (superficial dermis), an intermediate layer rich in nerve endings, which is separated from the epidermis by the dermal-epidermal junction (Figure 1); and b) the reticular dermis (deep and medium dermis), a dense connective tissue composed of a network of elastic fibers. The dermis harbors blood vessels, hair follicles, nerve endings, sweat, and sebaceous glands that support and nourish the epidermis, and protects the vascular network and nerve fibers (Figure 1). The dermis also harbors an abundance of different resident cell types, including fibroblasts that synthesize collagen/ECM essential for tissue elasticity, and histiocytes such as macrophages, lymphocytes and mast cells important in skin immune response.

1.4. The Hypodermis

Underneath the dermis lies the hypodermis or subcutaneous tissue (or subcutis) containing adipose tissue, blood vessels, nerves, and sometimes invaginations of epidermal appendages such as sweat glands, sebaceous glands, and hair follicles, enabling the hypodermis to insulate the body and serve as an energy reservoir as well [1,6] (Figure 1). As shown in (Figure 1), skin derivatives or appendages including the hair follicles, nails, sebaceous, sweat, and apocrine glands, also derived from embryonic ectoderm are absent in palmoplantar or load bearing skin sites such as the palms and soles [6]. Although the skin varies in thickness based on the anatomical site, age, and the presence and density of derivatives, the basic structure is maintained at all body sites; all three skin tissue compartments interact and communicate with each other through the secretion of immune-mediators, extracellular matrix proteins, growth factors and hormones [6,34]. Being a functional barrier, the skin is constantly in direct contact with the outside environment. Since healthy skin is a major component of our physical appearance, skin also plays an important role in our social and sexual relationships. Perturbation in the intricate organization due to acquired or inherited factors leads to several cutaneous diseases, most of which are chronic and recalcitrant to treatment.

2. Risk Factors Associated with Cutaneous Carcinogenesis

Two major risk factors are associated with the pathophysiology of many cutaneous carcinogenesis, including environmental (also termed modifiable) and genetic (also termed non-modifiable) risk factors [18,35]. The most common environmental risk factor or trigger for almost all skin cancer types is exposure to ultraviolet (UV) radiation [36], which can damage DNA in skin cells like keratinocytes and melanocytes, leading to tanned and sunburned skin [35]. As depicted in (Figure 2), UV radiation forms a portion of the electromagnetic (EM) spectrum that reaches the Earth from the sun. UV rays are situated between X-rays and visible light, and contain three major types, UVA, UVB, and UVC, at varying wavelengths ranging from 100 to 400 nanometers (nm) with varied skin penetrating properties [37]. UVA rays have the longest wavelengths (320–400 nm), followed by UVB rays with medium wavelengths (290–320 nm), and then UVC rays have the shortest wavelengths (100–280 nm). UVA rays are not absorbed by the atmosphere (Earth's ozone layer), so they are transmitted through and can penetrate deep into the middle layer of the skin via the basement membrane, where the melanocytes reside to the superficial dermis [38]. UVB rays are almost totally absorbed by the epidermis (Figure 2). UVC rays are mostly absorbed by the ozone layer and the atmosphere, to some extent, that is often dependent on the climatic conditions [36,39]. Thus, most of the UV rays that come in contact with the skin are UVA with a small amount of UVB [38]. Both UVA and UVB exposures can result in a tanned skin appearance [37], and overexposure to UVB radiation causes erythema, swelling, and pain, the characteristic signs of sunburn, which generally take several hours to develop.

Incident UV rays unto the skin can intermingle with numerous light-emitting skin layer specific molecules to elicit both desirable and undesirable effects, contingent upon the UV rays' exposure, sources, and wavelength. Desirable effects include priming the skin to synthesize vitamin D precursor as well as their intake, in view of treating various cutaneous diseases including cancers [40–42]. Undesirable effects of UV rays in skin include allergic and inflammatory diseases, immunosuppression, photo-aging, oxidative stress, carcinogenesis, and increased drug sensitivity [35,43–45]. The molecular mechanism of UV-induced skin cancers is associated with eliciting increased DNA damage signals, e.g., activation of the p53 pathway and induction of the apoptotic pathway, which profoundly alter cell physiology to mediate cell cycle arrest and activate DNA repair [44,45]. Interestingly, exposure of human keratinocytes to UVA and UVB results in activation of the phosphatidyl-inositiol 3-kinase

(PI3K) as well as phosphorylation of Akt at S473 by UVB and at Thr308 by UVA as well as increased phosphorylation of the mammalian or mechanistic target of rapamycin (mTOR) and p70 S6 kinase 1 (S6K1). Rapamycin pretreatment has been shown to suppress the expression of phosphorylated S6K1 upon exposure to UV radiation, and the silencing of Akt had no effect on its expression, an indication that exposure to UV radiation can activate the PI3K/Akt/mTOR-S6K1 pathway [46].

3. The PI3K/Akt/Mtor Signaling and Interrelations in Tissue Development

In multicellular organisms, several signaling pathways are associated with the regulation of gene expressions, thus contributing to the organized complex physiological processes critically involved in skin cell growth, proliferation, survival, and differentiation, as well as skin tissue development [47,48]. Consequently, alterations in these pathways can modulate protein synthesis, negatively impact skin cell growth and proliferation, and result in phenotypically diverse skin diseases [47–49]. Knowledge of the intracellular signals and mechanisms through which cells receive and integrate extracellular cues is important for the diagnosis and the development of novel and well-targeted therapeutic regimen for ensuing cutaneous malignancies. Amongst various signal transduction pathways, the PI3K/Akt/mTOR pathways [50,51] are the hub involved in a variety of physiologic functions linking growth factors, nutrients, and energy availability to lipid and protein synthesis, metabolism, cell growth, proliferation, survival, apoptosis, angiogenesis, and tissue development [52,53]. These pathways and associated components have been frequently observed to be deregulated in diverse cancers including the melanoma and non-melanoma skin cancers and are emerging as clinically relevant therapeutic targets [52,53].

4. Structure and Function of the mTOR Pathway

When talking about mTOR, we have to mention rapamycin. Rapamycin (sirolimus) is an antifungal antibiotic that was first isolated from the bacterial strain *Streptomyces hygroscopicus* NRRL 5491 in 1975 [52,54] in the soil of Rapa Nui Island (Easter Island) from which its name was derived [52]. In 1991, Hall laboratory first discovered target of rapamycin (TOR) in yeast [55,56]. Until mid-1990s, the mammalian counterpart (mTOR) was discovered by Sabatini and colleagues [57]. Rapamycin forms a complex with FK506-binding protein 12 (FKBP-12), and then the rapamycin-FKBP-12 complex binds to the FKBP-rapamycin-binding (FRB) domain of mTOR, inhibiting mTOR function [50]. Thus, mTOR is also termed FKBP-12-rapamycin-associated protein (FRAP), rapamycin and FKBP-12 target (RAFT1), rapamycin target 1 (RAPT 1), or sirolimus effector protein (SEP). mTOR belongs to the PI3K-related protein kinases (PIKKs) family with a C-terminus that shares strong homology to the PI3K catalytic domain (Figure 3). mTOR interacts with several proteins and forms at least two distinctive complexes, namely mTOR complex 1 (mTORC1) and 2 (mTORC2), with distinct kinase activities and cellular functions [46,50,57]. These complexes are large but have different sensitivities to rapamycin as well as different effectors. Both mTORC1 and mTORC2 share the following common components: Catalytic mTOR subunit, mammalian lethal with sec-13 protein8 (mLST8 or GβL), the negative regulator DEP domain containing mTOR-interacting protein (DEPTOR), and the Tti1/Tel2 complex (reviewed in Reference [50]). The mTORC1 discretely comprises the regulatory-associated protein of mTOR (Raptor), and another negative regulator, proline-rich Akt substrate 40 kDa (PRAS40). In addition to the above common components, the mTORC2 additionally contains the rapamycin-insensitive companion of mTOR (Rictor), the mammalian stress-activated MAP kinase-interacting protein 1 (mSin1), and protein observed with Rictor 1 and 2 (Proctor 1/2) (Figure 4) [46,50,57]. Both Raptor and mLST8 are positive regulators of mTORC1's activity and function, while PRAS40 and DEPTOR are both negative regulators of the mTORC1 [46,52,58]. Raptor serves as a scaffold for recruiting mTORC1 substrates, while mLST8 binds the mTOR kinase domain, and positively regulates its kinase activity. On the other hand, PRAS40 associates with mTOR via raptor to inhibit the activity of mTORC1, while DEPTOR functions as mTOR-interacting protein, to both mTORC1 and mTORC2, as a negative regulator of their activities [50,52].

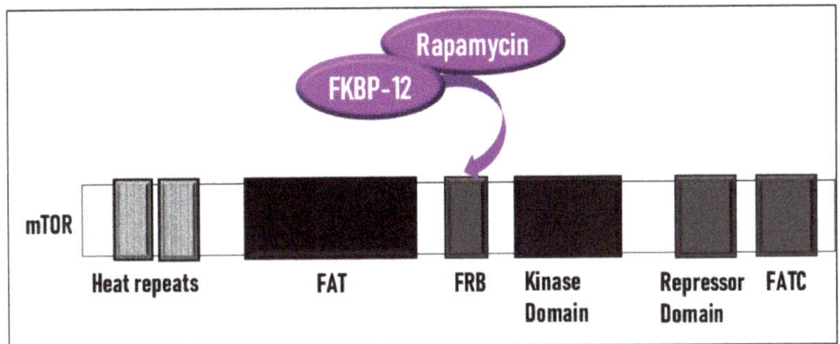

Figure 3. Schematic of the domain structure of mTOR showing the *N*-terminus with two tandem HEAT repeats, followed by a FAT domain (domain shared by PI3K-related protein kinases (PIKK) family members), an FRB domain (FKBP-12-rapamycin-binding site), a kinase catalytic domain, a repressor domain (RD), and a FAT C terminus domain located at the *C*-terminus of the protein. The FRB domain forms a deep hydrophobic cleft that serves as the high-affinity binding site for the inhibitory complex FKBP12-rapamycin [52].

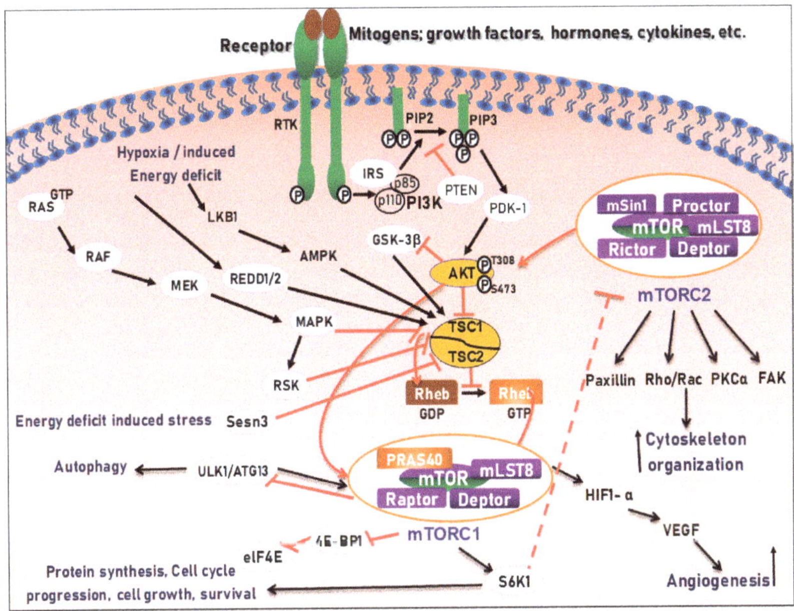

Figure 4. Schematic illustration of the PI3K/Akt/mTOR signaling pathway. Arrows represent activation, whereas bars represent inhibition [50]. Upon receptor activation, insulin receptor substrate (IRS) activates phosphatidylinositol 3-kinase (PI3K), which is in turn phosphorylated to generate phosphatidylinositol [3,4,5]-trisphosphate (PIP3). Phosphatase and tensin homolog (PTEN) can dephosphorylate PIP3 to regulate the pathway activity. AKT is activated through the binding of PIP3 to its amino terminal, pleckstrin homology (PH) domain (stripped), which then promotes the translocation of AKT to the plasma membrane, where the carboxyl terminal T308 is phosphorylated by phosphoinositide-dependent kinase-1 (PDK1), and S473 is phosphorylated by mTORC2. AKT regulates several cellular processes such as survival and cell proliferation, through a variety of downstream proteins like glycogen synthase kinase 3-beta (GSK-3β), Forkhead Box O (FOXO), amid others (not shown). AKT is able to directly phosphorylate and thus inactivates the 40 kDa proline-rich protein

(PRAS40), relieving the suppressive regulation on mTORC1 activity. Furthermore, AKT can phosphorylate and inactivate the tuberous sclerosis (TSC) tumor suppressor protein complex that acts as a GTPase-activating protein (GAP) for the RAS homolog enriched in brain (Rheb) small G protein to regulate its activity. Retention of the Rheb-GTP bound form activates mTOR, which is comprised of two main complexes that are associated with diverse proteins such as Raptor, mLST8, PRAS40 and Deptor for complex I (mTORC1), and Rictor, mLST8, Deptor, mSin1 and Protor for complex II (mTORC2). mTORC1 is regulated by a variety of environmental signals mediated via several proteins including REDD1/2 (regulated in development and DNA damage responses 1/2), AMP-activated protein kinase (AMPK), among others. mTORC1 phosphorylates downstream S6K1 (p70S6 Kinase 1) and modulates the eukaryotic initiation factor 4E-binding protein (4E-BP1), which discharges it from hindering eIF4E, and enabling 40S ribosomal subunit to be recruited to mRNAs, leading to the initiation of protein translation. S6K also phosphorylates ribosomal protein S6 that is also involved in translational regulation by the 40S ribosomal subunit. By contrast, the regulation of mTORC2 is still under investigation, but it is known to be regulated by growth factors. mTORC2 phosphorylates distinct groups of proteins, enabling the regulation of actin cytoskeleton and migration via activating protein kinase C α (PKC-α), small GTPases (Rhoa, Rac1 and Cdc42), and focal adhesion proteins, such as focal adhesion kinase (FAK) and paxillin. Essentially, the activation of the RAS-RAF-MEK-ERK-RSK pathway mediated by growth factor is another mechanism of regulated crosstalk with the PI3K/AKT/mTOR signaling pathway.

mTORC1 is sensitive to rapamycin, growth factors, energy (ATP), nutrients (amino acids), oxidative stress, and DNA damage, and regulates cell growth and proliferation by controlling protein/lipid/nucleotide synthesis, and lysosome biogenesis through mediating phosphorylation of S6K1 and eukaryotic initiation factor 4E (eIF4E) binding protein 1 (4E-BP1) (Figure 4) [50]. mTORC2 is only sensitive to prolonged (>24 h) rapamycin exposure in certain cases and growth factors and regulates cell survival and cytoskeletal organization in part by regulating phosphorylation of Akt, serum and glucocorticoid-inducible kinase 1 (SGK1), protein kinase Cα (PKCα) and focal adhesion proteins, as well as the activity of small GTPases [46,57,59]. Though the functions of the mTOR complexes remain to be unveiled, current data indicate that mTOR plays a central role in the regulation of cell growth, proliferation, differentiation, survival, autophagy, and motility, as well as angiogenesis and lymphangiogenesis [46,50,57,59].

5. Regulation of the PI3K/Akt/mTOR Pathways in Development and Carcinogenesis

To date, mTORC1 is known to be regulated by multiple pathways (as reviewed below), but how mTORC2 is regulated remains poorly understood. Currently PI3K is the only known upstream modulator of mTORC2, as overexpression of PI3K leads to upregulated mTORC2 activity (Figure 4) [46]. In response to the binding of a growth factor, the corresponding receptor, such as insulin-like growth factor receptor (IGFR), platelet-derived growth factor receptor (PDGFR), or epidermal growth factor receptor (EGFR) on the cell surface, is activated and signals to downstream molecules, leading to the activation of multiple pathways, including the PI3K-Akt and RAS-RAF-mitogen-activated protein kinase kinase (MEK)-extracellular-signal-regulated kinase (ERK)-ribosomal protein S6 kinase (RSK) pathways [57]. The activated PI3K catalyzes phosphatidylinositol [4,5]-bisphosphate (PIP2) to phosphatidylinositol [3,4,5]-triphosphate (PIP3), which is antagonized by the phosphatase and tensin homologue deleted on chromosome 10 (PTEN), a lipid and protein phosphatase (Figure 4). The PIP3 binds to the pleckstrin homology (PH) domain of the serine/threonine kinase, Akt, facilitating Akt docking to the cell membrane, where it is activated via phosphorylation by mTORC2 on S473 and by the phosphoinositide-dependent kinase 1 (PDK1) on T308 [46]. Akt can thus be positively regulated by PI3K, and negatively regulated by PTEN [46,50,57,59]. Therefore, loss of *PTEN* and/or mutations of *PIK3CA* result in constitutive activation of Akt/mTOR, which have been documented in various cancers [52].

Tuberous sclerosis complex 1 (TSC1 or hamartin), TSC2 (or tuberin), and TBC1D7 form a complex, acting as a GTPase-activating protein (GAP) for the Ras homolog enriched in brain (Rheb)

GTPase [46,50,57,59]. The GTP-bound form of Rheb interacts with mTORC1 to potently stimulate its kinase activity [46,50,57,59]. Being a Rheb GAP, the TSC1/2 complex negatively regulates mTORC1 by converting an active GTP-bound Rheb into an inactive GDP-bound state [50]. In response to growth factor stimulation, the activated Akt can phosphorylate TSC2 at S939 and T1462, preventing TSC2 from forming a complex with TSC1, so that the active (GTP-bound) Rheb state remains, leading to activation of mTORC1 [46,50,57,59] (Figure 4). Of note, through a TSC1/2-independent manner, Akt can also activate mTORC1 by phosphorylating PRAS40, triggering the dissociation of PRAS40 from raptor [50].

In fact, the TSC1/2 complex can transmit more signals to mTORC1 as well. In response to growth factor stimulation, the activated ERK1/2 and ribosomal S6 kinase 1 (RSK1) can directly phosphorylate TSC2 at S664/540 and at S1798, respectively, inhibiting the TSC1/2 complex and consequently activating mTORC1 [46,50,57,59]. In response to the pro-inflammatory cytokine, tumor necrosis factor-α (TNFα), IκB kinase β (IKKβ) is activated, which can phosphorylate TSC1 at S511/487, causing TSC1/2 inhibition and mTORC1 activation. Furthermore, the canonical Wnt signaling which inhibits glycogen synthase kinase 3β (GSK3-β) can also activate mTORC1 through TSC1/2, considering that GSK3-β is normally responsible for phosphorylation (S1371, S1375, S1379, and S1387) and activation of TSC2 [46,50,57,59].

Furthermore, the AMP-activated protein kinase (AMPK) and/or the regulated in development and DNA damage responses 1 (REDD1), a hypoxia-induced tumor suppressor, can activate the TSCl/2 complex, inhibiting the mTORC1 signaling pathway [46,50,57,59]. The other critically regulated cell growth/survival intracellular signaling pathway is the RAS, which is activated when complexed with GTP, and the conversion of GDP to GTP is inhibited by neurofibromatosis type 1 (NF1), a protein that causes neurofibromatosis type 1 when mutated [57]. As discussed above, downstream of activated RAS are RAF and MEK, which affect the activation of ERK1/2 and RSK1 [46,50,57,59]. The activated ERK1/2 and RSK1 can specifically phosphorylate TSC2, inhibiting the TSC1/2 complex and thereby activating mTORC1 (Figure 4). Consequently, both the PI3K-Akt and the RAS-RAF-MEK-ERK-RSK pathways unite to regulate the mTORC1 signaling [46,50,57,59].

Consequently, the activated mTORC1 further regulates energy metabolism, protein/lipid/nucleotide synthesis, lysosome biogenesis, autophagy, and angiogenesis through S6K1, 4E-BP1, lipin1, activating transcription factor 4 (ATF4), transcription factor EB (TFEB), Unc-51 like autophagy activating kinase 1 (ULK1), hypoxia-inducible factor 1α (HIF-1α), etc. (for details, see review in [53] and [50,57,59]. In particular, activated mTORC1 phosphorylates S6K1 and 4E-BP1, through an interaction between Raptor and a TOR signaling motif in S6K and 4E-BP1. On one hand, activated S6K1 then phosphorylates S6 (40S ribosomal protein S6), thereby improving mRNA translation. This, in turn, further stimulates translation and activates RNA polymerases I and III transcription factors, leading to the synthesis and assembly of ribosomes, tRNAs, and translation factors [57,59]. On the other hand, 4E-BP1 plays an inhibitory role in the initiation of translation by binding and inactivating the eukaryotic translation initiation factor 4E (eIF4E) [50,57,59]. When 4E-BP1 is phosphorylated by mTORC1, it will dissociate from eIF4E. Subsequently, the released eIF4E can bind to eIF4G and eIF4A, forming the eIF4F complex, which binds the 5′ cap of mRNAs and promotes eukaryotic translation initiation [46,57,59].

Unlike mTORC1, little is known vis-à-vis the upstream activators of the mTORC2 pathway. So far, mTORC2 is known to respond to cue from growth factors such as insulin, via direct links to ribosome in a PI3K-dependent fashion [46,50,57]. mTORC2 directly activates Akt via phosphorylation at its hydrophobic motif (S473) and SGK1 (S422), a kinase that control ion transport and growth [46,50,60]. Nonetheless, the loss of mTORC2 does not avert phosphorylation of some Akt targets like TSC2, although on the other hand it completely obliterates the activity of SGK1 [46,50]. Therefore, besides activating mTORC2 via endorsing its association with ribosomes, PI3K, also controls the activation of mTORC1 through the Akt-dependent TSC1/TSC2 inhibition, as described above. It has been proposed that PI3K promotes mTORC2 binding to ribosomes, which directly activates mTORC2; and that mTORC2 activates Akt through phosphorylation at S473 [50,57]. Although mTORC2 is known to be less responsive to rapamycin, prolonged (>24 h) treatment with rapamycin or rapalogs has been reported to inhibit mTORC2 assembly through disruption of the rictor-mTOR complex and consequently inhibiting Akt signaling [46,58]. Of note, there exists a direct phosphorylation of insulin receptor substrate 1 (IRS1) by S6K1, which promotes IRS1 degradation and PI3K/Akt down-regulation [46,50,58]. It has been found that treatment of cancer cells with rapamycin or rapalogs can cause the activation of PI3K/Akt through the S6K1-IRS negative feedback mechanism, reducing the apoptotic potential in cancer cells [46,50,58]. This has become one of the explanations for the unsatisfactory anticancer activity of rapalogs clinically.

In addition, mTORC2 regulates cell actin cytoskeleton and migration via activating protein kinase C α (PKC-α), small GTPases (Rhoa, Rac1, and Cdc42), and focal adhesion proteins, such as focal adhesion kinase (FAK) and paxillin. Therefore, mTORC2 is capable of regulating cell growth, proliferation, survival, and motility [46,58] (Figure 4).

6. Cutaneous Cancers Associated with Dysregulation of the PI3K/Akt/mTOR Pathways

Deregulation of the PI3K/Akt/mTOR pathways has been implicated in the pathogenesis of multiple solid human cancers including several skin cancers [46,52]. As a result, this has specifically stimulated the development of specific PI3K, Akt, and mTOR inhibitors for targeted cancer therapy (reviewed in References [46,52,57]). As mentioned above, skin cancers are partly caused by UVA and UVB exposure, which is also associated with the mTOR pathway deregulation [46,50,52,57]. Obligingly, abnormalities due to modifiable (UV) and non-modifiable (genetic) distresses in target genes or proteins of the intracellular networks regulating skin homeostasis, have been shown to result in a gamut of phenotypically varied, and overlapping cutaneous cancers. These cancers are characterized by tissue neoplastic and hyperplastic growth including but not limited to melanoma and non-melanoma skin cancers (basal and squamous cell carcinoma, Merkel cell carcinoma [46,52,57]).

Given that the molecular basis and targets of most skin cancers are well-understood, novel development and effective delivery of chemotherapeutic agents targeting the dysfunction towards maintaining skin tissue homeostasis and integrity are promising therapeutic strategies [46,50,61,62]. In this light, various synthetic small molecule compounds and naturally occurring nutraceuticals have been shown to modulate the activities of the PI3K/Akt/mTOR and may thus serve as novel therapies for these cutaneous cancers [46,50]. Below, we summarize the relevance of the PI3K/Akt/mTOR pathways and the effector molecules in diverse skin cancers and discuss the mechanistic role of several synthetic molecules and dietary phytochemicals in inhibiting these pathways as potential therapeutic approaches.

6.1. *Role of the PI3K/Akt/mTOR and their Targeting in Melanoma Skin Cancer*

Melanoma is a skin cancer derived from malignant transformation of epidermal melanocytes [63,64]. Melanoma is one of the two major forms of skin cancer including non-melanoma skin cancer and is the least common form of skin cancer accounting for only 1% of the total skin cancer incidences, and due to it metastatic potential, is the most aggressive skin malignancy accounting for ~75% of all skin cancer-related deaths worldwide [63–65]. The 2018 cancer facts and figures by American Cancer Society reported that about 178,560 cases of melanoma will be diagnosed in the United States. Among those, 87,290 cases will be in situ, or noninvasive and confined to the epidermis, while 91,270 cases will be invasive, or penetrating the epidermis into the dermis. Prognosis of an estimated 9320 people (5990 men and 3330 women) were expected to die of melanoma in 2018 (https://www.skincancer.org/skin-cancer-information/skin-cancer-facts#melanoma). This single fact, in addition to established risk factors for developing melanoma, is imposed by increased socioeconomic burden among melanoma patients with grim 5-years survival depending on metastasis site ranging 12–28% [63,64,66]. Primary benign melanomas are mostly initiated as horizontal lesions with plaque-like appearance in the epidermis, often called the radial growth phase (RGP), which often progresses to the vertical growth phase (VGP), an invasive phase that eventually metastasis to distant organs such as the lung. The transition from RGP to VGP has been reported to be associated with Akt activation, where heightened Akt/mTOR activities has been reported in about 70% of metastatic melanoma [63,67].

Using melanoma models, we recently reported that Akt acts as a molecular switch linked with elevated mTOR, S6K1, angiogenesis, and concomitant production of peroxides, which further nurture the aggressiveness of metastatic melanoma [68].

6.2. Targeting PI3K/Akt/mTOR and Associated Pathways with Chemotherapeutics, Biologic Drugs, Natural Products, and Synthetic Derivatives in Melanoma

Activation of the mTOR pathway has been suggested to be strongly associated with the pathogenesis of melanoma [63,64,66,69]. Constitutive activation of mTOR inhibits autophagic cell death and dysregulates the normal cell cycle [70]. Due to advances in knowledge of the molecular genetics of melanoma, novel agents targeting this signal transduction pathway have been developed, including rapalogs (everolimus, deforolimus, and temsirolimus) and mTOR kinase inhibitors [71].

6.2.1. Chemotherapeutic Small Molecules and Biologic Drugs

Several in vitro and in vivo preclinical studies have shown that dual PI3K/mTOR inhibitors have significant inhibitory activities against cell proliferation and activation of Akt, some of which are undergoing clinical trials in patients with selective mTOR mutations (reviewed in Reference [63]). Some synthetic small molecule compounds have been shown to mechanistically target the PI3K/Akt/mTOR and associated RAS/RAF/MEK/ERK or MAPK signaling pathway as promising treatments against metastatic melanoma [63]. Other agents including the BRAF inhibitors dabrafenib and vemurafenib, as well as trametinib, a MEK1/2 inhibitor in metastatic melanoma patients, have yielded extended survival [63]. Moreover, the combination treatment with dabrafenib and trametinib yielded better outcomes in the patients with metastatic melanoma than the individual drug treatment. Despite the rapid development of resistance to these agent treatments, combinatorial approach with additional compounds co-targeting the PI3K/Akt/mTOR (but mostly PI3K), MAPK, and other signal transduction pathways are under investigation at various clinical trials [63].

In addition, mutation of the serine/threonine kinase BRAF is found in almost 50% of the malignant melanoma patients; in more than 90% of the cases BRAF harbors V600E point mutation [72]. Mutation in BRAF activates the MAPK pathway, which is involved in cancer cell survival and proliferation. Inhibition of BRAF is also a promising approach in treating malignant melanoma. A number of small molecules have been introduced by far to inhibit BRAF, some of which are approved by the FDA for targeting BRAF mutated malignant melanoma. However, resistance to such molecules is quite common leading to therapeutic failure. A number of mechanisms contributing to escape the BRAF inhibition have been reported in several studies. One of the studies reported the involvement of hepatocyte growth factor (HGF) in acquisition of resistance against BRAF inhibitors through upregulation of c-MET and GAB1, leading to activation of the MAPK pathway [73]. Abnormal expression of long non-coding RNAs (lncRNAs) has also been implicated in the metastatic growth of cells in many cancer types. Activation of c-MET lncRNAs KCNQ1OT1 [74], or downregulation of tumor suppressor microRNA MiR-22 by MALAT1 [75] and miR-152–3p by HOTAIR [76] was also found to increase metastatic growth of melanoma cells [74]. Treating the melanoma cells with combination of BRAF inhibitor Vemurafenib and c-MET inhibitor AMG 337 [73] or siRNA exhibited therapeutic benefits in BRAF mutant malignant melanoma. Hersey et al. reported that the combination of small chemotherapeutic molecules and targeted biological therapies was clinically beneficial in distant metastatic disease [71].

Rapamycin, a specific mTORC1 inhibitor inhibits the cell growth and proliferation as shown in several melanoma cell lines [77,78]. Two other rapamycin analogs, everolimus and temsirolimus, also showed promising results in preclinical studies, inducing cytostatic tumor growth inhibition and decreasing angiogenic capillary perfusion. However, in a phase II clinical study, everolimus failed to demonstrate adequate efficacy in treating patients with metastatic melanoma; but its antiangiogenic role suggested potential utilization in combination therapy [79]. Another mTOR inhibitor, temsirolimus, in combination with chemotherapeutic agent temozolomide, exhibited significant decrease in tumor growth and increased apoptotic death in melanoma cells that showed resistance to BRAF inhibitor vemurafenib [80]. In a phase I clinical study, the combination of temsirolimus and an autophagy inhibitor hydroxychloroquine, accelerated the cell death in melanoma [81].

In a phase 2 trial from the Sarah Cannon Oncology Research Consortium, Hainsworth et al., reported that the combination of Bevacizumab and everolimus was tolerable and showed moderate activity in the treatment of patients with metastatic melanoma [82]. This is an indication that further investigation of compounds with such mechanisms of action, even in combination with inhibitors of secondary signaling pathways, are promising research areas [82]. VS-5584 is a low-molecular weight compound and a novel potent and highly selective dual PI3K and mTOR inhibitor that is well tolerated in animal models with good pharmacokinetic properties [67,83]. Since melanoma is highly resistant to conventional chemotherapeutics, by investigating the in vitro and in vivo anti-melanoma activity of VS-5584, Shao et al. reported a significant and simultaneous blockade of activated components of Akt/mTOR pathway as well as the downregulation of cyclin D1 expression in melanoma, indicating its utility as a potent PI3K/mTOR dual inhibitor [67,84–86]. Moreover, the activity of orally administered VS-5584 suppressed the growth of A375 melanoma xenograft in nude mice. Co-administration of ABT-737 (Bcl-2 inhibitor) with VS-5584 further enhanced the suppressive effects, suggesting a rationale for the clinical assessment of VS-5584 in melanoma patients as well as the development of ABT-737 and other target inhibitors in adjuvants settings [67].

SKLB-M8 is a derivative of millepachine (MIL, a novel chalcone with a 2,2-dimethylbenzopyran motif derived from Millettia pachycarpa Benth (*Leguminosae*), and a flavonoid-rich traditional Chinese medicine). The modified derivative of MIL, (*E*)-3-(3-amino-4-methoxyphenyl)-1-(5-methoxy-2,2-dimethyl-2*H*-chromen-8-yl) prop-2-en-1-one hydrochloride (SKLB-M8) [87,88] has been reported to have antitumor and especially anti-melanoma activity. Wang et al. reported that SKLB-M8 treatment inhibited the proliferation and the expression of cdc2, induced G2/M arrest, and elicited apoptosis via the down-regulation of activated Akt/mTOR signaling pathway in melanoma models as well as inhibited angiogenesis associated with inhibition of ERK1/2 phosphorylation (Table 1) [87,88].

Due to the superior benefit of multi-target strategy over single target approach, Oudart et al. employed the Type XIX collagen NC1 domain associated with basement membranes [NC1 (XIX)], a 19-amino acid peptide localized at the C-terminal end of the α1 (XIX) chain in multi-target identification in an anti-melanoma approach [89] (see Table 1). Using the tumor NC1 domain of collagen Type XIX [C1(XIX)], Oudart et al. were able to target the $\alpha v\beta 3$ integrin interaction and demonstrated the inhibition of migration, invasion, and PI3K/Akt/mTOR and FAK pathways in melanoma cells and preclinical melanoma model [89–91].

Table 1. List of reported lead synthetic compounds targeting mTOR, PI3K, and Akt signaling pathway in various skin cancers.

Lead Compound(s)	Protein Target(s)	Skin Cancer Type	Compound Structure	References
Everolimus (RAD-001)	mTOR	Melanoma, Basal Cell Carcinoma		[92,93]
Erufosine	mTOR	Oral Squamous Cell Carcinoma		[94]
GDC-0084	PI3K, mTOR, Akt	cutaneous Squamous Cell Carcinoma		[95]
Isoselenocyanate-4	Akt	Melanoma		[96]
MLN0128 (Sapanisertib)	mTOR	Melanoma Merkel Cell Carcinoma		[97–99]

Table 1. *Cont.*

Lead Compound(s)	Protein Target(s)	Skin Cancer Type	Compound Structure	References
NVP-BEZ235	PI3K, Akt, mTOR	Melanoma Merkel Cell Carcinoma		[100,101]
NC1 domain of collagen Type XIX [NC1(XIX)]	PI3K, Akt, mTOR, FAK	Melanoma		[89–91]
PBISe	Akt3	Invasive metastatic Melanoma	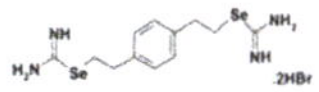	[102]
Rapamycin	PI3k, Akt, mTOR	Melanoma, Esophageal squamous cell carcinoma		[77,78,103]
SKLB-M8	Akt, mTOR	Melanoma		[87,88]
PI-103	PI3K, mTOR	Melanoma		[104]
Perifosine	Akt	Metastatic Melanoma		[105]
Tazarotene	IGFR, PI3K, Akt, mTOR	Basal cell carcinoma		[106,107]
Temsirolimus	mTOR	Metastatic melanoma		[108]
WYE-354	mTOR	Merkel cell carcinoma		[109]
VS-5584	PI3K and mTOR	Melanoma		[67,84–86]
Itraconazole	PI3K and mTOR	Melanoma Basal Cell Carcinoma		[110,111]
LY3023414	PI3K/mTOR	Cutaneous Basal Cell Carcinoma, cutaneous Squamous Cell Carcinoma		[112]
Ku-0063794	mTORC1 and mTORC2	BRAF-Mutant Melanoma in combination with MEK inhibitory agents Merkel cell carcinoma		[113]

Itraconazole is another FDA-approved azole compound that belongs to the antifungal drug family, which has been repurposed for the treatment of various cancers including melanoma [110,111]. Report by Liang et al., demonstrated an anti-melanoma effect of itraconazole and indicated the molecular mechanism to include the inhibition of the PI3K/mTOR, and Hedgehog/Wnt pathways [110].

Increasing evidence has identified the aberrant expression of miRNAs (microRNAs) in melanomagenesis including Uveal melanoma (UM) [114,115]. Jiang and Liu reported that miR-25 target the RNA-binding motif protein 47 (RBM47) and activated the PI3K/Akt/mTOR signaling pathway in melanoma cells [116]. Meng et al. recently demonstrated clinically that miR-138 is significantly upregulated in malignant melanoma patients by regulating the PI3K/Akt/mTOR autophagy pathway via PDK1 dependent expression [117]. Moreover, a study by Li et al. also reported that by targeting PIK3R3/AKT3, miR-224-5p suppressed proliferation, migration, and invasion in uveal melanoma (UM) cells, representing a therapeutic and diagnosis target for patients with UM [115]. Furthermore, Micevic et al. reported that the loss of the overexpressed DNA methyltransferase (DNMT3B) that plays a pro-tumorigenic role in human melanoma resulted in a dramatic suppression of melanoma formation in the Braf/Pten mouse melanoma model [118]. This loss also resulted in hypomethylation of the promoter of miR-196b and a subsequent increase in the expression of miR-196b, which directly targets

Rictor (mTORC2 component) to inhibit the activation of mTORC2, which is dire for formation and growth of melanoma. Thus, this study establishes DNMT3B as a regulator of melanoma development via its influence on mTORC2 signaling and suggests a new therapeutic target in melanoma [118].

Moreover, since aggressive and highly metastatic cutaneous melanoma involves the overexpression of Rictor, the major regulator of Akt phosphorylation, the effect of Rictor inhibition in melanoma models with specific accent on liver metastasis has been investigated. Schmidt et al. reported for the first time that mTORC2/Rictor play a critical role in melanoma liver metastasis via the interactions between cancer cells and cancer-associated hepatic stellate cells (HSC); inhibition of mTORC2/Rictor led to significant inhibition of Akt phosphorylation and motility of the cancer cells [119]. Additionally, Damsky et al. reported that mTORC1 activation blocked BrafV600E-induced growth arrest but was insufficient for the arrest of melanoma development, concluding that activation of both mTORC1/2 is required for Braf-induced melanomagenesis [120].

Rapamycin, its analogs and other protein kinase inhibitors have been investigated for targeting the main mTOR hub in melanoma cells. Ciołczyk-Wierzbicka et al. investigated the role of rapamycin (mTOR), everolimus (mTOR), U0126 (ERK1/2), LY294002 (PI3K), CHIR-99021 (GSK-3β) and others in human VGP (WM793) and metastatic (Lu1205) melanoma cells and observed their antiproliferative effects, in view of the crucial role of the PI3K/Akt/mTOR and ERK1/2 signaling pathways in melanoma progression [121] (Table 1). Temsirolimus (Torisel) targets multiple hallmarks of cancer to impede melanoma growth in vivo. An earlier study has shown that everolimus (RAD-001), an orally active rapamycin analog, may be potentially beneficial to treatment of metastatic melanoma (NCCTG-N0377, Alliance). However, a phase II trial reported that the single treatment with everolimus did not exhibit sufficient anticancer activity, suggesting that a combination with other drugs be considered in future clinical studies [79]. In a recent phase II study of everolimus against advanced melanoma patients with mTOR mutations, Si et al. reported that mTOR inhibitors had restricted activity in non-selected melanoma patient population, whereas a significant percentage of mTOR mutation melanoma patients responded better [92]. The study suggested that it may be possible for future prospective studies to identify suitable patients that would respond to mTOR inhibitors treatment (Clinical trial information: NCT01960829). Rao et al. recently conducted a two-stage Phase II multi-institutional trial evaluating everolimus (RAD-001) in the treatment of patients with metastatic melanoma by assessing progression free survival with baseline derived from historical controls (DOI: 10.1200/jco.2006.24.18_suppl.8043). They observed from interim analysis of 20 patients that RAD-001 was well tolerated with sufficient anti-metastatic melanoma activity. This result encouraged them to have open enrollment to the second stage of this trial.

PBISe (Se,Se′-1,4-phenylenebis(1,2-ethanediyl)bis-isoselenourea) is a selenium containing isosteric analog of the iNOS inhibitor PBIT [*S*,*S*′-1,4-phenylenebis(1,2-ethanediyl)bis-isothiourea], which has been utilized as a small molecule to treat systemic metastasis of melanoma [122]. Recently, Chung et al. showed that topical application of PBISe prevented cutaneous melanocytic lesion or melanoma development at 70–80% in reconstructed melanoma skin model and approximately 50% in melanoma tumor xenograft in mice via the downregulation of the activated Akt signaling, and a concomitant activation of ERK1/2 pathway. The data suggest that PBISe treatment concurrently targeting both pathways has potential to prevent cutaneous metastatic melanoma development in skin [102].

Perifosine, an alkylphosphocholine analog and Akt inhibitor, failed to demonstrate a good efficacy in a phase II clinical trial with various side effects and biochemical toxicity, suggesting no further development of this single compound for recurrent melanoma in human [105,123].

PI-103, a kinase inhibitor targeting Class I PI3K and both mTORC1 and mTORC2, has been shown to exhibit only modest anti-proliferative, cytotoxic effect in vitro and in vivo murine model as a single agent. However, Werzowa et al. showed that the combination of PI-103 and rapamycin in vitro (in melanoma cells) and in vivo (in a melanoma mouse model), synergistically induced apoptosis and suppressed Akt/S6 protein phosphorylation with superior efficacy against malignant melanoma [104].

6.2.2. Natural Plant-Derived Extracts, and Phytochemicals and their Synthetic Derivatives

Additionally, natural dietary phytochemicals have been proven to be effective in many cancer types including melanoma, and have generated encouraging anti-proliferation, anti-invasive, and anti-metastatic effects, which are often associated with their ability to target PI3K/Akt/mTOR and other signaling pathways involved in melanoma carcinogenesis (melanomagenesis). Since several of these natural phytochemicals have lower toxicities and side-effects at physiological attainable doses, exploration of their utility in a single basis or in combination or in adjuvant settings to this plethora of known anticancer targets are laudable therapeutic opportunity. Outlined below are a non-exhaustive list of examples that have proven useful in melanoma.

Acacetin (5,7-dihydroxy-4′-methoxyflavone) is a natural flavonoid derived from *Robinia p seudoacacia*, also termed black locust, with well documented antioxidant, anti-inflammatory, and anticancer properties [124,125]. Its anticancer effect has been associated with modulation of the PI3K/Akt/IKK and the MLK3/MKK3/6 and p38 MAP kinase pathways (reviewed in Reference [124]). Using cell-free, biophysical, computational, cell-based, and in vivo melanoma xenograft models, Jung et al. reported that acacetin inhibited PI3K activity, suppressed Akt phosphorylation and significantly regressed SK-MEL-28 melanoma tumor growth in vivo, suggesting an anti-melanoma agent [124].

Capsaicin (trans-8-methyl-*N*-vanillyl-6-nonenamide) is an active compound found in chili pepper that has been shown to inhibit migration and angiogenesis of B16-F10 melanoma cells in vitro; the mechanism of action was via the inhibition of the PI3K/Akt/Rac1 signal pathway [126].

Evodiamine is a major natural alkaloid component of *Evodiae fructus*. Evodiamine treatment of human melanoma A375-S2 induced cell death that was mediated through PI3K/Akt/caspase and Fas-L/NF-κB signaling pathways and this effect was synergized upon treatment with ubiquitin-proteasome inhibition [127].

Isoliquiritigenin (ISL) is a natural flavonoid with known ability to reprogram cancer cells with in vitro and in vivo anticancer activity. ISL treatment significantly inhibited A375 melanoma cell proliferation, induced G2/M cell cycle arrest, and up-regulated terminal melanocyte differentiation indicators, and significantly decreased melanoma cachexia through decreasing the protein expression levels of activated mTORC2-Akt-GSK3β signaling pathway components. Combined-treatment of ISL and Ku-0063794 (mTOR-specific inhibitor) synergistically inhibited proliferation, and increased melanocyte differentiation markers than ISL alone treatment [128].

Melittin, an amphiphilic small peptide containing 26 amino acid residues, is the major active ingredient from bee venom (BV) with known anti-inflammatory, antibacterial and anticancer activities. It has been demonstrated that BV and melittin have antimelanoma effects with mechanisms associated with the suppression of the activation of the PI3K/Akt/mTOR and MAPK signaling pathways. Moreover, the combination of melittin with temozolomide (TMZ; chemotherapeutic agent) significantly inhibited growth and invasion of melanoma cells compared to the individual single agents, strongly suggesting that melittin could be a potential anti-melanoma agent [129].

Panduratin A, a phytochemical isolated form the rhizome of *Boesenbergia rotunda* was reported in many studies to show effectiveness in several cancer types. In a recent study it has been shown to induce autophagic cell death in melanoma cells that are resistant to apoptosis-inducing chemotherapeutics. Induction of autophagy by panduratin A was found to be mediated by suppression of the mTOR signaling pathway [130].

Another multicomponent Chinese herbal preparation Compound Muniziqi granule (MNZQ), which contains 13 medicinal plants, has been used traditionally for the treatment of endocrine disorder-induced acne, chloasma, dysmenorrhea, menopausal syndrome, and melanoma. One of the components of the MNZQ is *Peganum harmala* plant seed extract, which contains β-carboline alkaloid, harmine. Harmine has been found to induce apoptosis and autophagic cell death in mouse B16 melanoma cells. The autophagy induced was due to the inhibition of multiple signaling pathways which include the Akt/mTOR and ERK1/2 signaling pathways [131].

Sinomenine (7,8-didehydro-4-hydroxy-3,7-dimethoxy-17-methylmorphinane-6-one), the main active component of the medicinal plant *Sinomenium acutum*, was also shown to induce apoptosis and decrease proliferation in B16 melanoma cells and mouse tumor xenografts. The apoptotic and antiproliferative effect of sinomenine is due to induction of autophagy by inhibiting the PI3K/Akt/mTOR signaling pathway [132].

Studies by Espona-Fiedler et al. demonstrated that the anti-melanoma effect of two small molecules members of the prodiginines family, prodigiosin and obatoclax, is related to inhibition of both mTORC1 and mTORC2. Of note, prodigiosin and obatoclax inhibited Akt phosphorylation at S473, despite no effect on T308 [133].

Fisetin (3,7,3′,4′-tetrahydroxyflavone), a dietary bioactive flavanol abundantly found in pigmented fruits and vegetables including apples, cucumbers, onions, persimmons, and strawberries, has been reported to possess pleiotropic effects in diverse human diseases including cancer [134–136]. Fisetin has been investigated for treatment cutaneous cancer, particularly melanoma, having constitutive activation of the Akt/mTOR signaling, due to mutations of PTEN, PIK3CA, or TSC1/2 [77,137,138]. Syed et al. reported that fisetin targeted several key melanomagenesis markers by diverse mechanisms, including inhibition of the Akt/RSK/mTOR/S6K axis, suggesting that fisetin is a potent anti-melanoma agent [68,139,140]. Pal et al. in a study of the effect of fisetin alone and in combination therapy with sorafenib, demonstrated that fisetin enhanced sorafenib-induced apoptosis and abrogated tumor growth in athymic nude mice xenografted with BRAF-mutated melanoma cells through the inhibition of the expressions of activated components of PI3K and MAPK pathways [141]. The data suggest that simultaneous inhibition of both PI3K and MAPK pathways by the combination of fisetin and sorafenib may be a better anti-melanoma therapeutic option.

Curcumin (diferuloylmethane), a polyphenolic active component of turmeric, derived from the rhizome of the plant *Curcuma longa*, has shown a wide variety of health benefits including anti-inflammatory, antioxidant and pro-apoptotic effects in diverse cancers by modulating multiple signal transduction pathways. By treating human melanoma cells with curcumin, Zhao et al., reported the induction of autophagy, and concomitant inhibition of proliferation and invasion via the suppression of activated components of the Akt/mTOR signaling pathway [142]. Another study by Rozzo et al., showed that an analog of curcumin (D6) significantly inhibited proliferation and induced apoptosis in melanoma cells involving the down-regulation of the PI3K/Akt and NF-κB pathway [143].

Resveratrol (*trans*-3,5,4′-trihydroxystilbene), a natural bioactive phenolic compound commonly found in pigmented fruits such as cranberries, grapes, and peanuts, has been shown to exhibit multifaceted biological and health beneficial effects by targeting multiple disease molecular markers [144,145]. Wang et al. showed that resveratrol treatment of B16 melanoma cells resulted in induction of autophagy in a mechanism involving the accumulation of ceramide and inhibition of Akt/mTOR pathway, suggesting a potential in treating melanoma [144,146]. Moreover, Bhattacharya et al. demonstrated that treatment with resveratrol reduced cell migration and invasion, and inactivated Akt/mTOR effectors in malignant melanoma and fibroblast cell lines [46,145].

Honokiol, a natural phenolic compound used for a long time in Chinese and Japanese traditional medicine, has recently been shown as a promising anticancer agent. Kaushik et al. reported that honokiol treatment resulted in the induction of cytotoxicity and cytostatic effects by inhibiting the Akt/mTOR and Notch signaling pathways in malignant melanoma cancer cells [63,147].

Epigallocatechin-3-gallate (EGCG), the most abundant catechin found in green tea (*Camellia sinensis*), has shown several health beneficial effects including anticancer activity (Table 2). In particular, EGCG possesses in vitro and in vivo pharmacological effects of EGCG on the migration and/or metastasis and on the management of melanoma, by inhibiting the PI3K and several signaling pathways like Reference [148].

Nexrutine^R, derived from *Phellodendron amurense*, is an inducer of oxidative stress, inhibiting antioxidant response (Table 2). Since melanoma cells exhibit heightened oxidative stress phenotype associated with increased protein damage, oxidized glutathione, reactive oxygen species (ROS), and KEAP1/NRF2 pathway activity, compared with normal melanocytes contributing to hyperactive proliferation and increased survival, agents targeting these can curtail the disease [149]. Hambright et al. reported Nexrutine^R treatment augmented the constitutively elevated oxidative stress markers, reduced proliferation, survival, and colony formation in melanoma cells, which was associated with selective inhibition of activated components of the PI3K/Akt/mTOR pathway [149,150].

Table 2. List of reported lead natural dietary compounds and extracts targeting mTOR, PI3K, and Akt signaling pathway in various skin cancers.

Lead Compound (s)	Protein Targets	Skin Cancer Type	Compound Structure	References
Acacetin	PI3K, Akt, mTOR	Malignant Melanoma		[124]
Bee Venom Melittin	PI3K, Akt, mTOR	Melanoma		[129]
Capsaicin	PI3K, Akt, Rac1	Melanoma		[126]
Curcumin	PI3K, Akt, mTOR	Melanoma		[142,143]
Epigallocatechin-3 (EGCG)	mTOR	Melanoma	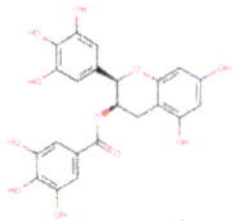	[148]
Evodiamine	PI3K, Akt	Melanoma		[127]
Fisetin	PI3K, Akt, mTOR	Melanoma		[68,139–141]
Isoliquiritigenin	mTORC2, Akt, GSK-3β	Melanoma cachexia		[128]
Harmine	Akt, mTOR and ERK1/2	Melanoma		[131]
Obatoclax	Akt, mTOR	Melanoma		[133]
Panduratin A	mTOR	Melanoma		[130]

Table 2. *Cont.*

Lead Compound (s)	Protein Targets	Skin Cancer Type	Compound Structure	References
Prodigiosin	Akt, mTOR	Melanoma		[133]
Resveratrol	Akt, mTOR	Melanoma		[46,144–146]
Sinomenine	PI3K, Akt, mTOR	Melanoma		[132]
Honokiol	mTOR	Melanoma, Oral squamous cell carcinoma		[63,147,151]
Nexrutine[R]	PI3K/Akt/mTOR	Melanoma		[149,150]

6.3. Targeting PI3K/Akt/mTOR for Treatment of Basal Cell Carcinoma

Keratinocyte carcinoma (KC) comprises basal cell carcinoma (BCC) and cutaneous squamous cell carcinoma (cSCC), which constitute the main forms of non-melanoma skin cancers (NMSC) [152,153]. BCC is the most common type of non-melanoma skin cancer among Caucasians or ethnic groups having blue or green eyes, blond or red hair, and light-colored skin exposed to the sunlight for prolonged periods of time [154–156]. BCC arises in the basal epidermal cell layers and constitutes up to 80% of skin cancers and nearly a third of all cancers diagnosed in the United States [157–159] with an incidence rate of up to 5% per year generating a total cost of about $400 million/year [160]. BCC is usually not life threatening, but if left untreated it can cause loss of function, and disfiguration [161–164]. The morphological features of BCC include the presence of a group of tumors in the dermal layer of the skin that are composed of cells having cellular components similar to undifferentiated basal epidermal cells. An important feature of BCC is the palisade arrangement of epidermal cells in the tumor periphery that separates the tumor from the surrounding stroma. These cells often give the tumor nodular shape or form a band or string surrounding it. Compared to their normal counterparts, the tumor cells have less cytoplasm and chromatin-rich nucleus, which renders more frequent mitotic division but at the same time apoptotic cell death accounts for slow progression of the tumor [165] discussed above, the primary risk factor for BCC being the direct exposure to the sunlight (UV-A and UV-B radiation) and depends on the rate, extent, and duration of exposure to UV irradiation. Other risk factors for BCC include immunosuppression, trauma, arsenic poisoning and other skin disorders such as Gorlin-Goltz syndrome or xeroderma pigmentosum [166,167]. The clinical presentation of BCC may appear in different morphological patterns such as nodular or cystic, superficial, infiltering, and sclerotic or pigmented, which differ in their site of occurrence as well. The nodular or cystic BCC usually occurs as solitary, shiny, red nodules on the face whereas the superficial BCC tends to occur in the trunk. Infiltering BCC is the most aggressive type of tumor often with less defined border [155,165]. Upregulation of the Hedgehog signaling has been found to be the primary mechanism of BCC development, though other non-canonical pathways such as WNT, NOTCH, p53 as well as the P13K/Akt/mTOR pathways have been implicated in the pathogenesis of BCC.

Po-Lin So et al. demonstrated that even a brief inhibition of the PI3K/Akt/mTOR pathway can sustain the inhibition of BCC carcinogenesis long after treatment is completed [106]. This suggested that short-term exposure of BCCs to PI3K inhibitors can result in chemoprevention or chemotherapy, and consequently, evades toxicity and side effects that often result from chronic treatment with other antitumor agents such as tazarotene [106]. Everolimus (inhibitor of mTORC1) shows antiproliferative activity against several types of neoplasia, particularly against BCC. An oral daily dose of 1.5–3 mg everolimus has demonstrated significant improvement in BCC patients with partial or complete recession of the disease [93]. Since the crosstalk among the different signaling pathways such as p53, WNT, Hedgehog, NOTCH along with the P13K/Akt/mTOR pathway exists in BCC, resistance to a particular pathway inhibitor is common. In this scenario, combination treatment with inhibitors of different signaling pathways has been proven beneficial. The PI3K inhibitor, buparlisib, in combination with the smoothened (SMO) inhibitor, erismodegib, in Hedgehog signaling pathway is currently under investigation for determining the efficacy of the treatment in BCC [166]. Furthermore, there is a considerable crosstalk among these pathways creating an intricate network of molecules which confers resistance to the drugs targeting a particular signaling [93]. GANT61 (inhibitor of GLI) in the Hedgehog pathway has been tried to establish an efficient mode of targeting BCC. However, the crosstalk between the PI3K and Hedgehog signaling has evolved the resistance against the inhibitors of Hedgehog signaling. Combination of PI3K/mTOR inhibitor, PI103 and GANT61, has been shown to synergistically overcome the resistance in a rhabdomyosarcoma model [166]. Another PI3K inhibitor buparlisib is under investigation to determine the effectiveness of the SMO inhibitor erismodegib in BCC [166]. Pharmacological doses of retinoids are reported to be effective in BCC carcinogenesis in human. The retinoid, tazarotene, has been shown to inhibit murine BCC through inhibition of IGFR/PI3K/Akt/mTOR signaling pathway [106,107].

Drug repurposing has led to the identification of novel anticancer drugs through the discovery of new pharmacological effect of the existing drugs with established therapeutic activity. Itraconazole was discovered as an antifungal agent in the 1980s, which in recent years has been discovered to have anticancer activity through inhibition of different signaling pathways such as Akt/mTOR, Hedgehog signaling, and Wnt/β-catenin signaling. Itraconazole has been found to inhibit the mTOR signaling by binding to the voltage sensitive anion channel in mitochondria and interfering with ATP production. Itraconazole showed a promising result in BCC treatment in recent clinical trials and is also being used in ongoing clinical trials for BCC [168]. Pharmacological doses of retinoids are reported to be effective in BCC carcinogenesis in human. The retinoic acid receptors (RAR) α, β, and γ in the endodermal layer of the skin bind the retinoic acid derived from endogenous conversion of retinoid to retinoic acid and retard BCC carcinogenesis. A typical example is tazarotene, a prodrug, when converted to tazarotenic acid, can potently bind and activate RAR β and γ receptors, thus exerting the anticancer effect [106,107].

6.4. *Targeting PI3K/Akt/mTOR for Treatment of Cutaneous Squamous Cell Carcinoma*

Cutaneous squamous cell carcinoma (cSCC) is the second most common form of non-melanoma skin cancer globally after BCC, accounting for almost 20% of NMSCs [169], and clinical presentation including but not limited to hyperkeratotic plaque, formation of nodular mass or ulceration on the skin, which may be associated with pain, pruritus, or bleeding [170]. Actinic Keratosis AK and Bowen's disease [171] are two premalignant forms of cSCC, which, if not treated well, result in malignant transformation. Although 95% of the cSCC are curable by surgical means, it is estimated that 20% of the skin cancer deaths are instigated by cSCC [172]. The cSCC can be metastasized to the surrounding dermis layer and to the local lymph nodes. Approximately 5% of the patients have been found to develop metastasis in the lymph nodes [173]. Exposure to ultraviolet radiation from the sunlight is the major cause of cSCC development. Other causes include occupational exposure to ionizing radiation and radiotherapy for the treatment of other skin conditions such as psoriasis. Immunosuppressive treatment followed by organ transplantation has also been found to be implicated with development of cSCC. Prolonged administration of immunosuppressant azathioprine was found to increase the risk of cSCC in patients after lung transplantation [174].

Overexpression of EGFR has been reported in cSCC, so targeting EGFR has been a promising therapeutic approach in treating this type of cancer. In a phase II clinical study EGFR tyrosine kinase inhibitor gefitinib showed a promising result in patients with aggressive cSCC of the head and neck [175]. Furthermore, cSCC has been reported to exhibit a higher level of mTOR activity compared to other non-melanoma skin cancers particularly BCC [176,177]. Although cSCC shows more aggressive behavior than BCC, interestingly it shows better response to mTOR inhibitors due to higher mTOR level in these cell types [178]. Elevated mTOR level was also observed in cSCC, compared to its premalignant forms AK and BD. Increased expression of cyclin-dependent kinase 2 (CDK2) was also observed in this type of cancer, suggesting its correlation with the Akt/mTOR signaling. This may act as another potential therapeutic target for treating cSCC in addition to targeting mTOR [179]. Since rapalogs (sirolimus, everolimus, etc.) have immunosuppressive effects and at the same time also possess antiproliferative activity, these mTOR inhibitors can be promising agents in treating patients with malignancies following organ transplantation. Rapalogs have been reported to be effective in posttransplant skin malignancies, especially in cSCCs [180]. In a recent study, a novel orally bioactive PI3K/mTOR dual inhibitor, LY3023414, has been found to inhibit human cSCC and is currently under phase I/II clinical trials for the treatment of patients with cSCC [112]. Einspahr et al. utilized reverse phase protein microarray analysis to study independent tissue sets of isolated and enriched epithelial cells squamous cell carcinoma (SCC) compared with actinic keratosis (AK) obtained by laser capture microdissection, and observed aberrant activation of MEK-ERK, EGFR, and mTOR pathways [181]. In another study, Chen et al. reported frequent constitutive activation of the Akt/mTOR pathway components in mostly malignant epidermal tumors, which also highly correlated with CDK2 expression, suggesting that this pathway induces the malignant transition through CDK2 in epidermal tumors [179].

The PI3K/Akt/mTOR signaling pathway has been implicated in the development of resistance to EGFR inhibitors in head and neck SCC. Therefore, everolimus (mTORC1 inhibitor) was combined with erlotinib (EGFR inhibitor) in a phase II clinical trial to treat head and neck SCC. Unfortunately, this combination treatment did not show any clinically significant benefits to the patients with metastatic cancer [182].

Erufosine (erucylphospho-*N*, *N*,*N*-trimethylpropylammonium), an alkylphosphocholine, showed promising anticancer effects in oral squamous cell carcinoma. The induction of apoptotic and autophagic cell death and the antiproliferative effect of erufosine resulted from its downregulatory effect on the mTOR signaling cascade [94].

Several small molecules have been developed to target mTOR signaling to treat cSCC. Some of these molecules have shown promising results in ex vivo experiments. GDC-0084, a novel small molecule showed potent inhibitory effect on both mTORC1 and mTORC2. It exhibited antiproliferative and cytotoxic effects on several established and primary cSCC cell lines. A clinical study in human subjects showed the good safety and tolerability as well as the complete shutdown of the PI3K-Akt-mTOR signaling cascade, presenting GDC-0084 as a potential therapeutic agent in treating cSCC [95]. Another PI3K/mTOR dual inhibitor, LY3023414, also showed potent cytotoxic activity in several cSCC cell lines as well as the in vivo tumor xenograft models. The high aqueous solubility and oral bioavailability positioned LY3023414 as a potential chemotherapeutic agent for treatment of cSCC. This small molecule is currently undergoing phase I and II clinical trials [112].

6.5. *Targeting PI3K/Akt/mTOR for Treatment of Merkel Cell Carcinoma*

Merkel cell carcinoma (MCC), first discovered by Toker in 1972, is a devastating nonmelanoma skin cancer of neuroendocrine origin, which contains neurosecretory granules with an increasing/grim prevalence of 1500 cases per year in the US alone. A five-year epidemiologic data revealed in Australia and other parts of the world that there is an annual increase of 8% in an age adapted incidence of MCC compared to only 3% corresponding increases in cutaneous melanoma. About 50% of MCC patients progressed to metastatic disease with an associated 46% mortality rate, which thus far exceeds that of melanoma [109]. In spite of this increase, approximately 11-fold greater cases of MCC have been reported in the patients with AIDS. The electron-dense neuroendocrine granule containing Merkel cells originate from epidermal stem cells [183]. MCC most commonly occurs in the sun-exposed areas of the body especially the head and neck, and the pathogenesis remains incompletely understood.

In 2008, Merkel cell polyomavirus (MCV) was discovered as a causative agent of MCC, suggesting that integration of the viral genetic material to the cell is responsible for the virus induced pathogenesis, and the molecular mechanism associated with the disease pathogenesis has been reported in diverse literatures, which involves p53, PTEN, Ras/MAPK, and PI3K/Akt. A recent report suggested the involvement of activated Akt/mTOR and its downstream effector molecules p-4E-BP1 (S65) and p-S6K in MCC cells [109], and a positive correlation between MCV specific T cell antigen and the translation initiation factor 4E-BP1 was testified to activate Akt/mTOR signaling in MCV positive tumor [184]. It was also shown that constitutive activation of 4E-BP1 that preserved from being phosphorylated upset the MCV specific T cell transformation activity, suggesting that 4E-BP1 inhibition (phosphorylation) is required for MCV transformation [109,184]. More studies have reported that Akt is hyper-phosphorylated in MCC regardless of the presence of MCV [109,184].

There are no real treatments for the MCC, even though the first-generation inhibitors of mTOR including rapamycin and other rapalogs are currently being investigated in different cancer types including some skin malignancies. Diminished efficacy of rapalogs due to feedback activation of Akt has instigated the development of the second-generation mTOR inhibitors that can target both mTORC1 and mTORC2. For instance, MLN0128, a compound that inhibits both mTORC1 and mTORC2, was able to potently suppress the growth of MCC both in vitro culture and in vivo in mouse xenograft models. MLN0128 is now at the verge of a dose escalation protocol in phase I clinical trial and undergoing a phase II clinical trial for the treatment of MCC [98,99].

WYE-354 is an allosteric mTOR inhibitor that has been reported to increase autophagy in primary human MCC cell line with over two-fold more efficacious than Ku-006394 in 24 h treatment regimen [109].

Another small molecular target of PI3K/Akt/mTOR pathway is NVP-BEZ235, which inhibits the activity of both PI3K and mTOR kinase by interacting with the ATP-binding site. It is an imidazoquinoline derivative with significant oral bioavailability. In different preclinical trials NVP-BEZ235 has been reported to be effective against osteosarcoma, glioblastoma, breast, prostate, and pancreatic cancer and is now in phase I clinical trial for solid tumors. This molecule has been found to inhibit proliferation and induce cell cycle arrest of MCC cells in culture. The mechanism of NVP-BEZ235 action against cancer cell proliferation was found to be associated with dual inhibition of PI3K/mTOR [101].

6.6. Targeting PI3K/Akt/mTOR for Treatment of Tuberous Sclerosis

Tuberous sclerosis complex (TSC) is an inconsistently expressed, mostly autosomal dominantly inherited neurocutaneous syndrome/disorder that affects the skin, brain, kidneys, eyes, and other organ systems known to affect persons of several races, including those from sub-Saharan Africa or of black, African, ancestry [185,186]. TSC exhibits a wide spectrum of clinical sequelae, pathologically characterized by benign, non invasive, tumor-like lesions or organs hamartomas). These typically occur in multiple organs including the central nervous system, kidneys, lungs, and skin (observed as hypomelanotic macules, confetti skin lesions, facial angiofibromas, shagreen patches, fibrous cephalic plaques, or ungual fibromas) [187–189]. At molecular levels, mutations of TSC genes modulate proliferation and differentiation of cells, causing the hamartomas, tumors, or altered neuronal polarity. It has been reported that most persons affected by TSC defects display distinctive phenotypes associated to attacks, including childhood spasms with autism related to variable intellectual disability [188,190]. It is commonly estimated that TSC is a rare condition with an average frequency of 1:6000 live births, and the prevalence ranges between 1:14,000 and 1:25,000 [189,191], and mutations in one of the two tumor suppressor genes, TSC1 (encoding hamartin) or TSC2 (encoding tuberin) occurs in more than 85% of TSC cases [192]. Under abnormal conditions, hamartin and tuberin are activated to avoid substrate usage through the biosynthesis courses facilitated by the mTORC1. Importantly, mutation of either TSC1 (on chromosome 9) or TSC2 (on chromosome 16) in patients with TSC leads to dysfunction of hamartin or tuberin. This causes the activation of mTOR signaling, leading to anomalies resulting from downstream kinase signaling cascade with deregulated cell cycle progression, transcription, translation, and metabolic control [193,194]. The stratification of mosaic TSC into subtypes corresponding to disease prognosis and severity, identified phenotypic distinctions between mosaic forms of TSC. They reported mosaic patients with; asymmetrically distributed facial angiofibromas, bilaterally symmetric facial angiofibromas, and germline TSC, involving both cutaneous and internal organs [186,195,196]. Some neonates are characterized by cardiac failure due to intracardiac rhabdomyomas, and often can show an age-dependent increase in the likelihood of developing renal angiomyolipomas. Central nervous system tumors are the principal basis of morbidity and mortality, while renal disease comes second as leading cause of early death due to TSC [189]. The pathophysiology and the clinical diagnostic and therapeutic management of TSC has recently been reviewed [191]. A high morbidity and mortality due to pulmonary involvement occurs predominantly in women. As explained above, both proteins form a functional complex that modulates the mTOR pathway, as such drugs that suppress mTOR are currently employed to treat TSC-related tumors and are being investigated as potential agents to alleviate other complications associated with TSC. Treatment of TSC patients possess severe burdens time wise and costly healthcare for both the patients, their families, as well as the healthcare system with no acceptable standard treatment approach that are clinically and economically effective [191]. A consensus was reached that a multidisciplinary strategy is required for ideal care of TSC patients [188]. Current treatment and care options are conservative including surgery, pharmacologic treatment with mTOR inhibitors, and recent proposals including biologics therapy (e.g., anti-EGFR antibody), as well as ultrasound guided percutaneous microwaves [191]. A recent review associated with meta-analysis including a total of 262 patients in 40 studies discussed the role of topical applications/indication of mTOR inhibitors and evaluation of their efficacy and safety in dermatologic conditions including TSC [197]. The study identified amongst 11 dermatological conditions that over 157 of the 262 patients frequently had mostly angiofibromas linked to TSC, and that topically applied mTOR inhibitors, such as sirolimus was more effective against angiofibromas than placebo and were well tolerated [197]. Importantly, everolimus (Afinitor) has been approved by FDA for treatment of certain types of brain and kidney tumors caused by TSC [198], and there is more hope in future to target this to clearing of cutaneous tumor-like lesion and systemic disease management.

7. Clinical Implications, Conclusions, and Future Prospects

The identification of disease molecular basis and targets for most cutaneous malignancies are well-understood and possesses diagnostic and therapeutic indications, as it will permit the development of novel, safer, cheaper, as well as effective delivery of chemotherapeutic, biologic, and natural dietary agents targeting the dysfunction towards maintaining skin tissue homeostasis and integrity. Given that the PI3K/Akt/mTOR signaling plays a crucial role in skin cancers, targeting this pathway with therapeutic nature-derived bioactive phytochemicals, biologic molecules, and synthetic small molecules alone or in various combinations is a promising strategy to treat skin cancers. So far, several of the aforementioned synthetic molecules have been widely employed as prescription agents in clinical trials. Most of these agents have known limitations as undesired side effects, and thus require the careful selection and development of more potent, cost effective, and safer remedies [199–201]. Natural phytochemicals and synthetic molecules have been extensively tested for their ability to remedy PI3K/Akt/mTOR-associated cutaneous disorders [202,203]. While the global market for nutraceuticals research was estimated to reach approximately $340 billion by 2024 (Variant Market Research, Pune, India), nothing is yet suggestive of the synthetics and their analogs. Recent technological innovations in in vitro and in vivo animal disease models coupled with high throughput drug screening will speed up anticancer drug discovery and development. Some inhibitors currently available are being tested in early-stage clinical trials, and their applications in the treatment of skin related malignancies warrant further testing.

8. Materials and Methods

A literature search was performed in PubMed using the following MESH terms ("melanoma" OR "skin" OR "cancer small molecule") AND ("mTOR, PI3K, Akt or mTOR, PI3K/Akt inhibitors" OR "phytochemical and skin cancer" OR "Merkel cell carcinoma and PI3K/Akt/mTOR") AND ("skin, cutaneous, cancer" OR "Biologics for mTOR" OR "cutaneous malignancies"). We then identified 307 reviews/original articles, observational studies and guidelines and selected relevant English-language publications based on their abstracts. Publications of all studies investigating the mechanism and the use of pathway inhibitors in dermatological malignancies were included, and the most references were targeted within the last 6 years except for very rare conditions or reports.

Funding: Research work related to this review topic is partially supported by a startup fund from the University of Louisiana at Monroe (ULM), College of Pharmacy (COP), a ULM COP Faculty Research Seed Grant (#5CALHN-260615), a Carson Research Scholar Award in Psoriasis from American Skin Association (ASA), a University of Wisconsin (UW)-Madison Skin Disease Research Center (SDRC) Pilot and Feasibility Research Award from NIH/NIAMS grant P30AR066524, and a Louisiana Biomedical Research Network (LBRN) Pilot Award from an IDeA NIH/NIGMS grant P20GM12345 to (JCC). RCNC is supported by Summer Research Internship Scholarship Award from POHOFI Inc. ALW is supported by a ULM Academic Innovation Center (AIC) Grant Award # R0018856.

Conflicts of Interest: The authors state no conflict of interest.

Abbreviations

PI3K	phosphatidyl-inositiol 3-kinase
BMZ	Bbaement membrane zone
mTOR	mammalian target of rapamycin
UV	ultraviolet
EGCG	Epigallocatechin-3-gallate
Akt	protein kinase B
NMSC	non-melanoma skin cancer
SCC	squamous cell carcinoma
BCC	basal cell carcinoma
MCC	Merkel cell carcinoma
K6	keratin 6
K14	keratin 14
EMT	epithelial-mesenchymal transition
CK15	cytokeratin 15
CK5	cytokeratin

References

1. Chamcheu, J.C.; Adhami, V.M.; Siddiqui, I.V.; Mukhtar, H. Cutaneous Cell-and Gene-Based Therapies for Inherited and Acquired Skin Disorders. In *Gene and Cell Therapy: Therapeutic Mechanisms and Strategies*, 4th ed.; Templeton, N.S., Ed.; CRC Press, Taylor and Francis Group: Boca Raton, FL, USA, 2015.
2. Gonzales, K.A.U.; Fuchs, E. Skin and Its Regenerative Powers: An Alliance between Stem Cells and Their Niche. *Dev. Cell* **2017**, *43*, 387–401. [CrossRef] [PubMed]
3. Magin, T.M.; Vijayaraj, P.; Leube, R.E. Structural and regulatory functions of keratins. *Exp. Cell Res.* **2007**, *313*, 2021–2032. [CrossRef] [PubMed]
4. Elias, P.M.; Eichenfield, L.F.; Fowler, J.F.; Horowitz, P.; McLeod, R.P. Update on the structure and function of the skin barrier: Atopic dermatitis as an exemplar of clinical implications. *Semin. Cutan. Med. Surg.* **2013**, *32*, S21–S24. [CrossRef] [PubMed]
5. Pullar, J.M.; Carr, A.C.; Vissers, M.C.M. The Roles of Vitamin C in Skin Health. *Nutrients* **2017**, *9*, 866. [CrossRef] [PubMed]
6. Segre, J.A. Epidermal barrier formation and recovery in skin disorders. *J. Clin. Investig.* **2006**, *116*, 1150–1158. [CrossRef]

7. Chamcheu, J.C.; Siddiqui, I.A.; Syed, D.N.; Adhami, V.M.; Liovic, M.; Mukhtar, H. Keratin gene mutations in disorders of human skin and its appendages. *Arch. Biochem. Biophys.* **2011**, *508*, 123–137. [CrossRef] [PubMed]
8. Elmets, C.A.; Katiyar, S. Green tea and skin cancer: Photoimmunology, angiogenesis and DNA repair. *J. Nutr. Biochem.* **2007**, *18*, 287–296.
9. Eckhart, L.; Lippens, S.; Tschachler, E.; Declercq, W. Cell death by cornification. *Biochim. Biophys. Acta* **2013**, *1833*, 3471–3480. [CrossRef]
10. Zhang, L.; Li, Y.; Yang, X.; Wei, J.; Zhou, S.; Zhao, Z.; Cheng, J.; Duan, H.; Jia, T.; Lei, Q.; et al. Characterization of Th17 and FoxP3^{+} Treg Cells in Paediatric Psoriasis Patients. *Scand. J. Immunol.* **2016**, *83*, 174–180. [CrossRef]
11. Ng, W.L.; Qi, J.T.Z.; Yeong, W.Y.; Naing, M.W. Proof-of-concept: 3D bioprinting of pigmented human skin constructs. *Biofabrication* **2018**, *10*, 25005. [CrossRef]
12. Tharmarajah, G.; Eckhard, U.; Jain, F.; Marino, G.; Prudova, A.; Urtatiz, O.; Fuchs, H.; De Angelis, M.H.; Overall, C.M.; Van Raamsdonk, C.D. Melanocyte development in the mouse tail epidermis requires the Adamts9 metalloproteinase. *Pigment. Cell Melanoma Res.* **2018**, *31*, 693–707. [CrossRef]
13. Serre, C.; Busuttil, V.; Botto, J.-M. Intrinsic and extrinsic regulation of human skin melanogenesis and pigmentation. *Int. J. Cosmet. Sci.* **2018**, *40*, 328–347. [CrossRef]
14. Bertolesi, G.E.; McFarlane, S. Seeing the light to change colour: An evolutionary perspective on the role of melanopsin in neuroendocrine circuits regulating light-mediated skin pigmentation. *Pigment. Cell Melanoma Res.* **2018**, *31*, 354–373. [CrossRef]
15. De Assis, L.V.M.; Moraes, M.N.; Magalhaes-Marques, K.K.; de Lauro Castrucci, A.M. Melanopsin and rhodopsin mediate UVA-induced immediate pigment darkening: Unravelling the photosensitive system of the skin. *Eur. J. Cell Biol.* **2018**, *97*, 150–162. [CrossRef]
16. Yamaguchi, Y.; Hearing, V.J. Physiological factors that regulate skin pigmentation. *BioFactors* **2009**, *35*, 193–199. [CrossRef]
17. Yamaguchi, Y.; Morita, A.; Maeda, A.; Hearing, V.J. Regulation of skin pigmentation and thickness by dickkopf 1 (DKK1). *J. Investig. Dermatol. Symp. Proc.* **2009**, *14*, 73–75. [CrossRef]
18. Perez-Sanchez, A.; Barrajon-Catalan, E.; Herranz-Lopez, M.; Micol, V. Nutraceuticals for Skin Care: A Comprehensive Review of Human Clinical Studies. *Nutrients* **2018**, *10*, 403. [CrossRef]
19. Sagi, Z.; Hieronymus, T. The Impact of the Epithelial–Mesenchymal Transition Regulator Hepatocyte Growth Factor Receptor/Met on Skin Immunity by Modulating Langerhans Cell Migration. *Front. Immunol.* **2018**, *9*, 517. [CrossRef]
20. Deckers, J.; Hammad, H.; Hoste, E. Langerhans Cells: Sensing the Environment in Health and Disease. *Front. Immunol.* **2018**, *9*, 93. [CrossRef]
21. Iwamoto, K.; Nümm, T.J.; Koch, S.; Herrmann, N.; Leib, N.; Bieber, T.; Stroisch, T.J. Langerhans and inflammatory dendritic epidermal cells in atopic dermatitis are tolerized toward TLR2 activation. *Allergy* **2018**, *73*, 2205–2213. [CrossRef]
22. Petersson, M.; Niemann, C. Stem cell dynamics and heterogeneity: Implications for epidermal regeneration and skin cancer. *Curr. Med. Chem.* **2012**, *19*, 5984–5992. [CrossRef]
23. Wang, L.-J.; Wang, Y.-L.; Yang, X. Progress in epidermal stem cells. *Yi Chuan* **2010**, *32*, 198–204. [CrossRef]
24. Flores, E.R.; Halder, G. Stem cell proliferation in the skin: Alpha-catenin takes over the hippo pathway. *Sci. Signal.* **2011**, *4*, pe34. [CrossRef]
25. Uzarska, M.; Porowińska, D.; Bajek, A.; Drewa, T. Epidermal stem cells—Biology and potential applications in regenerative medicine. *Postępy Biochem.* **2013**, *59*, 219–227.
26. Shen, Q.; Jin, H.; Wang, X. Epidermal Stem Cells and Their Epigenetic Regulation. *Int. J. Mol. Sci.* **2013**, *14*, 17861–17880. [CrossRef]
27. Lavker, R.M.; Sun, T.-T. Epidermal stem cells: Properties, markers, and location. *Proc. Natl. Acad. Sci. USA* **2000**, *97*, 13473–13475. [CrossRef]
28. Matsui, T.; Amagai, M. Dissecting the formation, structure and barrier function of the stratum corneum. *Int. Immunol.* **2015**, *27*, 269–280. [CrossRef]
29. Osawa, R.; Akiyama, M.; Shimizu, H. Filaggrin Gene Defects and the Risk of Developing Allergic Disorders. *Allergol. Int.* **2011**, *60*, 1–9. [CrossRef]

30. Sandilands, A.; Sutherland, C.; Irvine, A.D.; McLean, W.H.I. Filaggrin in the frontline: Role in skin barrier function and disease. *J. Cell Sci.* **2009**, *122*, 1285–1294. [CrossRef]
31. Koster, M.I. Making an epidermis. *Ann. N. Y. Acad. Sci.* **2009**, *1170*, 7–10. [CrossRef]
32. Labat-Robert, J. Information exchanges between cells and extracellular matrix. Influence of aging. *Biol. Aujourd'hui* **2012**, *206*, 103–109. [CrossRef]
33. Quan, T.; Fisher, G.J. Role of Age-Associated Alterations of the Dermal Extracellular Matrix Microenvironment in Human Skin Aging: A Mini-Review. *Gerontology* **2015**, *61*, 427–434. [CrossRef]
34. Watt, F.M. Mammalian skin cell biology: At the interface between laboratory and clinic. *Science* **2014**, *346*, 937–940. [CrossRef]
35. Gilchrest, B.A.; Geller, A.C.; Yaar, M.; Eller, M.S. The Pathogenesis of Melanoma Induced by Ultraviolet Radiation. *N. Engl. J. Med.* **1999**, *340*, 1341–1348. [CrossRef]
36. Watson, M.; Holman, D.M.; Maguire-Eisen, M. Ultraviolet Radiation Exposure and Its Impact on Skin Cancer Risk. *Semin. Oncol. Nurs.* **2016**, *32*, 241–254. [CrossRef]
37. Miyamura, Y.; Coelho, S.G.; Schlenz, K.; Batzer, J.; Smuda, C.; Choi, W.; Brenner, M.; Passeron, T.; Zhang, G.; Kolbe, L.; et al. The deceptive nature of UVA tanning versus the modest protective effects of UVB tanning on human skin. *Pigment Cell Melanoma Res.* **2011**, *24*, 136–147. [CrossRef]
38. D'Orazio, J.; Jarrett, S.; Amaro-Ortiz, A.; Scott, T. UV radiation and the skin. *Int. J. Mol. Sci.* **2013**, *14*, 12222–12248. [CrossRef]
39. McKenzie, R.L.; Aucamp, P.J.; Madronich, S.; Tourpali, K.; Bais, A.; Bernhard, G.; Ilyas, M. Ozone depletion and climate change: Impacts on UV radiation. *Photochem. Photobiol. Sci.* **2015**, *14*, 19–52. [CrossRef]
40. Holick, M.F. Sunlight, UV-radiation, vitamin D and skin cancer: How much sunlight do we need? *Adv. Exp. Med. Biol.* **2008**, *624*, 1–15.
41. Juzeniene, A.; Moan, J. Beneficial effects of UV radiation other than via vitamin D production. *Dermato-Endocrinology* **2012**, *4*, 109–117. [CrossRef]
42. Gupta, A.; Avci, P.; Dai, T.; Huang, Y.-Y.; Hamblin, M.R. Ultraviolet Radiation in Wound Care: Sterilization and Stimulation. *Adv. Wound Care* **2013**, *2*, 422–437. [CrossRef]
43. Slominski, A.; Wortsman, J. Neuroendocrinology of the skin. *Endocr. Rev.* **2000**, *21*, 457–487. [CrossRef]
44. Diffey, B.L. Solar ultraviolet radiation effects on biological systems. *Phys. Med. Boil.* **1991**, *36*, 299–328. [CrossRef]
45. Abeyama, K.; Eng, W.; Jester, J.V.; Vink, A.A.; Edelbaum, D.; Cockerell, C.J.; Bergstresser, P.R.; Takashima, A. A role for NF-kappaB-dependent gene transactivation in sunburn. *J. Clin. Investig.* **2000**, *105*, 1751–1759. [CrossRef]
46. Leo, M.S.; Sivamani, R.K. Phytochemical modulation of the Akt/mTOR pathway and its potential use in cutaneous disease. *Arch. Dermatol. Res.* **2014**, *306*, 861–871. [CrossRef]
47. Huang, S. Targeting mTOR signaling for cancer therapy. *Curr. Opin. Pharmacol.* **2003**, *3*, 371–377. [CrossRef]
48. Wullschleger, S.; Loewith, R.; Hall, M.N. TOR Signaling in Growth and Metabolism. *Cell* **2006**, *124*, 471–484. [CrossRef]
49. Guertin, D.A.; Sabatini, D.M. Defining the Role of mTOR in Cancer. *Cancer Cell* **2007**, *12*, 9–22. [CrossRef]
50. Laplante, M.; Sabatini, D.M. mTOR signaling in growth control and disease. *Cell* **2012**, *149*, 274–293. [CrossRef]
51. Vanhaesebroeck, B.; Stephens, L.; Hawkins, P. PI3K signalling: The path to discovery and understanding. *Nat. Rev. Mol. Cell Boil.* **2012**, *13*, 195–203. [CrossRef]
52. Pópulo, H.; Lopes, J.M.; Soares, P. The mTOR Signalling Pathway in Human Cancer. *Int. J. Mol. Sci.* **2012**, *13*, 1886–1918. [CrossRef]
53. Saxton, R.A.; Sabatini, D.M. mTOR Signaling in Growth, Metabolism, and Disease. *Cell* **2017**, *169*, 361–371. [CrossRef]
54. Sehgal, S.N.; Baker, H.; Vezina, C. Rapamycin (AY-22,989), a new antifungal antibiotic. II. Fermentation, isolation and characterization. *J. Antibiot.* **1975**, *28*, 727–732. [CrossRef]
55. Sabatini, D.M.; Erdjument-Bromage, H.; Lui, M.; Tempst, P.; Snyder, S.H. RAFT1: A mammalian protein that binds to FKBP12 in a rapamycin-dependent fashion and is homologous to yeast TORs. *Cell* **1994**, *78*, 35–43. [CrossRef]
56. Buerger, C. Epidermal mTORC1 Signaling Contributes to the Pathogenesis of Psoriasis and Could Serve as a Therapeutic Target. *Front. Immunol.* **2018**, *9*, 2786. [CrossRef]

57. Vahidnezhad, H.; Youssefian, L.; Uitto, J.; Youssefian, L. Molecular Genetics of the PI3K-AKT-mTOR Pathway in Genodermatoses: Diagnostic Implications and Treatment Opportunities. *J. Investig. Dermatol.* **2016**, *136*, 15–23. [CrossRef]
58. Huang, T.; Lin, X.; Meng, X.; Lin, M. Phosphoinositide-3 Kinase/Protein Kinase-B/Mammalian Target of Rapamycin Pathway in Psoriasis Pathogenesis. A Potential Therapeutic Target? *Acta Derm. Venereol.* **2014**, *94*, 371–379. [CrossRef]
59. Kezić, A.; Popovic, L.; Lalic, K. mTOR Inhibitor Therapy and Metabolic Consequences: Where Do We Stand? *Oxidative Med. Cell. Longev.* **2018**, *2018*, 1–8. [CrossRef]
60. Corti, F.; Nichetti, F.; Raimondi, A.; Niger, M.; Prinzi, N.; Torchio, M.; Tamborini, E.; Perrone, F.; Pruneri, G.; Di Bartolomeo, M.; et al. Targeting the PI3K/AKT/mTOR pathway in biliary tract cancers: A review of current evidences and future perspectives. *Cancer Treat. Rev.* **2019**, *72*, 45–55. [CrossRef]
61. Cragg, G.M.; Newman, D.J. Natural Products: A Continuing Source of Novel Drug Leads. *Biochim. Biophys. Acta* **2013**, *1830*, 3670–3695. [CrossRef]
62. Newman, D.J.; Cragg, G.M. Natural Products as Sources of New Drugs from 1981 to 2014. *J. Nat. Prod.* **2016**, *79*, 629–661. [CrossRef]
63. Strickland, L.R.; Pal, H.C.; Elmets, C.A.; Afaq, F. Targeting Drivers of Melanoma with Synthetic Small Molecules and Phytochemicals. *Cancer Lett.* **2015**, *359*, 20–35. [CrossRef]
64. Cronin, K.A.; Lake, A.J.; Scott, S.; Sherman, R.L.; Noone, A.-M.; Howlader, N.; Henley, S.J.; Anderson, R.N.; Firth, A.U.; Ma, J.; et al. Annual Report to the Nation on the Status of Cancer, part I: National cancer statistics. *Cancer* **2018**, *124*, 2785–2800. [CrossRef]
65. Liu, X.; Wu, J.; Qin, H.; Xu, J. The Role of Autophagy in the Resistance to BRAF Inhibition in BRAF-Mutated Melanoma. *Target. Oncol.* **2018**, *13*, 437–446. [CrossRef]
66. Wahid, M.; Jawed, A.; Mandal, R.K.; Dar, S.A.; Akhter, N.; Somvanshi, P.; Khan, F.; Lohani, M.; Areeshi, M.Y.; Haque, S. Recent developments and obstacles in the treatment of melanoma with BRAF and MEK inhibitors. *Crit. Rev. Oncol.* **2018**, *125*, 84–88. [CrossRef]
67. Shao, Z.; Bao, Q.; Jiang, F.; Qian, H.; Fang, Q.; Hu, X. VS-5584, a Novel PI3K-mTOR Dual Inhibitor, Inhibits Melanoma Cell Growth In Vitro and In Vivo. *PLoS ONE* **2015**, *10*, e0132655. [CrossRef]
68. Syed, D.N.; Chamcheu, J.-C.; Khan, M.I.; Sechi, M.; Lall, R.K.; Adhami, V.M.; Mukhtar, H. Fisetin inhibits human melanoma cell growth through direct binding to p70S6K and mTOR: Findings from 3-D melanoma skin equivalents and computational modeling. *Biochem. Pharmacol.* **2014**, *89*, 349–360. [CrossRef]
69. Bosbous, M.W.; Dzwierzynski, W.W.; Neuburg, M. Lentigo Maligna: Diagnosis and Treatment. *Clin. Plast. Surg.* **2010**, *37*, 35–46. [CrossRef]
70. Broussard, L.; Howland, A.; Ryu, S.; Song, K.; Norris, D.; Armstrong, C.A.; Song, P.I. Melanoma Cell Death Mechanisms. *Chonnam Med. J.* **2018**, *54*, 135–142. [CrossRef]
71. Hersey, P.; Bastholt, L.; Chiarion-Sileni, V.; Cinat, G.; Dummer, R.; Eggermont, A.M.M.; Espinosa, E.; Hauschild, A.; Quirt, I.; Robert, C.; et al. Small molecules and targeted therapies in distant metastatic disease. *Ann. Oncol.* **2009**, *20*, vi35–vi40. [CrossRef]
72. Yu, Y. A novel combination treatment against melanoma with NRAS mutation and therapy resistance. *EMBO Mol. Med.* **2018**, *10*, e8573. [CrossRef]
73. Caenepeel, S.; Cooke, K.; Wadsworth, S.; Huang, G.; Robert, L.; Moreno, B.H.; Parisi, G.; Cajulis, E.; Kendall, R.; Beltran, P.; et al. MAPK pathway inhibition induces MET and GAB1 levels, priming BRAF mutant melanoma for rescue by hepatocyte growth factor. *Oncotarget* **2017**, *8*, 17795–17809. [CrossRef]
74. Guo, B.; Zhang, Q.; Wang, H.; Chang, P.; Tao, K. KCNQ1OT1 promotes melanoma growth and metastasis. *Aging* **2018**, *10*, 632–644. [CrossRef]
75. Luan, W.; Li, L.; Shi, Y.; Bu, X.; Xia, Y.; Wang, J.; Djangmah, H.S.; Liu, X.; You, Y.; Xu, B. Long non-coding RNA MALAT1 acts as a competing endogenous RNA to promote malignant melanoma growth and metastasis by sponging miR-22. *Oncotarget* **2016**, *7*, 63901–63912. [CrossRef]
76. Luan, W.; Li, R.; Liu, L.; Ni, X.; Shi, Y.; Xia, Y.; Wang, J.; Lu, F.; Xu, B. Long non-coding RNA HOTAIR acts as a competing endogenous RNA to promote malignant melanoma progression by sponging miR-152-3p. *Oncotarget* **2017**, *8*, 85401–85414. [CrossRef]
77. Karbowniczek, M.; Spittle, C.S.; Morrison, T.; Wu, H.; Henske, E.P. mTOR Is Activated in the Majority of Malignant Melanomas. *J. Investig. Dermatol.* **2008**, *128*, 980–987. [CrossRef]

78. Molhoek, K.R.; Brautigan, D.L.; Slingluff, C.L. Synergistic inhibition of human melanoma proliferation by combination treatment with B-Raf inhibitor BAY43-9006 and mTOR inhibitor Rapamycin. *J. Transl. Med.* **2005**, *3*, 39. [CrossRef]
79. Vera Aguilera, J.; Rao, R.D.; Allred, J.B.; Suman, V.J.; Windschitl, H.E.; Kaur, J.S.; Maples, W.J.; Lowe, V.J.; Creagan, E.T.; Erickson, L.A.; et al. Phase II Study of Everolimus in Metastatic Malignant Melanoma (NCCTG-N0377, Alliance). *Oncologist* **2018**, *23*, 887-e94. [CrossRef]
80. Niessner, H.; Kosnopfel, C.; Sinnberg, T.; Beck, D.; Krieg, K.; Wanke, I.; Lasithiotakis, K.; Bonin, M.; Garbe, C.; Meier, F. Combined activity of temozolomide and the mTOR inhibitor temsirolimus in metastatic melanoma involves DKK1. *Exp. Dermatol.* **2017**, *26*, 598–606. [CrossRef]
81. Rangwala, R.; Chang, Y.C.; Hu, J.; Algazy, K.M.; Evans, T.L.; Fecher, L.A.; Schuchter, L.M.; Torigian, D.A.; Panosian, J.T.; Troxel, A.B.; et al. Combined MTOR and autophagy inhibition: Phase I trial of hydroxychloroquine and temsirolimus in patients with advanced solid tumors and melanoma. *Autophagy* **2014**, *10*, 1391–1402. [CrossRef]
82. Hainsworth, J.D.; Infante, J.R.; Spigel, D.R.; Peyton, J.D.; Thompson, D.S.; Lane, C.M.; Clark, B.L.; Rubin, M.S.; Trent, D.F.; Burris, H.A., III. Bevacizumab and everolimus in the treatment of patients with metastatic melanoma: A phase 2 trial of the Sarah Cannon Oncology Research Consortium. *Cancer* **2010**, *116*, 4122–4129. [CrossRef]
83. Kolev, V.N.; Wright, Q.G.; Vidal, C.M.; Ring, J.E.; Shapiro, I.M.; Ricono, J.; Weaver, D.T.; Padval, M.V.; Pachter, J.A.; Xu, Q. PI3K/mTOR dual inhibitor VS-5584 preferentially targets cancer stem cells. *Cancer Res.* **2015**, *75*, 446–455. [CrossRef]
84. Hutchinson, L. Skin cancer. Golden age of melanoma therapy. *Nat. Rev. Clin. Oncol.* **2015**, *12*, 1. [CrossRef]
85. Webster, R.M.; Mentzer, S.E. The malignant melanoma landscape. *Nat. Rev. Drug Discov. Engl.* **2014**, *13*, 491–492. [CrossRef]
86. Schadendorf, D.; Hauschild, A. Melanoma in 2013: Melanoma—The run of success continues. *Nat. Rev. Clin. Oncol.* **2014**, *11*, 75–76. [CrossRef]
87. Mi, C.; Ma, J.; Shi, H.; Li, J.; Wang, F.; Lee, J.J.; Jin, X. 4′,6-dihydroxy-4-methoxyisoaurone inhibits the HIF-1alpha pathway through inhibition of Akt/mTOR/p70S6K/4E-BP1 phosphorylation. *J. Pharmacol. Sci.* **2014**, *125*, 193–201. [CrossRef]
88. Wang, J.; Yang, Z.; Wen, J.; Ma, F.; Wang, F.; Yu, K.; Tang, M.; Wu, W.; Dong, Y.; Cheng, X.; et al. SKLB-M8 induces apoptosis through the AKT/mTOR signaling pathway in melanoma models and inhibits angiogenesis with decrease of ERK1/2 phosphorylation. *J. Pharmacol. Sci.* **2014**, *126*, 198–207. [CrossRef]
89. Oudart, J.-B.; Doue, M.; Vautrin, A.; Brassart, B.; Sellier, C.; Dupont-Deshorgue, A.; Monboisse, J.C.; Maquart, F.X.; Brassart-Pasco, S.; Ramont, L. The anti-tumor NC1 domain of collagen XIX inhibits the FAK/PI3K/Akt/mTOR signaling pathway through alphavbeta3 integrin interaction. *Oncotarget* **2016**, *7*, 1516–1528. [CrossRef]
90. Lee, B.Y.; Timpson, P.; Horvath, L.G.; Daly, R.J. FAK signaling in human cancer as a target for therapeutics. *Pharmacol. Ther.* **2015**, *146*, 132–149. [CrossRef]
91. Sulzmaier, F.J.; Jean, C.; Schlaepfer, D.D. FAK in cancer: Mechanistic findings and clinical applications. *Nat. Rev. Cancer* **2014**, *14*, 598–610. [CrossRef]
92. Si, L.; Xu, X.; Kong, Y.; Flaherty, K.T.; Chi, Z.; Cui, C.; Sheng, X.; Li, S.; Dai, J.; Yu, W.; et al. Major Response to Everolimus in Melanoma With Acquired Imatinib Resistance. *J. Clin. Oncol.* **2012**, *30*, e37–e40. [CrossRef]
93. Pierard-Franchimont, C.; Hermanns-Lê, T.; Paquet, P.; Herfs, M.; Delvenne, P.; Piérard, G.E. Hedgehog-and mTOR-targeted therapies for advanced basal cell carcinomas. *Future Oncol.* **2015**, *11*, 2997–3002. [CrossRef]
94. Kapoor, V.; Zaharieva, M.M.; Das, S.N.; Berger, M.R. Erufosine simultaneously induces apoptosis and autophagy by modulating the Akt–mTOR signaling pathway in oral squamous cell carcinoma. *Cancer Lett.* **2012**, *319*, 39–48. [CrossRef]
95. Ding, L.-T.; Zhao, P.; Yang, M.-L.; Lv, G.-Z.; Zhao, T.-L. GDC-0084 inhibits cutaneous squamous cell carcinoma cell growth. *Biochem. Biophys. Res. Commun.* **2018**, *503*, 1941–1948. [CrossRef]
96. Nguyen, N.; Sharma, A.; Nguyen, N.; Sharma, A.K.; Desai, D.; Huh, S.J.; Amin, S.; Meyers, C.; Robertson, G.P. Melanoma chemoprevention in skin reconstructs and mouse xenografts using isoselenocyanate-4. *Cancer Prev. Res.* **2011**, *4*, 248–258. [CrossRef]

97. Kannan, A.; Lin, Z.; Shao, Q.; Zhao, S.; Fang, B.; Moreno, M.A.; Vural, E.; Stack, B.C., Jr.; Suen, J.Y.; Kannan, K.; et al. Dual mTOR inhibitor MLN0128 suppresses Merkel cell carcinoma (MCC) xenograft tumor growth. *Oncotarget* **2016**, *7*, 6576–6592. [CrossRef]
98. Cassler, N.M.; Merrill, D.; Bichakjian, C.K.; Brownell, I. Merkel Cell Carcinoma Therapeutic Update. *Curr. Treat. Options Oncol.* **2016**, *17*, 36. [CrossRef]
99. Villani, A.; Fabbrocini, G.; Costa, C.; Annunziata, M.C.; Scalvenzi, M. Merkel Cell Carcinoma: Therapeutic Update and Emerging Therapies. *Dermatol. Ther.* **2019**, *9*, 209–222. [CrossRef]
100. Calero, R.; Morchon, E.; Martinez-Argudo, I.; Serrano, R. Synergistic anti-tumor effect of 17AAG with the PI3K/mTOR inhibitor NVP-BEZ235 on human melanoma. *Cancer Lett.* **2017**, *406*, 1–11. [CrossRef]
101. Lin, Z.; Mei, H.; Fan, J.; Yin, Z.; Wu, G. Effect of the dual phosphatidylinositol 3-kinase/mammalian target of rapamycin inhibitor NVP-BEZ235 against human Merkel cell carcinoma MKL-1 cells. *Oncol. Lett.* **2015**, *10*, 3663–3667. [CrossRef]
102. Chung, C.-Y.; Madhunapantula, S.V.; Desai, D.; Amin, S.; Robertson, G.P. Melanoma prevention using topical PBISe. *Cancer Prev. Res.* **2011**, *4*, 935–948. [CrossRef]
103. Hou, G.; Xue, L.; Lu, Z.; Fan, T.; Tian, F.; Xue, Y. An activated mTOR/p70S6K signaling pathway in esophageal squamous cell carcinoma cell lines and inhibition of the pathway by rapamycin and siRNA against mTOR. *Cancer Lett.* **2007**, *253*, 236–248. [CrossRef]
104. Werzowa, J.; Koehrer, S.; Strommer, S.; Cejka, D.; Fuereder, T.; Zebedin, E.; Wacheck, V. Vertical Inhibition of the mTORC1/mTORC2/PI3K Pathway Shows Synergistic Effects against Melanoma In Vitro and In Vivo. *J. Investig. Dermatol.* **2011**, *131*, 495–503. [CrossRef]
105. Chong, K.; Daud, A.; Ortiz-Urda, S.; Arron, S.T. Cutting Edge in Medical Management of Cutaneous Oncology. *Semin. Cutan. Med. Surg.* **2012**, *31*, 140–149. [CrossRef]
106. So, P.-L.; Wang, G.Y.; Wang, K.; Chuang, M.; Chiueh, V.C.; Kenny, P.A.; Epstein, E.H. PI3K-AKT signaling is a downstream effector of retinoid prevention of murine basal cell carcinogenesis. *Cancer Prev. Res.* **2014**, *7*, 407–417. [CrossRef]
107. Wu, C.-S.; Chen, G.-S.; Lin, P.-Y.; Pan, I.-H.; Wang, S.-T.; Lin, S.H.; Yu, H.-S.; Lin, C.-C. Tazarotene Induces Apoptosis in Human Basal Cell Carcinoma via Activation of Caspase-8/t-Bid and the Reactive Oxygen Species-Dependent Mitochondrial Pathway. *DNA Cell Boil.* **2014**, *33*, 652–666. [CrossRef]
108. Velho, T.R. Metastatic melanoma—A review of current and future drugs. *Drugs Context* **2012**, *2012*, 1–17. [CrossRef]
109. Lin, Z.; McDermott, A.; Shao, L.; Kannan, A.; Morgan, M.; Stack, B.C., Jr.; Moreno, M.; Davis, D.A.; Cornelius, L.A.; Gao, L. Chronic mTOR activation promotes cell survival in Merkel cell carcinoma. *Cancer Lett.* **2014**, *344*, 272–281. [CrossRef]
110. Liang, G.; Liu, M.; Wang, Q.; Shen, Y.; Mei, H.; Li, D.; Liu, W. Itraconazole exerts its anti-melanoma effect by suppressing Hedgehog, Wnt, and PI3K/mTOR signaling pathways. *Oncotarget* **2017**, *8*, 28510–28525. [CrossRef]
111. Head, S.A.; Shi, W.; Zhao, L.; Gorshkov, K.; Pasunooti, K.; Chen, Y.; Deng, Z.; Li, R.-J.; Shim, J.S.; Tan, W.; et al. Antifungal drug itraconazole targets VDAC1 to modulate the AMPK/mTOR signaling axis in endothelial cells. *Proc. Natl. Acad. Sci. USA* **2015**, *112*, E7276–E7285. [CrossRef]
112. Zou, Y.; Ge, M.; Wang, X. Targeting PI3K-AKT-mTOR by LY3023414 inhibits human skin squamous cell carcinoma cell growth in vitro and in vivo. *Biochem. Biophys. Res. Commun.* **2017**, *490*, 385–392. [CrossRef]
113. Sweetlove, M.; Wrightson, E.; Kolekar, S.; Rewcastle, G.W.; Baguley, B.C.; Shepherd, P.R.; Jamieson, S.M.F. Inhibitors of pan-PI3K Signaling Synergize with BRAF or MEK Inhibitors to Prevent BRAF-Mutant Melanoma Cell Growth. *Front. Oncol.* **2015**, *5*, 135. [CrossRef]
114. Polini, B.; Carpi, S.; Romanini, A.; Breschi, M.C.; Nieri, P.; Podestà, A. Circulating cell-free microRNAs in cutaneous melanoma staging and recurrence or survival prognosis. *Pigment. Cell Melanoma Res.* **2019**, *32*, 486–499. [CrossRef]
115. Li, J.; Liu, X.; Li, C.; Wang, W. miR-224-5p inhibits proliferation, migration, and invasion by targeting PIK3R3/AKT3 in uveal melanoma. *J. Cell. Biochem.* **2019**, *120*, 12412–12421. [CrossRef]
116. Jiang, Q.-Q.; Liu, W.-B. miR-25 Promotes Melanoma Progression by regulating RNA binding motif protein 47. *Med. Sci.* **2018**, *34*, 59–65. [CrossRef]

117. Meng, F.; Zhang, Y.; Li, X.; Wang, J. Clinical significance of miR-138 in patients with malignant melanoma through targeting of PDK1 in the PI3K/AKT autophagy signaling pathway. *Oncol. Rep.* **2017**, *38*, 1655–1662. [CrossRef]
118. Micevic, G.; Muthusamy, V.; Damsky, W.; Theodosakis, N.; Liu, X.; Meeth, K.; Wingrove, E.; Santhanakrishnan, M.; Bosenberg, M. DNMT3b modulates melanoma growth by controlling levels of mTORC2 component RICTOR. *Cell Rep.* **2016**, *14*, 2180–2192. [CrossRef]
119. Schmidt, K.M.; Dietrich, P.; Hackl, C.; Guenzle, J.; Bronsert, P.; Wagner, C.; Fichtner-Feigl, S.; Schlitt, H.J.; Geissler, E.K.; Hellerbrand, C.; et al. Inhibition of mTORC2/RICTOR Impairs Melanoma Hepatic Metastasis12. *Neoplasia* **2018**, *20*, 1198–1208. [CrossRef]
120. Damsky, W.; Micevic, G.; Meeth, K.; Muthusamy, V.; Curley, D.P.; Santhankrishnan, M.; Erdélyi, I.; Platt, J.T.; Huang, L.; Theodosakis, N.; et al. mTORC1 activation blocks BrafV600E-induced growth-arrest, but is insufficient for melanoma formation. *Cancer Cell* **2015**, *27*, 41–56. [CrossRef]
121. Ciołczyk-Wierzbicka, D.; Gil, D.; Laidler, P. Treatment of melanoma with selected inhibitors of signaling kinases effectively reduces proliferation and induces expression of cell cycle inhibitors. *Med. Oncol.* **2017**, *35*, 7. [CrossRef]
122. Sikora, A.G.; Gelbard, A.; Davies, M.A.; Sano, D.; Ekmekcioglu, S.; Kwon, J.; Hailemichael, Y.; Jayaraman, P.; Myers, J.N.; Grimm, E.A.; et al. Targeted inhibition of inducible nitric oxide synthase inhibits growth of human melanoma in vivo and synergizes with chemotherapy. *Clin. Cancer Res.* **2010**, *16*, 1834–1844. [CrossRef]
123. Ernst, D.S.; Eisenhauer, E.; Wainman, N.; Davis, M.; Lohmann, R.; Baetz, T.; Bélanger, K.; Smylie, M. Phase II Study of Perifosine in Previously Untreated Patients with Metastatic Melanoma. *Investig. New Drugs* **2005**, *23*, 569–575. [CrossRef]
124. Jung, S.K.; Kim, J.E.; Lee, S.-Y.; Lee, M.H.; Byun, S.; Kim, Y.A.; Lim, T.G.; Reddy, K.; Huang, Z.; Bode, A.M.; et al. The P110 subunit of PI3-K is a therapeutic target of acacetin in skin cancer. *Carcinogenesis* **2014**, *35*, 123–130. [CrossRef]
125. Chien, S.-T.; Lin, S.-S.; Wang, C.-K.; Lee, Y.-B.; Chen, K.-S.; Fong, Y.; Shih, Y.W. Acacetin inhibits the invasion and migration of human non-small cell lung cancer A549 cells by suppressing the p38alpha MAPK signaling pathway. *Mol. Cell. Biochem.* **2011**, *350*, 135–148. [CrossRef]
126. Shin, D.-H.; Kim, O.-H.; Jun, H.-S.; Kang, M.-K. Inhibitory effect of capsaicin on B16-F10 melanoma cell migration via the phosphatidylinositol 3-kinase/Akt/Rac1 signal pathway. *Exp. Mol. Med.* **2008**, *40*, 486–494. [CrossRef]
127. Wang, C.; Li, S.; Wang, M. Evodiamine-induced human melanoma A375-S2 cell death was mediated by PI3K/Akt/caspase and Fas-L/NF-kappaB signaling pathways and augmented by ubiquitin-proteasome inhibition. *Toxicol. In Vitro* **2010**, *24*, 898–904. [CrossRef]
128. Chen, X.-Y.; Li, D.-F.; Han, J.-C.; Wang, B.; Dong, Z.-P.; Yu, L.-N.; Pan, Z.H.; Qu, C.J.; Chen, Y.; Sun, S.G.; et al. Reprogramming induced by isoliquiritigenin diminishes melanoma cachexia through mTORC2-AKT-GSK3beta signaling. *Oncotarget* **2017**, *8*, 34565–34575.
129. Lim, H.N.; Baek, S.B.; Jung, H.J. Bee Venom and Its Peptide Component Melittin Suppress Growth and Migration of Melanoma Cells via Inhibition of PI3K/AKT/mTOR and MAPK Pathways. *Molecules* **2019**, *24*, 929. [CrossRef]
130. Lai, S.-L.; Mustafa, M.R.; Wong, P.-F.; Mr, M. Panduratin A induces protective autophagy in melanoma via the AMPK and mTOR pathway. *Phytomedicine* **2018**, *42*, 144–151. [CrossRef]
131. Zou, N.; Wei, Y.; Li, F.; Yang, Y.; Cheng, X.; Wang, C. The inhibitory effects of compound Muniziqi granule against B16 cells and harmine induced autophagy and apoptosis by inhibiting Akt/mTOR pathway. *BMC Complement. Altern. Med.* **2017**, *17*, 517. [CrossRef]
132. Sun, Z.; Zheng, L.; Liu, X.; Xing, W.; Liu, X. Sinomenine inhibits the growth of melanoma by enhancement of autophagy via PI3K/AKT/mTOR inhibition. *Drug Des. Dev. Ther.* **2018**, *12*, 2413–2421. [CrossRef]
133. Espona-Fiedler, M.; Soto-Cerrato, V.; Hosseini, A.; Lizcano, J.M.; Guallar, V.; Quesada, R.; Gao, T.; Pérez-Tomás, R. Identification of dual mTORC1 and mTORC2 inhibitors in melanoma cells: Prodigiosin vs. obatoclax. *Biochem. Pharmacol.* **2012**, *83*, 489–496. [CrossRef]
134. Liu-Smith, F.; Meyskens, F. Molecular mechanisms of flavonoids in melanin synthesis and the potential for the prevention and treatment of melanoma. *Mol. Nutr. Food Res.* **2016**, *60*, 1264–1274. [CrossRef]

135. Pal, H.C.; Hunt, K.M.; Diamond, A.; Elmets, C.A.; Afaq, F. Phytochemicals for the Management of Melanoma. *Mini-Rev. Med. Chem.* **2016**, *16*, 953–979.
136. Adhami, V.M.; Syed, D.N.; Khan, N.; Mukhtar, H. Dietary flavonoid fisetin: A novel dual inhibitor of PI3K/Akt and mTOR for prostate cancer management. *Biochem. Pharmacol.* **2012**, *84*, 1277–1281. [CrossRef]
137. Syed, D.N.; Mukhtar, H. Botanicals for the prevention and treatment of cutaneous melanoma. *Pigment. Cell Melanoma Res.* **2011**, *24*, 688–702. [CrossRef]
138. Govindarajan, B.; Sligh, J.E.; Vincent, B.J.; Li, M.; Canter, J.A.; Nickoloff, B.J.; Rodenburg, R.J.; Smeitink, J.A.; Oberley, L.; Zhang, Y.; et al. Overexpression of Akt converts radial growth melanoma to vertical growth melanoma. *J. Clin. Investig.* **2007**, *117*, 719–729. [CrossRef]
139. Pearce, L.R.; Alton, G.R.; Richter, D.T.; Kath, J.C.; Lingardo, L.; Chapman, J.; Hwang, C.; Alessi, D.R. Characterization of PF-4708671, a novel and highly specific inhibitor of p70 ribosomal S6 kinase (S6K1). *Biochem. J.* **2010**, *431*, 245–255. [CrossRef]
140. Sechi, M.; Lall, R.K.; Afolabi, S.O.; Singh, A.; Joshi, D.C.; Chiu, S.-Y.; Mukhtar, H.; Syed, D.N. Fisetin targets YB-1/RSK axis independent of its effect on ERK signaling: Insights from in vitro and in vivo melanoma models. *Sci. Rep.* **2018**, *8*, 15726. [CrossRef]
141. Pal, H.C.; Baxter, R.D.; Hunt, K.M.; Agarwal, J.; Elmets, C.A.; Athar, M.; Afaq, F. Fisetin, a phytochemical, potentiates sorafenib-induced apoptosis and abrogates tumor growth in athymic nude mice implanted with BRAF-mutated melanoma cells. *Oncotarget* **2015**, *6*, 28296–28311. [CrossRef]
142. Zhao, G.; Han, X.; Zheng, S.; Li, Z.; Sha, Y.; Ni, J.; Sun, Z.; Qiao, S.; Song, Z. Curcumin induces autophagy, inhibits proliferation and invasion by downregulating AKT/mTOR signaling pathway in human melanoma cells. *Oncol. Rep.* **2016**, *35*, 1065–1074. [CrossRef]
143. Rozzo, C.; Fanciulli, M.; Fraumene, C.; Corrias, A.; Cubeddu, T.; Sassu, I.; Cossu, S.; Nieddu, V.; Galleri, G.; Azara, E.; et al. Molecular changes induced by the curcumin analogue D6 in human melanoma cells. *Mol. Cancer* **2013**, *12*, 37. [CrossRef]
144. Wang, M.; Yu, T.; Zhu, C.; Sun, H.; Qiu, Y.; Zhu, X.; Li, J. Resveratrol Triggers Protective Autophagy Through the Ceramide/Akt/mTOR Pathway in Melanoma B16 Cells. *Nutr. Cancer* **2014**, *66*, 435–440. [CrossRef]
145. Bhattacharya, S.; Darjatmoko, S.R.; Polans, A.S. Resveratrol modulates the malignant properties of cutaneous melanoma through changes in the activation and attenuation of the antiapoptotic protooncogenic protein Akt/PKB. *Melanoma Res.* **2011**, *21*, 180–187. [CrossRef]
146. Aggarwal, B.B.; Bhardwaj, A.; Aggarwal, R.S.; Seeram, N.P.; Shishodia, S.; Takada, Y. Role of resveratrol in prevention and therapy of cancer: Preclinical and clinical studies. *Anticancer Res.* **2004**, *24*, 2783–2840.
147. Kaushik, G.; Ramalingam, S.; Subramaniam, D.; Rangarajan, P.; Protti, P.; Rammamoorthy, P.; Anant, S.; Mammen, J.M.V. Honokiol induces cytotoxic and cytostatic effects in malignant melanoma cancer cells. *Am. J. Surg.* **2012**, *204*, 868–873. [CrossRef]
148. Prieto, J.M.; Alqathama, A. Natural products with therapeutic potential in melanoma metastasis. *Nat. Prod. Rep.* **2015**, *32*, 1170–1182.
149. Hambright, H.G.; Meng, P.; Kumar, A.P.; Ghosh, R. Inhibition of PI3K/AKT/mTOR axis disrupts oxidative stress-mediated survival of melanoma cells. *Oncotarget* **2015**, *6*, 7195–7208. [CrossRef]
150. Gong, J.; Munoz, A.R.; Chan, D.; Ghosh, R.; Kumar, A.P. STAT3 down regulates LC3 to inhibit autophagy and pancreatic cancer cell growth. *Oncotarget* **2014**, *5*, 2529–2541. [CrossRef]
151. Huang, K.-J.; Kuo, C.-H.; Chen, S.-H.; Lin, C.-Y.; Lee, Y.-R. Honokiol inhibits in vitro and in vivo growth of oral squamous cell carcinoma through induction of apoptosis, cell cycle arrest and autophagy. *J. Cell. Mol. Med.* **2018**, *22*, 1894–1908. [CrossRef]
152. Surdu, S. Non-melanoma skin cancer: Occupational risk from UV light and arsenic exposure. *Rev. Environ. Health* **2014**, *29*, 255–264. [CrossRef]
153. Burton, K.A.; Ashack, K.A.; Khachemoune, A. Cutaneous Squamous Cell Carcinoma: A Review of High-Risk and Metastatic Disease. *Am. J. Clin. Dermatol.* **2016**, *17*, 491–508. [CrossRef]
154. Cohen, B.J.; Cohen, E.S.; Cohen, P.R. Basal Cell Carcinoma: A Patient and Physician's Experience. *Dermatol. Ther.* **2018**, *8*, 329–337. [CrossRef]
155. Wong, C.S.M.; Strange, R.C.; Lear, J.T. Basal cell carcinoma. *BMJ* **2003**, *327*, 794–798. [CrossRef]
156. Marzuka, A.G.; Book, S.E. Basal Cell Carcinoma: Pathogenesis, Epidemiology, Clinical Features, Diagnosis, Histopathology, and Management. *Yale J. Boil. Med.* **2015**, *88*, 167–179.

157. Rogers, H.W.; Weinstock, M.A.; Feldman, S.R.; Coldiron, B.M. Incidence Estimate of Nonmelanoma Skin Cancer (Keratinocyte Carcinomas) in the U.S. Population, 2012. *JAMA Dermatol.* **2015**, *151*, 1081–1086. [CrossRef]
158. Eisemann, N.; Waldmann, A.; Geller, A.C.; Weinstock, M.A.; Volkmer, B.; Greinert, R.; Breitbart, E.W.; Katalinic, A. Non-Melanoma Skin Cancer Incidence and Impact of Skin Cancer Screening on Incidence. *J. Investig. Dermatol.* **2014**, *134*, 43–50. [CrossRef]
159. Guy, G.P.; Thomas, C.C.; Thompson, T.; Watson, M.; Massetti, G.M.; Richardson, L.C. Vital Signs: Melanoma Incidence and Mortality Trends and Projections—United States, 1982–2030. *MMWR. Morb. Mortal. Wkly. Rep.* **2015**, *64*, 591–596.
160. Rigel, D.S. Cutaneous ultraviolet exposure and its relationship to the development of skin cancer. *J. Am. Acad. Dermatol.* **2008**, *58*, S129–S132. [CrossRef]
161. Tarallo, M.; Cigna, E.; Frati, R.; Delfino, S.; Innocenzi, D.; Fama, U.; Corbianco, A.; Scuderi, N. Metatypical basal cell carcinoma: A clinical review. *J. Exp. Clin. Cancer Res.* **2008**, *27*, 65. [CrossRef]
162. Athar, M.; Li, C.; Kim, A.L.; Spiegelman, V.S.; Bickers, D.R. Sonic Hedgehog Signaling in Basal Cell Nevus Syndrome. *Cancer Res.* **2014**, *74*, 4967–4975. [CrossRef]
163. Noubissi, F.K.; Kim, T.; Kawahara, T.N.; Aughenbaugh, W.D.; Berg, E.; Longley, B.J.; Athar, M.; Spiegelman, V.S. Role of CRD-BP in the growth of human Basal Cell Carcinoma Cells. *J. Investig. Dermatol.* **2014**, *134*, 1718–1724. [CrossRef]
164. Chamcheu, J.C.; Rady, I.; Chamcheu, R.-C.N.; Siddique, A.B.; Bloch, M.B.; Mbeumi, S.B.; Babatunde, A.S.; Uddin, M.B.; Noubissi, F.K.; Jurutka, P.W.; et al. Graviola (*Annona muricata*) Exerts Anti-Proliferative, Anti-Clonogenic and Pro-Apoptotic Effects in Human Non-Melanoma Skin Cancer UW-BCC1 and A431 Cells In Vitro: Involvement of Hedgehog Signaling. *Int. J. Mol. Sci.* **2018**, *19*, 1791. [CrossRef]
165. Kasper, M.; Jaks, V.; Hohl, D.; Toftgård, R. Basal cell carcinoma—Molecular biology and potential new therapies. *J. Clin. Investig.* **2012**, *122*, 455–463. [CrossRef]
166. Bakshi, A.; Chaudhary, S.C.; Rana, M.; Elmets, C.A.; Athar, M. Basal cell carcinoma pathogenesis and therapy involving hedgehog signaling and beyond. *Mol. Carcinog.* **2017**, *56*, 2543–2557. [CrossRef]
167. Alter, M.; Hillen, U.; Leiter, U.; Sachse, M.; Gutzmer, R. Current diagnosis and treatment of basal cell carcinoma. *J. Dtsch. Dermatol. Ges.* **2015**, *13*, 863–875. [CrossRef]
168. Tsubamoto, H.; Ueda, T.; Inoue, K.; Sakata, K.; Shibahara, H.; Sonoda, T. Repurposing itraconazole as an anticancer agent. *Oncol. Lett.* **2017**, *14*, 1240–1246. [CrossRef]
169. Alam, M.; Ratner, D. Cutaneous squamous-cell carcinoma. *N. Engl. J. Med.* **2001**, *344*, 975–983. [CrossRef]
170. Lansbury, L.; Bath-Hextall, F.; Perkins, W.; Stanton, W.; Leonardi-Bee, J. Interventions for non-metastatic squamous cell carcinoma of the skin: Systematic review and pooled analysis of observational studies. *BMJ* **2013**, *347*, f6153. [CrossRef]
171. Agamia, N.F.; Abdallah, D.M.; Sorour, O.; Mourad, B.; Younan, D.N. Skin expression of mammalian target of rapamycin and forkhead box transcription factor O1, and serum insulin-like growth factor-1 in patients with acne vulgaris and their relationship with diet. *Br. J. Dermatol.* **2016**, *174*, 1299–1307. [CrossRef]
172. Gurney, B.; Newlands, C. Management of regional metastatic disease in head and neck cutaneous malignancy.1. Cutaneous squamous cell carcinoma. *Br. J. Oral Maxillofac. Surg.* **2014**, *52*, 294–300. [CrossRef]
173. Azimi, A.; Kaufman, K.L.; Ali, M.; Arthur, J.; Kossard, S.; Fernandez-Penas, P. Differential proteomic analysis of actinic keratosis, Bowen's disease and cutaneous squamous cell carcinoma by label-free LC–MS/MS. *J. Dermatol. Sci.* **2018**, *91*, 69–78. [CrossRef]
174. Voß, M.; Plasmeijer, E.I.; Van Bemmel, B.C.; Van Der Bij, W.; Klaver, N.S.; Erasmus, M.E.; De Bock, G.H.; Verschuuren, E.A.; Rácz, E. Azathioprine to mycophenolate mofetil transition and risk of squamous cell carcinoma after lung transplantation. *J. Hear. Lung Transplant.* **2018**, *37*, 853–859. [CrossRef]
175. Lewis, C.M.; Glisson, B.S.; Feng, L.; Wan, F.; Tang, X.; Wistuba, I.I.; El-Naggar, A.K.; Rosenthal, D.I.; Chambers, M.S.; Lustig, R.A.; et al. A Phase II Study of Gefitinib for Aggressive Cutaneous Squamous Cell Carcinoma of the Head and Neck. *Clin. Cancer Res.* **2012**, *18*, 1435–1446. [CrossRef]
176. Karayannopoulou, G.; Euvrard, S.; Kanitakis, J. Differential expression of p-mTOR in cutaneous basal and squamous cell carcinomas likely explains their different response to mTOR inhibitors in organ-transplant recipients. *Anticancer Res.* **2013**, *33*, 3711–3714.

177. Wu, N.; Du, Z.; Zhu, Y.; Song, Y.; Pang, L.; Chen, Z. The Expression and Prognostic Impact of the PI3K/AKT/mTOR Signaling Pathway in Advanced Esophageal Squamous Cell Carcinoma. *Technol. Cancer Res. Treat.* **2018**, *17*, 1533033818758772. [CrossRef]
178. Euvrard, S.; Claudy, A.; Kanitakis, J. Skin Cancers after Organ Transplantation. *N. Engl. J. Med.* **2003**, *348*, 1681–1691. [CrossRef]
179. Chen, S.-J.; Nakahara, T.; Takahara, M.; Kido, M.; Dugu, L.; Uchi, H.; Takeuchi, S.; Tu, Y.-T.; Moroi, Y.; Furue, M. Activation of the mammalian target of rapamycin signalling pathway in epidermal tumours and its correlation with cyclin-dependent kinase 2. *Br. J. Dermatol.* **2009**, *160*, 442–445. [CrossRef]
180. Monaco, A.P. The Role of mTOR Inhibitors in the Management of Posttransplant Malignancy. *Transplant.* **2009**, *87*, 157–163. [CrossRef]
181. Einspahr, J.G.; Calvert, V.; Alberts, D.S.; Curiel-Lewandrowski, C.; Warneke, J.; Krouse, R.; Stratton, S.P.; Liotta, L.; Longo, C.; Pellicani, G.; et al. Functional protein pathway activation mapping of the progression of normal skin to squamous cell carcinoma. *Cancer Prev. Res.* **2012**, *5*, 403–413. [CrossRef]
182. Massarelli, E.; Lin, H.; Ginsberg, L.E.; Tran, H.T.; Lee, J.J.; Canales, J.R.; Williams, M.D.; Blumenschein, G.R.; Lu, C.; Heymach, J.V.; et al. Phase II trial of everolimus and erlotinib in patients with platinum-resistant recurrent and/or metastatic head and neck squamous cell carcinoma. *Ann. Oncol.* **2015**, *26*, 1476–1480. [CrossRef]
183. Schrama, D.; Ugurel, S.; Becker, J.C. Merkel cell carcinoma: Recent insights and new treatment options. *Curr. Opin. Oncol.* **2012**, *24*, 141–149. [CrossRef]
184. Shuda, M.; Kwun, H.J.; Feng, H.; Chang, Y.; Moore, P.S. Human Merkel cell polyomavirus small T antigen is an oncoprotein targeting the 4E-BP1 translation regulator. *J. Clin. Investig.* **2011**, *121*, 3623–3634. [CrossRef]
185. Fahmy, M.D.; Gupta, A.; Padilla, R.J.; Segura, A.; Brookes, C.D. Desmoplastic fibroma associated with tuberous sclerosis: Case report and literature review. *Oral Surg. Oral Med. Oral Pathol. Oral Radiol.* **2019**, *128*, e92–e99. [CrossRef]
186. Ekure, E.N.; Addissie, Y.A.; Sokunbi, O.J.; Kruszka, P.; Muenke, M.; Adeyemo, A.A. Tuberous sclerosis in a patient from Nigeria. *Am. J. Med. Genet. Part A* **2019**, *179*, 1423–1425. [CrossRef]
187. Curatolo, P.; Bombardieri, R.; Jozwiak, S. Tuberous sclerosis. *Lancet* **2008**, *372*, 657–668. [CrossRef]
188. Islam, M.P.; Roach, E.S. Tuberous sclerosis complex. In *Handbook of Clinical Neurology*; Oxford University Press: Oxford, UK, 2015; Volume 132, pp. 97–109.
189. Northrup, H.; Koenig, M.K.; Pearson, D.A.; Au, K.S. *Tuberous Sclerosis Complex*; Adam, M.P., Ardinger, H.H., Pagon, R.A., Wallace, S.E., Bean, L.J.H., Stephens, K., Eds.; Oxford University Press: Seattle, WA, USA, 1993.
190. Wei, C.-C.; Sheu, J.-N.; Liu, J.-T.; Yang, S.-H.; Chou, I.-C.; Tsai, J.-D. Trend of seizure remission in patients with tuberous sclerosis complex: A retrospective medical review. *J. Chin. Med. Assoc.* **2018**, *81*, 724–728. [CrossRef]
191. Volpi, A.; Sala, G.; Lesma, E.; Labriola, F.; Righetti, M.; Alfano, R.M.; Cozzolino, M. Tuberous sclerosis complex: New insights into clinical and therapeutic approach. *J. Nephrol.* **2019**, *32*, 355–363. [CrossRef]
192. Crino, P.B.; Nathanson, K.L.; Henske, E.P. The tuberous sclerosis complex. *N. Engl. J. Med.* **2006**, *355*, 1345–1356. [CrossRef]
193. Curatolo, P.; Moavero, R. mTOR Inhibitors in Tuberous Sclerosis Complex. *Curr. Neuropharmacol.* **2012**, *10*, 404–415. [CrossRef]
194. Adil, A.; Singh, A.K. *Neurofibromatosis Type 1 (Von Recklinghausen)*; StatPearls Publishing: Treasure Island, FL, USA, 2019.
195. Treichel, A.M.; Hamieh, L.; Nathan, N.R.; Tyburczy, M.E.; Wang, J.-A.; Oyerinde, O.; Raiciulescu, S.; Julien-Williams, P.; Jones, A.M.; Gopalakrishnan, V.; et al. Phenotypic distinctions between mosaic forms of tuberous sclerosis complex. *Genet. Med.* **2019**, 1. [CrossRef]
196. Giannikou, K.; Lasseter, K.D.; Grevelink, J.M.; Tyburczy, M.E.; Dies, K.A.; Zhu, Z.; Hamieh, L.; Wollison, B.M.; Thorner, A.R.; Ruoss, S.J.; et al. Low-level mosaicism in tuberous sclerosis complex: Prevalence, clinical features, and risk of disease transmission. *Genet. Med.* **2019**, 1. [CrossRef]
197. Leducq, S.; Giraudeau, B.; Tavernier, E.; Maruani, A. Topical use of mammalian target of rapamycin inhibitors in dermatology: A systematic review with meta-analysis. *J. Am. Acad. Dermatol.* **2019**, *80*, 735–742. [CrossRef]
198. Combes, F.P.; Baneyx, G.; Coello, N.; Zhu, P.; Sallas, W.; Yin, H.; Nedelman, J. Population pharmacokinetics–pharmacodynamics of oral everolimus in patients with seizures associated with tuberous sclerosis complex. *J. Pharmacokinet. Pharmacodyn.* **2018**, *45*, 707–719. [CrossRef]

199. Weidinger, S.; Novak, N. Atopic dermatitis. *Lancet* **2016**, *387*, 1109–1122. [CrossRef]
200. Boehncke, W.-H.; Schon, M.P. Psoriasis. *Lancet* **2015**, *386*, 983–994. [CrossRef]
201. Desmet, S.J.; De Bosscher, K. Glucocorticoid receptors: Finding the middle ground. *J. Clin. Investig.* **2017**, *127*, 1136–1145. [CrossRef]
202. Das, L.; Bhaumik, E.; Raychaudhuri, U.; Chakraborty, R. Role of nutraceuticals in human health. *J. Food Sci. Technol.* **2012**, *49*, 173–183. [CrossRef]
203. Proksch, E.; Schunck, M.; Zague, V.; Segger, D.; Degwert, J.; Oesser, S. Oral intake of specific bioactive collagen peptides reduces skin wrinkles and increases dermal matrix synthesis. *Skin Pharmacol. Physiol.* **2014**, *27*, 113–119. [CrossRef]

Article

HDAC Inhibition Counteracts Metastatic Re-Activation of Prostate Cancer Cells Induced by Chronic mTOR Suppression

Jasmina Makarević [1], Jochen Rutz [1], Eva Juengel [1,2], Sebastian Maxeiner [1], Jens Mani [1], Stefan Vallo [1], Igor Tsaur [1,2], Frederik Roos [1], Felix K.-H. Chun [1] and Roman A. Blaheta [1,*]

1 Department of Urology, Goethe-Universität Frankfurt, D-60590 Frankfurt am Main, Germany; jmakarevic@air-net.de (J.M.); Jochen.Rutz@kgu.de (J.R.); Eva.Juengel@unimedizin-mainz.de (E.J.); SebastianMaxeiner@gmx.de (S.M.); Jens.Mani@kgu.de (J.M.); Stefan.Vallo@kgu.de (S.V.); Igor.Tsaur@unimedizin-mainz.de (I.T.); Frederik.Roos@kgu.de (F.R.); Felix.Chun@kgu.de (F.K.-H.C.)

2 Department of Urology and Pediatric Urology, University Medical Center Mainz, D-55131 Mainz, Germany

* Correspondence: blaheta@em.uni-frankfurt.de; Tel.: +49-69-6301-7109; Fax: +49-69-6301-7108

Received: 3 July 2018; Accepted: 30 August 2018; Published: 1 September 2018

Abstract: This study was designed to investigate whether epigenetic modulation by histone deacetylase (HDAC) inhibition might circumvent resistance towards the mechanistic target of rapamycin (mTOR) inhibitor temsirolimus in a prostate cancer cell model. Parental (par) and temsirolimus-resistant (res) PC3 prostate cancer cells were exposed to the HDAC inhibitor valproic acid (VPA), and tumor cell adhesion, chemotaxis, migration, and invasion were evaluated. Temsirolimus resistance was characterized by reduced binding of $PC3^{res}$ cells to endothelium, immobilized collagen, and fibronectin, but increased adhesion to laminin, as compared to the parental cells. Chemotaxis, migration, and invasion of $PC3^{res}$ cells were enhanced following temsirolimus re-treatment. Integrin α and β receptors were significantly altered in $PC3^{res}$ compared to $PC3^{par}$ cells. VPA significantly counteracted temsirolimus resistance by down-regulating tumor cell–matrix interaction, chemotaxis, and migration. Evaluation of integrin expression in the presence of VPA revealed a significant down-regulation of integrin $\alpha5$ in $PC3^{res}$ cells. Blocking studies demonstrated a close association between $\alpha5$ expression on $PC3^{res}$ and chemotaxis. In this in vitro model, temsirolimus resistance drove prostate cancer cells to become highly motile, while HDAC inhibition reversed the metastatic activity. The VPA-induced inhibition of metastatic activity was accompanied by a lowered integrin $\alpha5$ surface level on the tumor cells.

Keywords: mTOR; histone deacetylase; prostate cancer; integrins; adhesion; invasion

1. Introduction

Prostate cancer (PCa) remains a leading cause of death in men worldwide [1]. Once metastasized, PCa is difficult to treat and though androgen suppression prolongs survival it is not curative. In the last decade, several promising cytotoxic and immunological agents, bone-seeking radionuclides, and next-generation androgen receptor axis targeted compounds have been introduced to optimize treatment [2]. Still, no matter which drug is employed, resistance develops over time, leading to aggressive tumor re-growth and disappointing survival rates. To improve patient outcome, innovative and novel treatment concepts are needed, with targeted therapies increasingly gaining importance.

Due to the importance of the mechanistic target of rapamycin (mTOR) pathway in tumor development and progression, mTOR inhibitors have been designed or are under development for cancer treatment [3]. The mTOR inhibitors everolimus and temsirolimus have already been approved for the treatment of metastatic renal cell carcinoma (temsirolimus, everolimus), mantle cell lymphoma (temsirolimus), breast cancer (everolimus), and pancreatic neuroendocrine tumors (everolimus) [4].

Ciccarese et al. recently reported that about a quarter of cases of localized PCa display activating mutations of the mTOR pathway. In castration-resistant PCa, the mTOR pathway was most frequently mutated [5]. A tissue microarray study has shown that mTOR is up-regulated in PCa of all stages and grades [6]. It is thought, therefore, that targeting this pathway could lead to improved patient survival and therapeutic efficacy [7].

Nevertheless, it must be emphasized that treatment with mTOR inhibitors alone may not exert long-lasting anti-tumor effects. Rather, chronic use of either temsirolimus or everolimus in vitro has been associated with negative feedback loops leading to mitotic progression and accelerated cell proliferation [8,9]. In a clinical trial, mono-treatment with temsirolimus was not as efficient in castration-resistant PCa patients as had been expected. The trial was therefore prematurely stopped [10]. Earlier studies, have shown that epigenetic modulation by a histone deacetylase (HDAC) inhibitor might improve an mTOR inhibitor-based regime. Indeed, a single molecule inhibitor targeting both HDAC activity and Akt–mTOR signaling has recently been developed. Greater tumor growth inhibition and pro-apoptotic activity have been observed than when employing single-target Akt/mTOR or HDAC inhibitors in vitro and in vivo [11].

Combining an mTOR inhibitor with an HDAC inhibitor has been shown to exert synergistic effects on B-cell acute lymphoblastic leukemia, compared to isolated drug treatment [12]. Other groups have also demonstrated that treating prostate cancer cells with this drug combination has a synergistic blocking effect on tumor proliferation [13]. A phase I study on patients with renal cell carcinoma revealed beneficial effects of a combined mTOR–HDAC inhibitor protocol [14]. Counteracting drug resistance encountered when treating different cancers by means of pharmacologic interaction with epigenetic machinery has been shown possible in vitro, in vivo, and in clinical studies [9,15–17]. The present study was therefore designed to evaluate the effects of HDAC inhibition on the metastatic and invasive behavior of temsirolimus (TEM)-resistant prostate cancer cells.

2. Materials and Methods

2.1. Cell Culture

The human prostate tumor cell line PC3 was obtained from DSMZ (Braunschweig, Germany). The tumor cells were grown and subcultured in RPMI 1640 medium (Gibco/Invitrogen, Karlsruhe, Germany) supplemented with 10% fetal calf serum (FCS), 2% HEPES (2-[4-(2-hydroxyethyl) piperazin-1-yl]ethanesulfonic acid) buffer (1 M, pH 7.4), 2% glutamine, and 1% penicillin/streptomycin at 37 °C in a humidified, 5% CO_2 incubator. Temsirolimus-resistant sublines were developed over 12 months of continuous exposure to temsirolimus (Torisel®, LC Laboratories, Woburn, MA, USA), starting at 1 nm/mL and increasing stepwise to 10 µm/mL ($PC3^{res}$). Control cells remained untreated ($PC3^{par}$). The cell doubling times of 22.07 h ($PC3^{par}$) and 24.05 h ($PC3^{res}$) were calculated according to the following formula:

$$\text{Duration of culture} = \frac{\ln(2)}{\ln(\text{final cell number}) - \ln(\text{initial cell number})}$$

Human endothelial cells (human umbilical vein endothelial cells; HUVECs) were isolated from human umbilical veins and harvested by enzymatic treatment with dispase (Gibco/Invitrogen). They were grown in Medium 199 (M199; Biozol, Munich, Germany), supplemented with 10% FCS, 10% pooled human serum, 20 µg/mL endothelial cell growth factor (Boehringer, Mannheim, Germany), 0.1% heparin, 100 ng/mL gentamycin, and 20 mm HEPES buffer (pH 7.4). Subcultures from passages 2–6 were selected for experimental use.

2.2. Drugs

Temsirolimus (TEM) was dissolved in dimethylsulfoxide (DMSO) as a 10-mm stock solution and stored as aliquots at −20 °C. Prior to experiments, TEM was diluted in cell culture medium to a final concentration of 10 µm/mL. Control cell cultures received cell culture medium without TEM. This treatment procedure lasted for at least 1 year. To exclude toxic effects of the compounds, cell viability was determined by trypan blue (Gibco/Invitrogen, Karlsruhe, Germany). Valproic acid (VPA; G.L. Pharma GmbH, Lannach, Austria), which served as a HDAC inhibitor prototype, was used at a final concentration of 1 mm. $PC3^{par}$ or $PC3^{res}$ cells were pre-treated for 3 days with VPA before adhesion and chemotaxis experiments were performed.

The response to therapeutic TEM concentrations (drug re-treatment) was also investigated. Preparation for TEM re-treatment was carried out by incubating the $PC3^{res}$ cells for three days with TEM-free medium. Subsequently, 10 nm of TEM was applied to the $PC3^{res}$ and $PC3^{par}$ cells, and cell cultures were then subjected to the below assays.

2.3. Tumor Cell Binding to HUVECs

To analyze tumor cell adhesion, HUVECs were transferred to six-well multiplates (Falcon Primaria; BD Biosciences, Franklin Lakes, NJ, USA) in complete HUVEC medium. When confluency was reached, $PC3^{par}$ or $PC3^{res}$ cells were detached from the culture flasks by accutase treatment (PAA Laboratories, Cölbe, Germany) and 0.5×10^6 cells were then added to the HUVEC monolayer for 30, 60, or 120 min. Subsequently, non-adherent tumor cells were washed off using warmed (37 °C) Medium 199. The remaining cells were fixed with 1% glutaraldehyde. Adherent tumor cells (appearing as rounded, light cells) were counted in five different fields (5×0.25 mm^2) using a phase contrast microscope and the mean cellular adhesion rate was calculated.

2.4. Attachment to Extracellular Matrix Components

Six-well plates were coated with collagen G (extracted from calfskin, consisting of 90% collagen type I and 10% collagen type III (Seromed) diluted to 400 µg/mL in phosphate buffered saline (PBS)), laminin (derived from the Engelbreth–Holm–Swarm mouse tumor, diluted to 50 µg/mL in PBS; BD Biosciences), or fibronectin (derived from human plasma; diluted to 50 µg/mL in PBS; BD Biosciences) overnight. Unspecific cell binding was evaluated by culture plates treated with poly-D-lysine (Nunc, Wiesbaden, Germany). Plastic dishes served as the background control. Plates were washed with 1% bovine serum albumin (BSA) in PBS to block nonspecific cell adhesion. Then, 0.5×10^6 tumor cells were added to each well for 60 min. Subsequently, non-adherent cells were fixed with 1% glutaraldehyde and counted microscopically. The mean cellular adhesion rate, defined by the number of cells which adhered to the coated wells minus the number of cells which adhered to the non-coated wells (background), was calculated from five different observation fields (5×0.25 mm^2).

2.5. Tumor Cell Motility (Chemotaxis), Migration, and Invasion

Serum-induced chemotactic movement was examined using six-well Transwell chambers (Greiner, Frickenhausen. Germany) with 8-µm pores. Either 0.5×10^6 $PC3^{par}$ or $PC3^{res}$ cells per mL were placed in an upper chamber in serum-free medium. To evaluate cell migration, Transwell chambers were pre-coated with collagen (400 µg/mL). Cell invasion was investigated by coating the Transwell chambers with collagen (400 µg/mL), which were then overlaid with HUVECs. The lower chamber contained 10% serum. After 20 h of incubation, the upper surface of the Transwell membrane was gently wiped with a cotton swab to remove non-migrating cells. Cells that had moved to the lower surface of the membrane were stained using hematoxylin and counted microscopically. The mean chemotaxis, migration, or invasion rate was calculated from five different observation fields (5×0.25 mm^2).

2.6. Integrin Surface Expression

Integrin surface expression was compared on PC3^{par} and PC3^{res} cells, and on PC3^{par} and PC3^{res} cells treated with VPA. Tumor cells were washed in blocking solution (PBS, 0.5% BSA) and then incubated for 60 min at 4 °C with phycoerythrin (PE)-conjugated monoclonal antibodies directed against the following integrin subtypes: anti-α1 (IgG1; clone SR84, dilution 1:1000), anti-α2 (IgG2a; clone 12F1-H6, dilution 1:250), anti-α3 (IgG1; clone C3II.1, dilution 1:1000), anti-α4 (IgG1; clone 9F10, dilution 1:200), anti-α5 (IgG1; clone IIA1, dilution 1:5000), anti-α 6 (IgG2a; clone GoH3, dilution 1:200), anti-β1 (IgG1; clone MAR4, dilution 1:2500), anti-β3 (IgG1; clone VI-PL2, dilution 1:2500), or anti-β4 (IgG2a; clone 439-9B, dilution 1:250; all BD Biosciences). Integrin surface expression was then measured using FACscan (BD Biosciences; FL-2H (log) channel histogram analysis; 1×10^4 cells per scan) and expressed as mean fluorescence units (MFU). A mouse IgG1-PE (MOPC-21) or IgG2a-PE (G155-178; all: BD Biosciences) was used as an isotype control.

2.7. Western Blot Analysis

To explore the integrin proteins in PC3^{par} and PC3^{res} cells, tumor cell lysates were applied to a 7% polyacrylamide gel and electrophoresed for 90 min at 100 V. The protein was then transferred to nitrocellulose membranes (1 h, 100 V). After blocking with non-fat dry milk for 1 h, the membranes were incubated overnight with the monoclonal antibodies listed above (unconjugated). Additionally, integrin-related signaling was investigated using anti-integrin-linked kinase (ILK; clone 3, dilution 1:1000) and anti-phospho-specific focal adhesion kinase (FAK; pY397; clone 18, dilution 1:1000) antibodies (all: BD Biosciences).

To evaluate the target specificity of TEM and VPA, the following monoclonal antibodies were employed: Anti Akt, anti-phospho Akt (pAkt; clone 104A282, both: mouse IgG1, dilution 1:500, BD Biosciences), and anti-acetylated H3 (aH3; rabbit IgG, clone Y28, dilution 1:500, Epitomics, USA). Horseradish peroxidase (HRP) -conjugated goat-anti-mouse IgG (Upstate Biotechnology, Lake Placid, NY, USA; dilution 1:5000) served as the secondary antibody. Membranes were briefly incubated with enhanced chemiluminescence (ECL) detection reagent (ECLTM, Amersham/GE Healthcare, München, Germany) to visualize the proteins and then analyzed by the Fusion FX7 system (Peqlab, Erlangen, Germany). β-actin (1:1000; Sigma, Taufenkirchen, Germany) served as the internal control. Gimp 2.8 software was used to perform pixel density analysis of the protein bands and to calculate the ratio of protein intensity/β-actin intensity.

2.8. Real-Time (RT)-qPCR

RT-qPCR was done in triplicate. cDNA synthesis was performed using 3 µg of total RNA per sample according to the manufacturer's protocol by AffinityScript QPCR cDNA Synthesis Kit (Stratagene, Amsterdam, the Netherlands). Quantitative gene expression analysis by RT-PCr was performed by the Mx3005p (Stratagene) using SYBER-Green Super Array (SABioscience Corporation, Valencia, CA, USA) and SuperArray primer sets: *GPDH* (NM_002046.3, Hs.592355), *integrin* α1 (ITGA1, NM_181501, Hs.644652), *integrin* α2 (ITGA2, NM_02203, Hs.482077), *integrin* α3 (ITGA3, NM_002204, Hs.265829), *integrin* α4 (ITGA4, NM_000885, Hs. 694732), *integrin* α5 (ITGA5, NM_002205, Hs. 505654), *integrin* α6 (ITGA6, NM_000210, Hs.133397), *integrin* β1 (ITGB1, NM_002211, Hs.643813), *integrin* β3 (ITGB3, NM_000212, HS218040), and *integrin* β4 (ITGB4, NM_000213, Hs.632226; all SABioscience Corporation). Calculation of the relative expression of each gene was done by the ΔΔCt method in the analysis program from SABioscience Corporation. The housekeeping gene, *GAPDH*, was used for normalization.

2.9. Blocking Studies

To determine whether integrin α2, α5, and β1 impact metastatic spread, drug-sensitive or -resistant cells were incubated for 60 min with 10 µg/mL function-blocking anti-integrin α2 (clone P1E6) mouse mAb, anti-integrin α5 (clone P1D6) mouse mAb, or anti-integrin β1 (clone 6S6) mouse mAB (all: from Millipore, Burlington, MA, USA), respectively. Controls were incubated with cell culture medium alone. Subsequently, tumor cell adhesion to immobilized collagen, as well as chemotaxis and migration, were evaluated as described above.

2.10. Statistics

All experiments were performed 3–6 times. Statistical significance was determined with the Wilcoxon–Mann-Whitney-U-test. Differences were considered statistically significant at a p-value less than 0.05.

3. Results

3.1. Adhesion Characteristics

More $PC3^{par}$ cells adhered to HUVECs over time than did $PC3^{res}$ cells (Figure 1A). The application of 10 nm of TEM resulted in a diminished attachment rate of $PC3^{par}$ but not of $PC3^{res}$ cells. Tumor cell interaction with an extracellular matrix revealed lower adhesion of $PC3^{res}$ cells to immobilized collagen (Figure 1B) or fibronectin (Figure 1C), but an increased adhesion to laminin (Figure 1D), compared to the parental cells. Treating the tumor cells with TEM (10 nm) suppressed binding of the drug-sensitive ($PC3^{par}$) but not the drug-resistant ($PC3^{res}$) cells to collagen, fibronectin, and laminin.

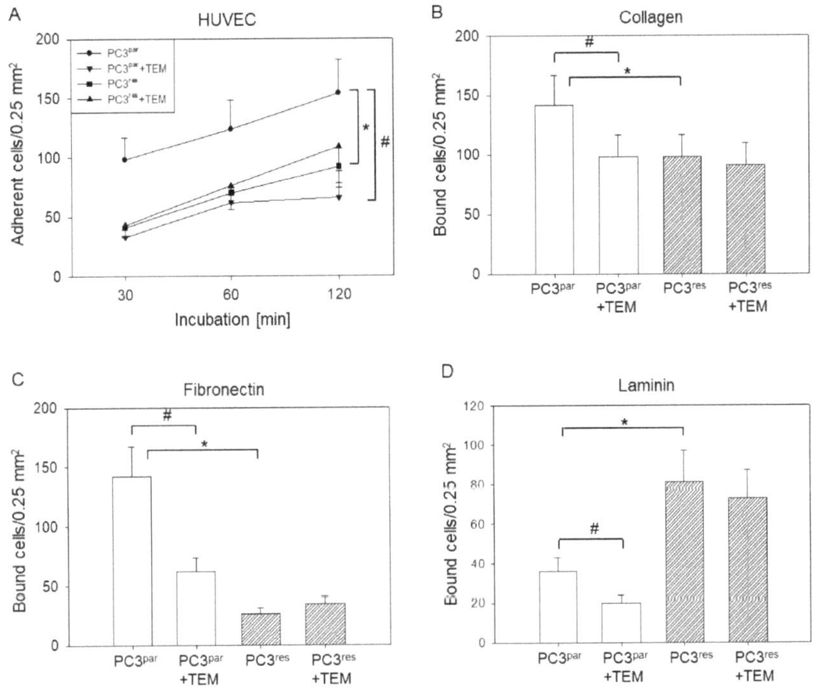

Figure 1. Adhesion of temsirolimus (TEM)-resistant ($PC3^{res}$) and TEM-sensitive ($PC3^{par}$) prostate cancer cells. (**A**) time-dependent PC3 adhesion to human vein endothelial cells (HUVEC); (**B**) binding to immobilized collagen; (**C**) binding to immobilized fibronectin; (**D**) binding to immobilized laminin. * indicates significant difference to the temsirolimus-free control ($PC3^{par}$). # indicates significant difference between cells exposed to 10 nm of TEM and untreated cells.

3.2. Tumor Motility, Migration, and Invasion

Chemotaxis was significantly elevated in $PC3^{res}$ versus $PC3^{par}$ cells (Figure 2A), whereas motile spreading through a collagen matrix (migration, Figure 2B) or through HUVECs layered onto collagen (invasion, Figure 2C) was diminished. TEM blocked migration and invasion of $PC3^{par}$ but not of $PC3^{res}$ cells. Rather, chemotaxis, migration, and invasion of $PC3^{res}$ cells were even enhanced following TEM re-treatment.

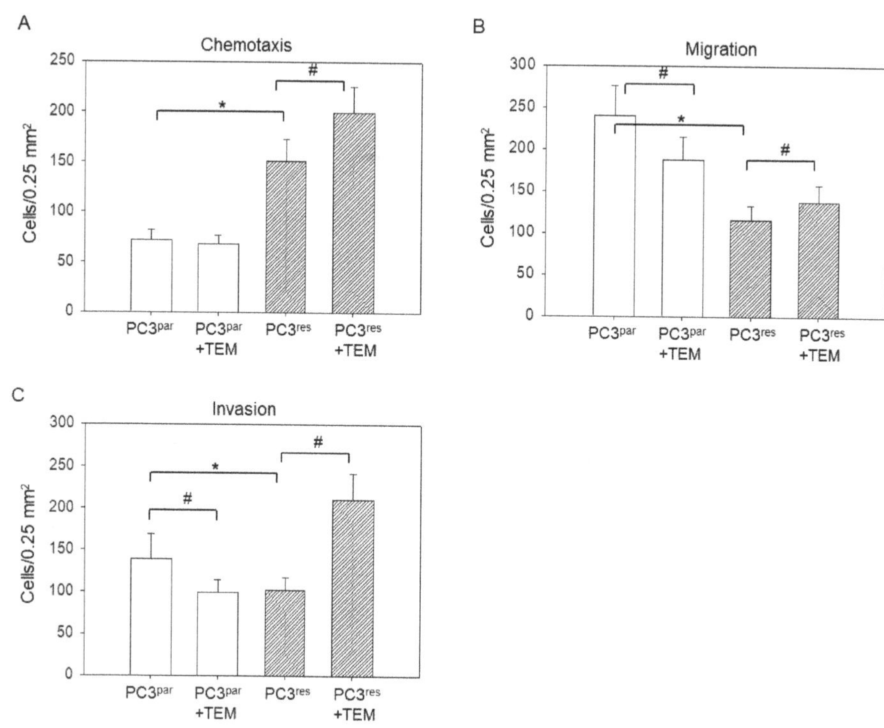

Figure 2. Motility of temsirolimus (TEM)-resistant ($PC3^{res}$) and TEM-sensitive ($PC3^{par}$) prostate cancer cells. Mean values were calculated from five counts. (**A**) PC3 chemotaxis; (**B**) PC3 migration; (**C**) PC3 invasion. * indicates significant difference to the temsirolimus-free control ($PC3^{par}$). # indicates significant difference between the treated and untreated PC3 sublines with 10 nm TEM.

3.3. Integrin Expression Pattern in $PC3^{par}$ and $PC3^{res}$ Cells

The integrin subtypes α2, α3, α6, β1, and β4 were strongly expressed, α1 and α5 were moderately expressed, and β3 was expressed to a low extent in $PC3^{par}$ cells (Figure 3). The α4 integrin subtype was not detectable by flow cytometry, neither on $PC3^{par}$ nor on $PC3^{res}$ cells (data not shown). $PC3^{res}$ cells were characterized by distinct differences in the integrin-expression pattern, compared to controls. The α2, α3, β1, and β4 subtypes were distinctly elevated, whereas α5 was nearly lost on the $PC3^{res}$ cell membrane. β3 was also found to be (moderately) up-regulated in the resistant cell population. No significant differences in expression were seen with respect to integrin α1 or α6 ($PC3^{par}$ versus $PC3^{res}$).

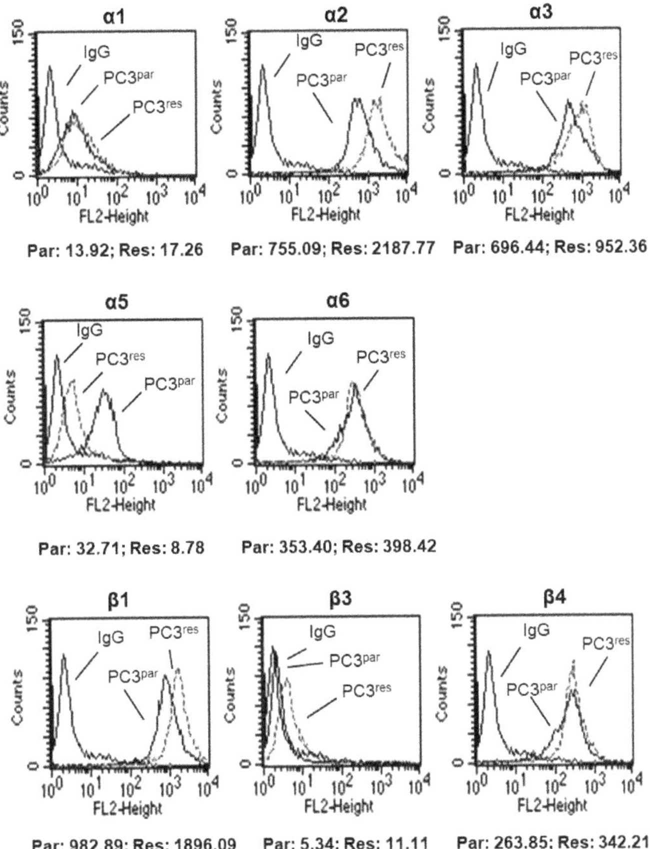

Figure 3. Flow activated cell sorting (FACS) analysis of integrin α and β subtype expression on temsirolimus (TEM)-resistant ($PC3^{res}$) and TEM-sensitive ($PC3^{par}$) prostate cancer cells. Mean fluorescence values are shown below the histograms (Par = PC3 parental cells, Res = PC3-resistant cells). One of three independent experiments.

Western blotting demonstrated enhanced protein content of integrins α2, α3, β1, and β4 in $PC3^{res}$ cells, as compared to the control cell line ($PC3^{par}$). The α5 integrin was considerably suppressed in $PC3^{res}$ cells. No differences were seen with integrin α6. The integrin members α1 and β3 were not detectable. FAK was slightly diminished, whereas pFAK and ILK were equally expressed in $PC3^{par}$ and $PC3^{res}$ cells (Figure 4A).

α5 *integrin* mRNA was expressed in $PC3^{res}$ at a very low level compared to the $PC3^{par}$ cells (Figure 4B). The mRNA of the other integrin subtypes displayed no significant differences between the sensitive and resistant cells.

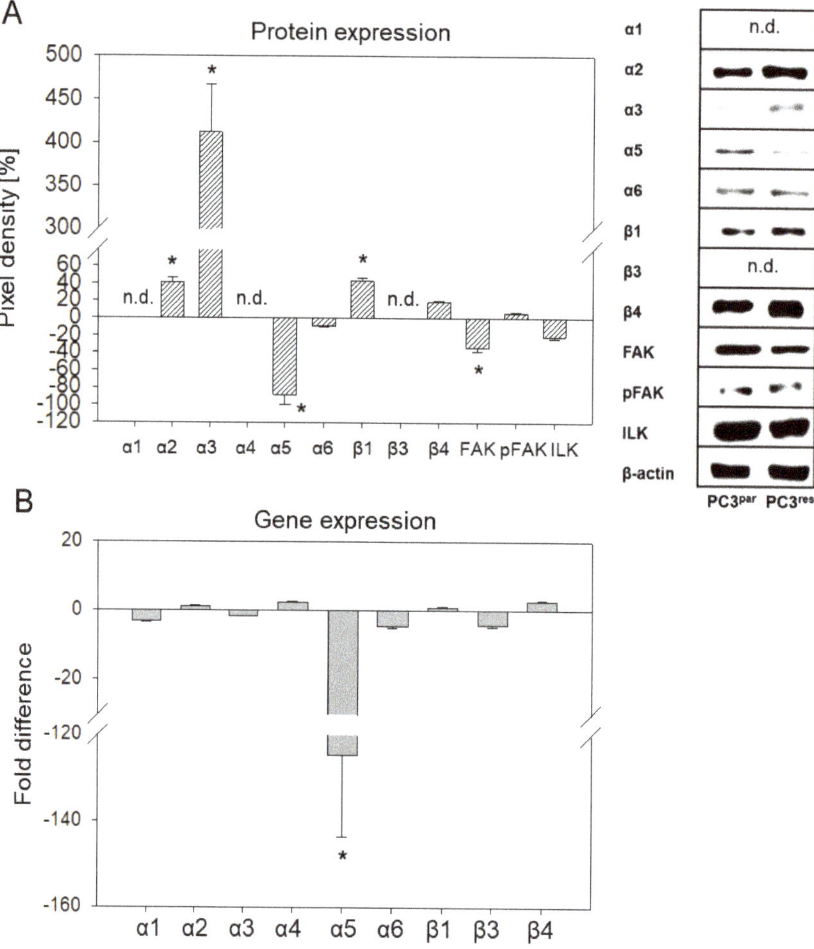

Figure 4. (**A**) Integrin protein level in temsirolimus-resistant (PC3^{res}) versus temsirolimus-sensitive (PC3^{par}) cells quantified by pixel density analysis. Integrins were evaluated three times, integrin-linked kinase (ILK), focal adhesion kinase (FAK), and phosphorylated focal adhesion kinase (pFAK) four times. Representative Western blots are shown on the right panel. (**B**) The integrin gene expression pattern in PC3^{res} versus PC3^{par} cells. Values are given as fold difference to PC3^{par} cells. * indicates a significant difference.

3.4. Blocking Studies

Blocking studies were carried out to investigate the function of α2 and β1 integrins, which were strongly elevated in PC3^{res} compared to PC3^{par}, and to explore the mode of action of integrin α5, which was distinctly diminished in the resistant cell population.

Blocking α2 or β1 significantly down-regulated adhesion, chemotactic movement, and migration of both PC3^{res} and PC3^{par} cells. The effect of receptor blockade on both cell sublines was similar, excepting chemotaxis, where β1 influenced PC3^{par} cells more efficiently than PC3^{res} cells (Figure 5). Blockade of integrin α5 differentially altered cell behavior. Adhesion of PC3^{par} to collagen was drastically reduced, while adhesion of PC3^{res} was only moderately diminished. Migration of PC3^{res}

and PC3^{par} increased to a similar extent. However, chemotaxis of PC3^{par} was up-regulated, whereas activity of PC3^{res} was down-regulated.

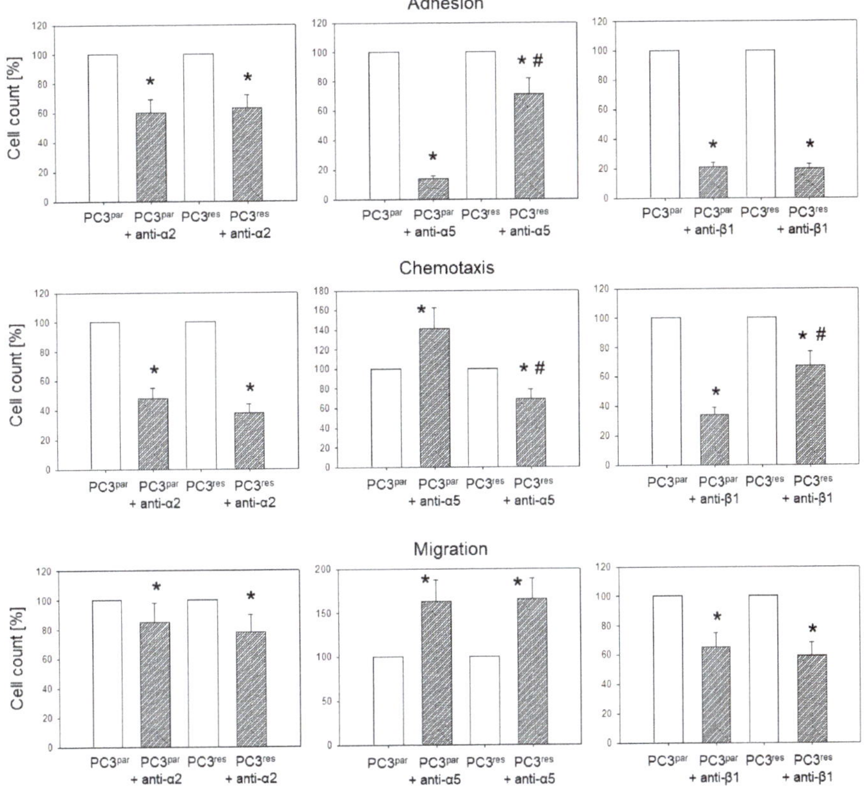

Figure 5. Influence of integrin α2, α5, or β1 blockade on PC3 adhesion, chemotaxis, and migration. Values are shown as percentage difference to their respective 100% controls. * indicates significant difference between the PC3 control subline and the PC3 subline treated with the function-blocking antibody. # indicates significant difference between temsirolimus-sensitive (PC3^{par}) and temsirolimus-resistant (PC3^{res}) cells whose integrin subtype was blocked.

3.5. Influence of VPA on Adhesion, Chemotaxis, Migration, and Integrin Expression of PC3^{par} and PC3^{res} Cells

VPA significantly down-regulated tumor cell binding to immobilized collagen, fibronectin, or matrigel of both PC3^{par} and PC3^{res} cells, as compared to the untreated controls (Figure 6). The same was true with respect to tumor cell attachment to HUVECs. Chemotactic movement and migration were also diminished when VPA was applied to drug-sensitive or drug-resistant tumor cells (Figure 7A,B). Integrin expression in the presence of VPA revealed a significant down-regulation of α5 in both PC3^{par} and PC3^{res} cells. Figure 7C depicts percentage difference of integrin expression level in VPA-treated cells, compared to the controls set to 100%. Figure 7D shows that VPA also acts on pAkt expression in both PC3^{par} and PC3^{res} cells. VPA did not induce toxic effects, as has been demonstrated by the trypan dye exclusion test (data not shown). Since VPA serves as an HDAC inhibitor, this was proved by staining VPA-treated PC3 cells with an anti-acetylated histone H3 (aH3) antibody. Pixel density analysis demonstrated an increase of aH3 to 205% (PC3^{par}) and 199% (PC3^{res}), as compared to PC3^{par} and PC3^{res} cells not treated with VPA (set to 100%).

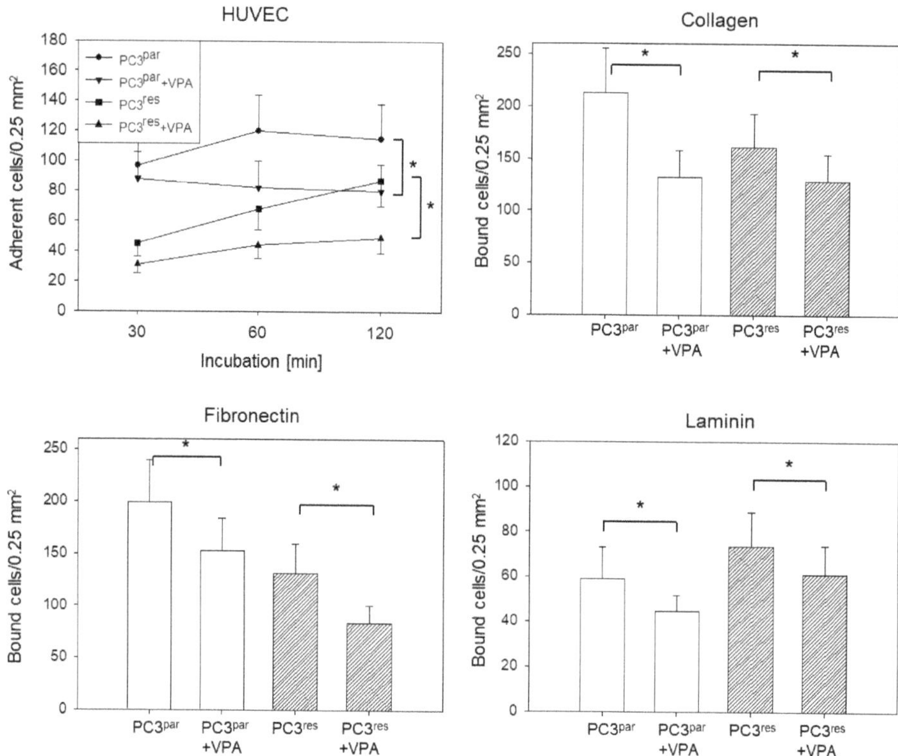

Figure 6. Adhesion of temsirolimus (TEM)-resistant (PC3^{res}) versus TEM-sensitive (PC3^{par}) prostate cancer cells in the presence of valproic acid (VPA). The figure depicts time-dependent PC3 adhesion to human umbilical vein endothelial cells (HUVEC), binding to immobilized collagen, fibronectin, or laminin. * indicates significant difference to controls not treated with VPA.

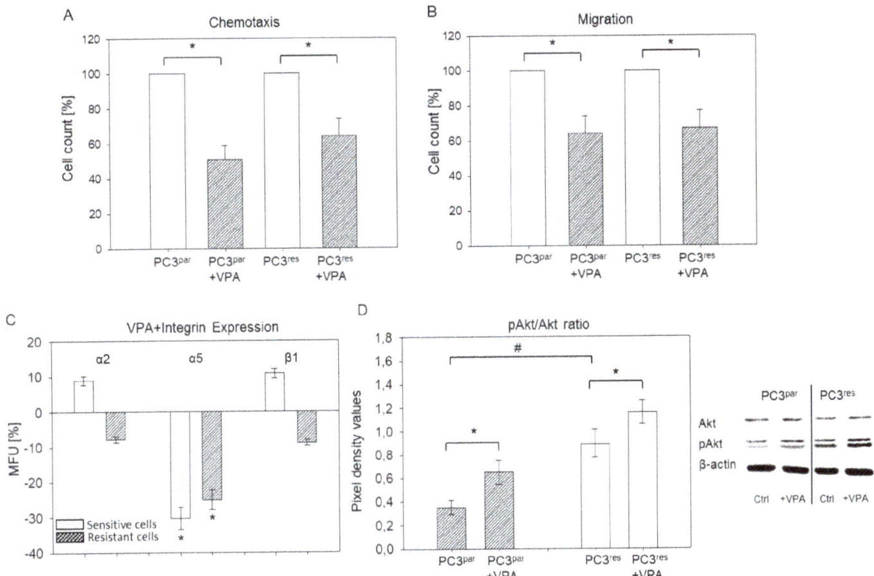

Figure 7. (**A**,**B**). Chemotactic movement and migration of PC3^{res} versus PC3^{par} cells treated with valproic acid (VPA). Values are given as percentage difference to their respective 100% controls. * indicates significant difference to controls not treated with VPA. (**C**). Influence of VPA on integrin α2, α5, or β1 expression. Mean fluorescence units (MFU) are shown as percentage difference to the respective 100% controls (not treated with VPA). (**D**) Influence of VPA on Akt expression. Akt and pAkt levels were quantified by Western blotting and pixel density analysis. Pixel density values of the pAkt/Akt ratio and representative Western blots are shown (Ctrl = control). * indicates significant difference between the PC3 control subline and the PC3 subline treated with VPA. # indicates significant difference between PC3^{par} and PC3^{res} cells.

4. Discussion

Prostate cancer cell adhesion, chemotaxis, and invasion were altered after TEM resistance developed. Tumor–HUVEC and tumor–matrix interactions were significantly blocked by TEM application to the drug-sensitive cells but not to the resistant PC3 cells. Binding of PC^{res} to immobilized laminin was considerably elevated compared to PC^{par} binding, corroborating studies from Liberio et al. where an increased laminin attachment of prostate cancer cells was associated with increased cell mobility [18]. Conversely, adhesion of PC3^{res} to endothelium, as well as to collagen and fibronectin, was significantly decreased, as compared to PC3^{par} cells. Earlier studies have shown that a loss of fibronectin binding promotes invasion by facilitating the detachment of cancer cells from the tumor mass [19]. Apparently, TEM resistance diminishes the (firm) contact of prostate cancer cells to the vascular wall and the underlying matrix proteins, collagen and fibronectin, while strengthening the laminin contact. These mechanisms serve as a prerequisite to metastatic progression.

The cell migration (through collagen) and invasion assay (through collagen and HUVECs) carried out in the present investigation could contradict the hypothesis that TEM-induced-resistance enhances motile crawling, since the number of counted PC3^{res} cells was lower than that for PC3^{par} cells. This could to be due to inhibited cell adhesion of PC3^{res} cells to collagen or HUVECs, compared to PC3^{par} cells, and consequently lead to a lower number of migrating cells. In fact, treating PC3^{res} cells with TEM significantly increased their migration and invasion potential, and chemotaxis strongly increased during resistance development, particularly when PC3 were re-treated with therapeutically relevant TEM concentrations, further activating their invasive behavior.

TEM resistance, therefore, may not only be coupled to "non-responsiveness" but rather to re-activation of the cellular motor machinery, leading to a highly invasive tumor phenotype. Chronic treatment of prostate cancer cells with another mTOR inhibitor, everolimus, has also been associated with increased metastatic activity [20]. The same increase in metastatic activity has been observed in renal cell carcinoma cells with acquired resistance towards TEM [21]. It is not yet clear whether mTOR represents the pivotal element triggering invasion. However, Caino and Altieri recently pointed to the close correlation between mTOR signaling and the crawling velocity of tumor cells [22]. Neuroblastoma and breast carcinoma models have provided evidence that mTOR activates pathways relevant to motility regulation [23,24].

Accelerated invasion was accompanied by a change in the integrin expression pattern. Both cell surface and cytoplasmic protein levels of α2- and β1- integrin increased, whereas α5 decreased in $PC3^{res}$ versus $PC3^{par}$ cells. Analysis of the integrin coding genes demonstrated a significant down-regulation of α5, indicating that this integrin subtype has been modified on a transcriptional level.

The role of α2 in prostate cancer is controversial. Current and former functional blocking studies have demonstrated a positive association between α2 expression and adhesion and migration properties [20,25]. In contrast, Ramirez et al. postulated that α2 integrin may act as a tumor suppressor and loss of it may induce bone metastasis [26]. However, this postulate was exclusively related to the expression of the α2 encoding gene. Since no change in the α2 gene was observed during the present investigation, an epigenetic mechanism behind α2 up-regulation under chronic drug treatment may be assumed. It is concluded that elevated α2 may promote cancer metastatic potential through activation of integrin-mediated signaling pathways. Accordingly, Van Slambrouck et al. showed that a high α2 expression level is necessary to maintain a highly invasive phenotype of prostate cancer cells [27]. A clinical investigation has presented evidence of elevated α2 in skeletal metastases, compared to primary prostate cancer lesions [28].

The integrin subtype β1 was similarly modified to α2 during resistance development. This may be expected, since β1 interacts with α2 to form a functional unit [28,29]. Still, the effect of β1 is controversial. Experiments on PC3 prostate cancer cells have shown that β1-integrin may stimulate tumor growth or invasion by dynamically regulating inside-out-signaling, while reduced β1-signaling may promote metastatic dissemination [30]. On the other hand, knockdown of β1 in the same cell line (PC3) significantly inhibits tumor cell migration [31], and overexpression is particularly apparent in human bone metastases [32], underpinning β1's role to drive invasive processes. Interestingly, the β1-integrin has recently been reported to control skeletal metastasis by crosstalking with the Akt–mTOR pathway [33,34], which might explain why chronic TEM application is coupled to both increased tumor invasiveness and increased β1-integrin expression. The Akt–mTOR pathway was not investigated in the present study. However, chronic TEM exposure has previously been shown to re-activate Akt-mTOR signaling [17]. It might, therefore, be likely that integrin β1 becomes up-regulated due to direct communication with the Akt–mTOR pathway, allowing tumor cells to begin the invasive cascade.

Tesfay et al. have concluded from their studies that autophagy loss may be coupled to intracellular β1-integrin accumulation caused by lacking lysosomal integrin degradation [35]. Dower and coworkers assumed that integrin-mediated regulation of cell migration and invasion might depend on endosome-mediated trafficking and integrin degradation [36]. Therefore, the higher protein levels of β1, along with α2, α3, and β4, in $PC3^{res}$ cells might be due to disrupted autophagy induced by chronic mTOR inhibition. Still this is hypothetical and requires further investigation. In fact, an integrin β1 protein increase combined with an increased α3 protein level has been considered to activate prostate cancer cell migration and invasion by regulating downstream signaling [37], whereby α3 might be a responsible triggering factor in controlling cell motility [38].

Integrin α5 was not only down-regulated in $PC3^{res}$ cells but was also found to be reduced in TEM-resistant renal cell carcinoma cells [21] or in prostate cancer cells with acquired resistance to everolimus [20]. This might reflect a generalized response to chronic mTOR-blockade, although the relevance of this behavior is not totally clear. Examination of 157 prostate cancer cases has revealed a negative correlation between α5 expression and Gleason score, pathological stage, lymph node metastasis, and prostate-specific antigen level [39], indicating that loss of α5 is associated with tumor progression. Based on an earlier study, it was hypothesized that resistance causes a functional switch of α5, driving the tumor cells from adhesiveness to invasiveness [20]. In fact, the present investigation demonstrates that α5 strongly regulates adhesion of TEM-sensitive tumor cells, while playing a major role in controlling migration during drug resistance. Since integrin α5 serves as a fibronectin receptor, this can explain why reduced α5 expression on $PC3^{res}$ was associated with cell detachment from the fibronectin matrix.

Attention should also be paid to the elevated surface expression of the laminin receptor integrin α3 in $PC3^{res}$ (versus $PC3^{par}$) cells, which correlated with enhanced tumor binding to laminin. Although the relevance of α3 on metastatic progression of prostate cancer has not been fully elucidated, in vitro experiments have recently demonstrated that α3 may promote survival of laminin-adherent tumor cells [35] and loss of α3 may impair adhesion to laminin [40]. Resistance development to the mTOR inhibitor TEM might, therefore, be accompanied by both a functional and quantitative modulation of the integrin subtype α5, associated with an α3 elevation, leading to an adhesive switch from fibronectin to laminin.

HDAC inhibition significantly blocked tumor cell adhesion and invasion of both TEM-sensitive and TEM-resistant PC3 cells. This finding is important, since it indicates that chemoresistant tumor patients may profit from HDAC targeting. We did not investigate histone acetylation in this study but have demonstrated earlier that VPA acetylates histones H3 and H4 in TEM-resistant bladder cancer [17] and renal cancer cells driven to everolimus non-responsiveness [9], indicating its role as an epigenetic modulator. Indeed, VPA has been reported to revert epithelial-mesenchymal-transition of mTOR inhibitor resistant tumor cells by targeting histone deacetylases [41].

VPA induced a strong integrin α5 decrease in both $PC3^{res}$ and $PC3^{par}$ cells, which was associated with an equally strong decrease in chemotaxis. The loss of α5 induced by TEM, on the other hand, was associated with increased chemotaxis. To account for the opposing chemotaxic effect the two drugs exert, it is important to remember that different signaling pathways are modified by TEM and VPA. Pan et al. recently pointed out that regulation of tumor cell behavior by integrins depends on how the receptors affect cell signaling [42]. TEM resistance has been shown to be coupled to a strong phosphorylation of Akt and the mTOR complex Raptor, along with c-Met activation [17,43,44]. This is relevant, since c-Met has recently been demonstrated to displace α5 integrin from β1 integrin during resistance development. This α5 loss is associated with increased invasion [45].

VPA has been shown to act on histone acetylation as its primary target, thereby reverting epithelial–mesenchymal transition and blocking tumor cell migration, both associated with α5 reduction [46,47]. Therefore, it may be assumed that invasion blocking by VPA, used here as the prototype HDAC inhibitor, is at least partially caused by suppression of α5 integrin expression (although further experiments are required for confirmation). The clinical importance of this has been demonstrated by Siva et al. who have recently identified α5 as a clinically relevant target for treating prostate cancer [48]. A peptide targeting activated α5β1 integrin has been developed, inducing complete regression of intramuscular prostate tumors in mice [49]. VPA use (in disorders for which a VPA indication exists) has been associated with a decreased cancer risk [50,51], and targeting HDAC is thought to be beneficial in treating prostate cancer, particularly in its advanced state [52,53]. In good accordance with our data, HDAC inhibition suppressed epithelial–mesenchymal plasticity and stemness activity observed in mesenchymal-like tumor cells resistant to Akt-mTOR pathway inhibitors [54]. This would support the role of HDAC inhibitors as a supplement to an mTORinhibitor based regimen.

VPA unexpectedly caused up-regulation of pAkt in both $PC3^{res}$ and $PC3^{par}$ cells. VPA has already been proven to block tumor cell growth and proliferation, even in chemo-resistant cells [17,55].

Therefore, elevation of pAkt by VPA may not be associated with increased growth activity of PC3 cells. Shi et al. presented evidence of reduced pAkt expression in PC3 cells caused by the HDAC inhibitor suberoylanilide hydroxamic acid (SAHA) [56]. In the same cell line, the HDAC inhibitor MS-275 (Entinostat) exerted only a marginal effect [57], and applying panobinostat (LBH589) did not lead to altered pAkt expression in PC3 cells [58]. Ellis et al. has shown that treating PC3 cells with panobinostat for 24 h down-regulated pAkt, whereas it was up-regulated with 48 h treatment [59]. This is of interest since VPA was also applied long-term in the present study (72 h) and likewise showed up-regulation. pAkt expression may, therefore, depend on the drug exposure time. Yan et al. has observed reduced pAkt expression in the presence of SAHA or panobinostat [60]. However, their results stem from the use of C4-2 cells, which are androgen receptor-positive, whereas PC3 cells are not. The communication between HDAC and Akt therefore seems complex and may depend on the tumor cell line, the androgen receptor status, and the applied HDAC inhibitor. Ongoing studies have been planned to deal with the Akt–mTOR cascade in more detail, to explore whether Akt up-regulation due to VPA exposure reflects a negative feedback loop or an unspecific epi-phenomenon.

Based on an in vitro model, evidence is presented here indicating that epigenetic suppression of the integrin $\alpha5$ subtype could provide a novel strategy to combat progressive prostate cancer. Verification through animal studies would be the next step with regard to use of the HDAC-mTOR inhibitor combination. Although this combination has already been demonstrated to delay resistance in renal cell cancer cells, the protocol must be tailored to the prostate model. Finally, integrin $\alpha5$ expression must be evaluated in tissue specimens derived from tumor patients.

Author Contributions: Conceptualization, E.J. and R.A.B.; Formal analysis, J.M. (Jens Mani) and S.V.; Investigation, J.M. (Jasmina Makarevic), J.R., S.M. and R.A.B.; Supervision, F.K.-H.C. and R.A.B.; Writing—Original draft, J.M. (Jasmina Makarevic), I.T. and R.A.B.; Writing—Review and Editing, F.R. and F.K.-H.C.

Funding: This research was funded by the "Monika Kutzner foundation", Berlin, and the "Vereinigung von Freunden und Förderern der Goethe-Universität Frankfurt".

Acknowledgments: We would like to thank Karen Nelson for critically reading the manuscript.

Conflicts of Interest: The authors declare no conflict of interest.

References

1. Siegel, R.L.; Miller, K.D.; Jemal, A. Cancer statistics, 2018. *CA Cancer J. Clin.* **2018**, *68*, 7–30. [CrossRef] [PubMed]
2. Ritch, C.R.; Cookson, M.S. Advances in the management of castration resistant prostate cancer. *BMJ* **2016**, *355*, i4405. [CrossRef] [PubMed]
3. Calimeri, T.; Ferreri, A.J.M. m-TOR inhibitors and their potential role in haematological malignancies. *Br. J. Haematol.* **2017**, *177*, 684–702. [CrossRef] [PubMed]
4. Pinto-Leite, R.; Arantes-Rodrigues, R.; Sousa, N.; Oliveira, P.A.; Santos, L. mTOR inhibitors in urinary bladder cancer. *Tumour Biol.* **2016**, *37*, 11541–11551. [CrossRef] [PubMed]
5. Ciccarese, C.; Massari, F.; Iacovelli, R.; Fiorentino, M.; Montironi, R.; Di Nunno, V.; Giunchi, F.; Brunelli, M.; Tortora, G. Prostate cancer heterogeneity: Discovering novel molecular targets for therapy. *Cancer Treat. Rev.* **2017**, *54*, 68–73. [CrossRef] [PubMed]
6. Yan, G.; Ru, Y.; Wu, K.; Yan, F.; Wang, Q.; Wang, J.; Pan, T.; Zhang, M.; Han, H.; Li, X.; et al. GOLM1 promotes prostate cancer progression through activating PI3K-AKT-mTOR signaling. *Prostate* **2018**, *78*, 166–177. [CrossRef] [PubMed]
7. Statz, C.M.; Patterson, S.E.; Mockus, S.M. mTOR Inhibitors in Castration-Resistant Prostate Cancer: A Systematic Review. *Target Oncol.* **2017**, *12*, 47–59. [CrossRef] [PubMed]
8. Tsaur, I.; Makarević, J.; Hudak, L.; Juengel, E.; Kurosch, M.; Wiesner, C.; Bartsch, G.; Harder, S.; Haferkamp, A.; Blaheta, R.A. The cdk1-cyclin B complex is involved in everolimus triggered resistance in the PC3 prostate cancer cell line. *Cancer Lett.* **2011**, *313*, 84–90. [CrossRef] [PubMed]

9. Juengel, E.; Nowaz, S.; Makarevi, J.; Natsheh, I.; Werner, I.; Nelson, K.; Reiter, M.; Tsaur, I.; Mani, J.; Harder, S.; et al. HDAC-inhibition counteracts everolimus resistance in renal cell carcinoma in vitro by diminishing cdk2 and cyclin A. *Mol. Cancer* **2014**, *13*, 152. [CrossRef] [PubMed]
10. Armstrong, A.J.; Shen, T.; Halabi, S.; Kemeny, G.; Bitting, R.L.; Kartcheske, P.; Embree, E.; Morris, K.; Winters, C.; Jaffe, T.; et al. A phase II trial of temsirolimus in men with castration-resistant metastatic prostate cancer. *Clin. Genitourin. Cancer* **2013**, *11*, 397–406. [CrossRef] [PubMed]
11. Qian, C.; Lai, C.J.; Bao, R.; Wang, D.G.; Wang, J.; Xu, G.X.; Atoyan, R.; Qu, H.; Yin, L.; Samson, M.; et al. Cancer network disruption by a single molecule inhibitor targeting both histone deacetylase activity and phosphatidylinositol 3-kinase signaling. *Clin. Cancer Res.* **2012**, *18*, 4104–4113. [CrossRef] [PubMed]
12. Beagle, B.R.; Nguyen, D.M.; Mallya, S.; Tang, S.S.; Lu, M.; Zeng, Z.; Konopleva, M.; Vo, T.T.; Fruman, D.A. mTOR kinase inhibitors synergize with histone deacetylase inhibitors to kill B-cell acute lymphoblastic leukemia cells. *Oncotarget* **2015**, *6*, 2088–2100. [CrossRef] [PubMed]
13. Thelen, P.; Krahn, L.; Bremmer, F.; Strauss, A.; Brehm, R.; Loertzer, H. Synergistic effects of histone deacetylase inhibitor in combination with mTOR inhibitor in the treatment of prostate carcinoma. *Int. J. Mol. Med.* **2013**, *31*, 339–346. [CrossRef] [PubMed]
14. Zibelman, M.; Wong, Y.N.; Devarajan, K.; Malizzia, L.; Corrigan, A.; Olszanski, A.J.; Denlinger, C.S.; Roethke, S.K.; Tetzlaff, C.H.; Plimack, E.R. Phase I study of the mTOR inhibitor ridaforolimus and the HDAC inhibitor vorinostat in advanced renal cell carcinoma and other solid tumors. *Investig. New Drugs* **2015**, *33*, 1040–1047. [CrossRef] [PubMed]
15. Dembla, V.; Groisberg, R.; Hess, K.; Fu, S.; Wheler, J.; Hong, D.S.; Janku, F.; Zinner, R.; Piha-Paul, S.A.; Ravi, V.; et al. Outcomes of patients with sarcoma enrolled in clinical trials of pazopanib combined with histone deacetylase, mTOR, Her2, or MEK inhibitors. *Sci. Rep.* **2017**, *7*, 15963. [CrossRef] [PubMed]
16. Booth, L.; Roberts, J.L.; Sander, C.; Lee, J.; Kirkwood, J.M.; Poklepovic, A.; Dent, P. The HDAC inhibitor AR42 interacts with pazopanib to kill trametinib/dabrafenib-resistant melanoma cells in vitro and in vivo. *Oncotarget* **2017**, *8*, 16367–16386. [CrossRef] [PubMed]
17. Juengel, E.; Najafi, R.; Rutz, J.; Maxeiner, S.; Makarevic, J.; Roos, F.; Tsaur, I.; Haferkamp, A.; Blaheta, R.A. HDAC inhibition as a treatment concept to combat temsirolimus-resistant bladder cancer cells. *Oncotarget* **2017**, *8*, 110016–110028. [CrossRef] [PubMed]
18. Liberio, M.S.; Sadowski, M.C.; Soekmadji, C.; Davis, R.A.; Nelson, C.C. Differential effects of tissue culture coating substrates on prostate cancer cell adherence, morphology and behavior. *PLoS ONE* **2014**, *9*, e112122. [CrossRef] [PubMed]
19. Jia, D.; Entersz, I.; Butler, C.; Foty, R.A. Fibronectin matrix-mediated cohesion suppresses invasion of prostate cancer cells. *BMC Cancer* **2012**, *12*, 94. [CrossRef] [PubMed]
20. Tsaur, I.; Makarević, J.; Juengel, E.; Gasser, M.; Waaga-Gasser, A.M.; Kurosch, M.; Reiter, M.; Wedel, S.; Bartsch, G.; Haferkamp, A.; et al. Resistance to the mTOR-inhibitor RAD001 elevates integrin α2- and β1-triggered motility, migration and invasion of prostate cancer cells. *Br. J. Cancer* **2012**, *107*, 847–855. [PubMed]
21. Juengel, E.; Makarević, J.; Reiter, M.; Mani, J.; Tsaur, I.; Bartsch, G.; Haferkamp, A.; Blaheta, R.A. Resistance to the mTOR inhibitor temsirolimus alters adhesion and migration behavior of renal cell carcinoma cells through an integrin α5- and integrin β3-dependent mechanism. *Neoplasia* **2014**, *16*, 291–300. [CrossRef] [PubMed]
22. Caino, M.C.; Altieri, D.C. Cancer cells exploit adaptive mitochondrial dynamics to increase tumor cell invasion. *Cell Cycle* **2015**, *14*, 3242–3247. [CrossRef] [PubMed]
23. Joly, M.M.; Williams, M.M.; Hicks, D.J.; Jones, B.; Sanchez, V.; Young, C.D.; Sarbassov, D.D.; Muller, W.J.; Brantley-Sieders, D.; Cook, R.S. Two distinct mTORC2-dependent pathways converge on Rac1 to drive breast cancer metastasis. *Breast Cancer Res.* **2017**, *19*, 74. [CrossRef] [PubMed]
24. Hua, Z.; Gu, X.; Dong, Y.; Tan, F.; Liu, Z.; Thiele, C.J.; Li, Z. PI3K and MAPK pathways mediate the BDNF/TrkB-increased metastasis in neuroblastoma. *Tumour Biol.* **2016**, *37*, 16227–16236. [CrossRef] [PubMed]
25. Van Slambrouck, S.; Jenkins, A.R.; Romero, A.E.; Steelant, W.F. Reorganization of the integrin α2 subunit controls cell adhesion and cancer cell invasion in prostate cancer. *Int. J. Oncol.* **2009**, *34*, 1717–1726. [CrossRef] [PubMed]
26. Ramirez, N.E.; Zhang, Z.; Madamanchi, A.; Boyd, K.L.; O'Rear, L.D.; Nashabi, A.; Li, Z.; Dupont, W.D.; Zijlstra, A.; Zutter, M.M. The α2β1 integrin is a metastasis suppressor in mouse models and human cancer. *J. Clin. Investig.* **2011**, *121*, 226–237. [CrossRef] [PubMed]

27. Van Slambrouck, S.; Hilkens, J.; Bisoffi, M.; Steelant, W.F. AsialoGM1 and integrin α2β1 mediate prostate cancer progression. *Int. J. Oncol.* **2009**, *35*, 693–699. [CrossRef] [PubMed]
28. Sottnik, J.L.; Daignault-Newton, S.; Zhang, X.; Morrissey, C.; Hussain, M.H.; Keller, E.T.; Hall, C.L. Integrin α2β1 (α2β1) promotes prostate cancer skeletal metastasis. *Clin. Exp. Metastasis* **2013**, *30*, 569–578. [CrossRef] [PubMed]
29. Lee, S.H.; Hatakeyama, S.; Yu, S.Y.; Bao, X.; Ohyama, C.; Khoo, K.H.; Fukuda, M.N.; Fukuda, M. Core3 *O*-glycan synthase suppresses tumor formation and metastasis of prostate carcinoma PC3 and LNCaP cells through down-regulation of α2β1 integrin complex. *J. Biol. Chem.* **2009**, *284*, 17157–17169. [CrossRef] [PubMed]
30. Pollan, S.G.; Huang, F.; Sperger, J.M.; Lang, J.M.; Morrissey, C.; Cress, A.E.; Chu, C.Y.; Bhowmick, N.A.; You, S.; Freeman, M.R.; et al. Regulation of inside-out β1-integrin activation by CDCP1. *Oncogene* **2018**, *37*, 2817–2836. [CrossRef] [PubMed]
31. Kurozumi, A.; Goto, Y.; Matsushita, R.; Fukumoto, I.; Kato, M.; Nishikawa, R.; Sakamoto, S.; Enokida, H.; Nakagawa, M.; Ichikawa, T.; et al. Tumor-suppressive microRNA-223 inhibits cancer cell migration and invasion by targeting ITGA3/ITGB1 signaling in prostate cancer. *Cancer Sci.* **2016**, *107*, 84–94. [CrossRef] [PubMed]
32. Jin, J.K.; Tien, P.C.; Cheng, C.J.; Song, J.H.; Huang, C.; Lin, S.H.; Gallick, G.E. Talin1 phosphorylation activates β1 integrins: A novel mechanism to promote prostate cancer bone metastasis. *Oncogene* **2015**, *34*, 1811–1821. [CrossRef] [PubMed]
33. Yu, M.; Wang, J.; Muller, D.J.; Helenius, J. In PC3 prostate cancer cells ephrin receptors crosstalk to β1-integrins to strengthen adhesion to collagen type I. *Sci. Rep.* **2015**, *5*, 8206. [CrossRef] [PubMed]
34. Virtakoivu, R.; Pellinen, T.; Rantala, J.K.; Perälä, M.; Ivaska, J. Distinct roles of AKT isoforms in regulating β1-integrin activity, migration, and invasion in prostate cancer. *Mol. Biol. Cell* **2012**, *23*, 3357–3369. [CrossRef] [PubMed]
35. Tesfay, L.; Schulz, V.V.; Frank, S.B.; Lamb, L.E.; Miranti, C.K. Receptor tyrosine kinase Met promotes cell survival via kinase-independent maintenance of integrin α3β1. *Mol. Biol. Cell* **2016**, *27*, 2493–2504. [CrossRef] [PubMed]
36. Dower, C.M.; Wills, C.A.; Frisch, S.M.; Wang, H.G. Mechanisms and context underlying the role of autophagy in cancer metastasis. *Autophagy* **2018**, *4*, 1–19. [CrossRef] [PubMed]
37. Kurozumi, A.; Goto, Y.; Matsushita, R.; Fukumoto, I.; Kato, M.; Nishikawa, R.; Sakamoto, S.; Enokida, H.; Nakagawa, M.; Ichikawa, T.; et al. Tumor-suppressive microRNA-223 inhibits cancer cell migration and invasion by targeting ITGA3/ITGB1 signaling in prostate cancer. *Cancer Sci.* **2016**, *10*, 84–94. [CrossRef] [PubMed]
38. Das, L.; Anderson, T.A.; Gard, J.M.; Sroka, I.C.; Strautman, S.R.; Nagle, R.B.; Morrissey, C.; Knudsen, B.S.; Cress, A.E. Characterization of Laminin Binding Integrin Internalization in Prostate Cancer Cells. *J. Cell. Biochem.* **2017**, *118*, 1038–1049. [CrossRef] [PubMed]
39. Drivalos, A.; Chrisofos, M.; Efstathiou, E.; Kapranou, A.; Kollaitis, G.; Koutlis, G.; Antoniou, N.; Karanastasis, D.; Dimopoulos, M.A.; Bamias, A. Expression of α5-integrin, α7-integrin, E-cadherin, and *N*-cadherin in localized prostate cancer. *Urol. Oncol.* **2016**, *34*, 165.e11–165.e18. [CrossRef] [PubMed]
40. Varzavand, A.; Drake, J.M.; Svensson, R.U.; Herndon, M.E.; Zhou, B.; Henry, M.D.; Stipp, C.S. Integrin α3β1 regulates tumor cell responses to stromal cells and can function to suppress prostate cancer metastatic colonization. *Clin. Exp. Metastasis* **2013**, *30*, 541–552. [CrossRef] [PubMed]
41. Juengel, E.; Dauselt, A.; Makarević, J.; Wiesner, C.; Tsaur, I.; Bartsch, G.; Haferkamp, A.; Blaheta, R.A. Acetylation of histone H3 prevents resistance development caused by chronic mTOR inhibition in renal cell carcinoma cells. *Cancer Lett.* **2012**, *324*, 83–90. [CrossRef] [PubMed]
42. Pan, B.; Guo, J.; Liao, Q.; Zhao, Y. β1 and β3 integrins in breast, prostate and pancreatic cancer: A novel implication. *Oncol. Lett.* **2018**, *15*, 5412–5416. [CrossRef] [PubMed]
43. Imura, Y.; Yasui, H.; Outani, H.; Wakamatsu, T.; Hamada, K.; Nakai, T.; Yamada, S.; Myoui, A.; Araki, N.; Ueda, T.; et al. Combined targeting of mTOR and c-MET signaling pathways for effective management of epithelioid sarcoma. *Mol. Cancer* **2014**, *13*, 185. [CrossRef] [PubMed]
44. Etnyre, D.; Stone, A.L.; Fong, J.T.; Jacobs, R.J.; Uppada, S.B.; Botting, G.M.; Rajanna, S.; Moravec, D.N.; Shambannagari, M.R.; Crees, Z.; et al. Targeting c-Met in melanoma: Mechanism of resistance and efficacy of novel combinatorial inhibitor therapy. *Cancer Biol. Ther.* **2014**, *15*, 1129–1141. [CrossRef] [PubMed]

45. Jahangiri, A.; Nguyen, A.; Chandra, A.; Sidorov, M.K.; Yagnik, G.; Rick, J.; Han, S.W.; Chen, W.; Flanigan, P.M.; Schneidman-Duhovny, D.; et al. Cross-activating c-Met/β1 integrin complex drives metastasis and invasive resistance in cancer. *Proc. Natl. Acad. Sci. USA* **2017**, *114*, E8685–E8694. [CrossRef] [PubMed]
46. Makarević, J.; Tawanaie, N.; Juengel, E.; Reiter, M.; Mani, J.; Tsaur, I.; Bartsch, G.; Haferkamp, A.; Blaheta, R.A. Cross-communication between histone H3 and H4 acetylation and Akt-mTOR signalling in prostate cancer cells. *J. Cell. Mol. Med.* **2014**, *18*, 1460–1466. [CrossRef] [PubMed]
47. Wedel, S.; Hudak, L.; Seibel, J.M.; Makarević, J.; Juengel, E.; Tsaur, I.; Wiesner, C.; Haferkamp, A.; Blaheta, R.A. Impact of combined HDAC and mTOR inhibition on adhesion, migration and invasion of prostate cancer cells. *Clin. Exp. Metastasis* **2011**, *28*, 479–491. [CrossRef] [PubMed]
48. Siva, A.C.; Kirkland, R.E.; Lin, B.; Maruyama, T.; McWhirter, J.; Yantiri-Wernimont, F.; Bowdish, K.S.; Xin, H. Selection of anti-cancer antibodies from combinatorial libraries by whole-cell panning and stringent subtraction with human blood cells. *J. Immunol. Methods* **2008**, *330*, 109–119. [CrossRef] [PubMed]
49. Veine, D.M.; Yao, H.; Stafford, D.R.; Fay, K.S.; Livant, D.L. A D-amino acid containing peptide as a potent, noncovalent inhibitor of α5β1 integrin in human prostate cancer invasion and lung colonization. *Clin. Exp. Metastasis* **2014**, *31*, 379–393. [CrossRef] [PubMed]
50. Salminen, J.K.; Tammela, T.L.; Auvinen, A.; Murtola, T.J. Antiepileptic drugs with histone deacetylase inhibition activity and prostate cancer risk: A population-based case-control study. *Cancer Causes Control* **2016**, *27*, 637–645. [CrossRef] [PubMed]
51. Kang, H.; Gillespie, T.W.; Goodman, M.; Brodie, S.A.; Brandes, M.; Ribeiro, M.; Ramalingam, S.S.; Shin, D.M.; Khuri, F.R.; Brandes, J.C. Long-term use of valproic acid in US veterans is associated with a reduced risk of smoking-related cases of head and neck cancer. *Cancer* **2014**, *120*, 1394–1400. [CrossRef] [PubMed]
52. Motawi, T.K.; Darwish, H.A.; Diab, I.; Helmy, M.W.; Noureldin, M.H. Combinatorial strategy of epigenetic and hormonal therapies: A novel promising approach for treating advanced prostate cancer. *Life Sci.* **2018**, *198*, 71–78. [CrossRef] [PubMed]
53. Xu, Q.; Liu, X.; Zhu, S.; Hu, X.; Niu, H.; Zhang, X.; Zhu, D.; Nesa, E.U.; Tian, K.; Yuan, H. Hyper-acetylation contributes to the sensitivity of chemo-resistant prostate cancer cells to histone deacetylase inhibitor Trichostatin A. *J. Cell. Mol. Med.* **2018**, *22*, 1909–1922. [CrossRef] [PubMed]
54. Ruscetti, M.; Dadashian, E.L.; Guo, W.; Quach, B.; Mulholland, D.J.; Park, J.W.; Tran, L.M.; Kobayashi, N.; Bianchi-Frias, D.; Xing, Y.; et al. HDAC inhibition impedes epithelial-mesenchymal plasticity and suppresses metastatic, castration-resistant prostate cancer. *Oncogene* **2016**, *35*, 3781–3795. [CrossRef] [PubMed]
55. Tsaur, I.; Hudak, L.; Makarević, J.; Juengel, E.; Mani, J.; Borgmann, H.; Gust, K.M.; Schilling, D.; Bartsch, G.; Nelson, K.; et al. Intensified antineoplastic effect by combining an HDAC-inhibitor, an mTOR-inhibitor and low dosed interferon alpha in prostate cancer cells. *J. Cell. Mol. Med.* **2015**, *19*, 1795–1804. [CrossRef] [PubMed]
56. Shi, X.Y.; Ding, W.; Li, T.Q.; Zhang, Y.X.; Zhao, S.C. Histone Deacetylase (HDAC) Inhibitor, Suberoylanilide Hydroxamic Acid (SAHA), Induces Apoptosis in Prostate Cancer Cell Lines via the Akt/FOXO3a Signaling Pathway. *Med. Sci. Monit.* **2017**, *23*, 5793–5802. [CrossRef] [PubMed]
57. Chen, C.S.; Wenig, S.C.; Tseng, P.H.; Lin, H.P.; Chen, C.S. Histone acetylation-independent effect of histone deacetylase inhibitors on Akt through the reshuffling of protein phosphatase 1 complexes. *J. Biol. Chem.* **2005**, *280*, 38879–38887. [CrossRef] [PubMed]
58. Verheul, H.M.; Salumbides, B.; Van Erp, K.; Hammers, H.; Qian, D.Z.; Sanni, T.; Atadja, P.; Pili, R. Combination strategy targeting the hypoxia inducible factor-1α with mammalian target of rapamycin and histone deacetylase inhibitors. *Clin. Cancer Res.* **2008**, *14*, 3589–3597. [CrossRef] [PubMed]
59. Ellis, L.; Ku, S.Y.; Ramakrishnan, S.; Lasorsa, E.; Azabdaftari, G.; Godoy, A.; Pili, R. Combinatorial antitumor effect of HDAC and the PI3K-Akt-mTOR pathway inhibition in a Pten defecient model of prostate cancer. *Oncotarget* **2013**, *4*, 2225–2236. [CrossRef] [PubMed]
60. Yan, Y.; An, J.; Yang, Y.; Wu, D.; Bai, Y.; Cao, W.; Ma, L.; Chen, J.; Yu, Z.; He, Y.; et al. Dual inhibition of AKT-mTOR and AR signaling by targeting HDAC3 in PTEN- or SPOP-mutated prostate cancer. *EMBO Mol. Med.* **2018**, *10*, e8478. [CrossRef] [PubMed]

Article

Eribulin Synergistically Increases Anti-Tumor Activity of an mTOR Inhibitor by Inhibiting pAKT/pS6K/pS6 in Triple Negative Breast Cancer

Wei Wen [1], Emily Marcinkowski [1], David Luyimbazi [1], Thehang Luu [2], Quanhua Xing [1], Jin Yan [1], Yujun Wang [1], Jun Wu [3], Yuming Guo [3], Dylan Tully [1], Ernest S. Han [1], Susan E. Yost [2], Yuan Yuan [2] and John H. Yim [1,*]

1 Division of Surgery, Beckman Research Institute, City of Hope National Medical Center, 1500 East Duarte Rd., Duarte, CA 91010, USA
2 Department of Medical Oncology and Molecular Therapy, Beckman Research Institute, City of Hope National Medical Center, 1500 East Duarte Rd., Duarte, CA 91010, USA
3 Department of Comparative Medicine, Beckman Research Institute, City of Hope National Medical Center, 1500 East Duarte Rd., Duarte, CA 91010, USA
* Correspondence: jyim@coh.org; Tel.: +1-626-218-7100

Received: 30 June 2019; Accepted: 28 August 2019; Published: 30 August 2019

Abstract: Unlike other breast cancer subtypes, patients with triple negative breast cancer (TNBC) have poor outcomes and no effective targeted therapies, leaving an unmet need for therapeutic targets. Efforts to profile these tumors have revealed the PI3K/AKT/mTOR pathway as a potential target. Activation of this pathway also contributes to resistance to anti-cancer agents, including microtubule-targeting agents. Eribulin is one such microtubule-targeting agent that is beneficial in treating taxane and anthracycline refractory breast cancer. In this study, we compared the effect of eribulin on the PI3K/AKT/mTOR pathway with other microtubule-targeting agents in TNBC. We found that the phosphorylation of AKT was suppressed by eribulin, a microtubule depolymerizing agent, but activated by paclitaxel, a microtubule stabilizing agent. The combination of eribulin and everolimus, an mTOR inhibitor, resulted in an increased reduction of p-S6K1 and p-S6, a synergistic inhibition of cell survival in vitro, and an enhanced suppression of tumor growth in two orthotopic mouse models. These findings provide a preclinical foundation for targeting both the microtubule cytoskeleton and the PI3K/AKT/mTOR pathway in the treatment of refractory TNBC.

Keywords: TNBC; eribulin; PI3K/AKT/mTOR; everolimus; combination; synergy

---

1. Introduction

Triple negative breast cancer (TNBC) accounts for 12–17% of all breast cancers, and is characterized by a poor overall and relapse-free survival [1]. Unlike hormone receptor positive tumors and tumors with Her2-neu overexpression, patients with TNBC have worse outcomes after chemotherapy and have an unmet need for targeted therapy [2,3]. As such, efforts to profile TNBC tumors have identified the PI3K/AKT/mTOR pathway as a potential therapeutic target [4–6].

The PI3K/AKT/mTOR pathway is a key signal transduction pathway that mediates cellular responses to growth factors [7–9]. This pathway affects many cellular functions, including cell survival, cell proliferation, and apoptosis [10]. Patients with TNBC often have high levels of AKT expression and activation of the PI3K/AKT/mTOR pathway [2,4,11,12]. Treatment targeting the PI3K/AKT/mTOR pathway in patients with alterations in the PI3K/AKT/mTOR pathway resulted in significantly better outcomes in both treatment naïve and previously treated patients [2,12]. However, an mTOR inhibitor, when used alone, can induce increased levels of p-AKT via a negative feedback loop, leading to resistance of cells to mTOR inhibitors [13,14]. Novel combinations are urgently needed to effectively target PI3K pathway alterations in patients that progress on therapy or fail to respond [9,10,15,16].

Microtubule-targeting agents have been used with success to treat TNBC [1,5,17]. Eribulin mesylate (E7389), a synthetic macrocyclic analogue of the marine sponge natural product halichondrin B [18,19], suppresses mitosis by directly binding to microtubule ends, resulting in the inhibition of microtubule growth and formation of tubulin aggregates. This leads to abnormal mitotic spindles that cannot pass the metaphase/anaphase checkpoint, effectively inducing G2-M cell cycle arrest [20–26]. Eribulin has demonstrated a potent anti-tumor activity against a wide range of tumor cells both in vitro and in vivo. Eribulin can also be combined effectively with other anticancer agents [27,28]. Eribulin has been approved for the treatment of TNBC in heavily pretreated patients [29].

Breast cancer cells can eventually become resistant to targeted therapy or chemotherapy, despite initial response to the treatment [30–32]. Activation of the PI3K/AKT/mTOR pathway contributes to the resistance to anti-cancer agents, including microtubule-targeting agents. In this study, we investigated the effect of eribulin and a mTOR inhibitor, either alone or in combination, on the PI3K/AKT/mTOR pathway and tumor growth in TNBC. Our results demonstrate that eribulin, unlike paclitaxel, potently decreases the expression of p-AKT in TNBC. Dual treatment of eribulin and the mTOR inhibitor results in a synergistic suppression of cell survival in a number of TNBCs in vitro and an enhanced suppression of tumor growth in two TNBC mouse models.

2. Materials and Methods

2.1. Reagents

Eribulin was kindly provided by Eisai Co. Ltd. (Tokyo, Japan). Everolimus (RAD001), BKM120, and BEZ235 were kindly provided by Novartis (Basel, Switzerland). Antibodies against p-AKT (S473), AKT, p-S6K, S6K, p-S6, S6, and GAPDH were obtained from Cell Signaling Technology (Danvers, MA, USA). Anti-β-actin was obtained from EMD Millipore (Billerica, MA, USA).

2.2. Cell Viability Assays

The human breast cell lines BT549, Hs578T, MDA-MB-231, and MDA-MB-468 were obtained from American Type Culture Collection (Rockville, MD, USA). Cells were cultured in RPMI 1640 medium (Mediatech Inc., Manassas, VA, USA) for BT549 and 4T1 cells, in 1:1 DMEM/F12 (Gibco, Thermo Scientific, Waltham, MA, USA) for MDA-MB-468 cells, and in DMEM medium (Mediatech Inc., Manassas, VA, USA) for MDA-MB-231 and Hs578 cells. Culture media were supplemented with 10% fetal bovine serum (Atlanta Biologicals, Norcross, GA, USA) and 1% penicillin/streptomycin (Gibco, Thermo Scientific, Waltham, MA, USA).

Cells (4000 per well) were plated in a 96-well plate format in 100 μL growth medium. Cells were treated with dimethyl sulfoxide (DMSO) or drugs the next day at the indicated concentrations and incubated for an additional 3 days. Viable cells were determined by the 3-(4,5-dimethyl-thiazol-2-yl)-2,5-diphenyltetrazolium bromide (MTT) assay (Promega, Madison, WI, USA) [33]. After treatment, the media were removed and MTT dye was added to each well and incubated for 4 h according to the manuscripter's instruction. The resulting formazan crystals were dissolved in DMSO after removal of the media. Absorbance was read at 570 nm. The half maximal inhibitory concentration IC_{50} was determined using the Calcusyn software package (Biosoft, Ferguson, MO, USA).

2.3. Combination Index (CI)

The Chou-Talalay method was used to calculate the combination index (CI) with the Calcusyn software package (Biosoft, Ferguson, MO, USA) [34]. $CI < 1$ indicates synergism, $CI > 1$ indicates antagonism, and $CI = 1$ indicates an additive effect.

2.4. Western Blot

Cells were treated with DMSO or drugs in complete medium at the indicated concentrations and times, washed with cold phosphaste-bufferd saline (PBS), and lysed in radioimmuoprecipitation assay (RIPA) lysis buffer (Thermo Scientific, Waltham, MA, USA) containing Halt phosphotase and protease inhibitors (Thermo Scientific, Waltham, MA, USA). Equal amounts of protein were separated by sodium dodecyl sulfate (SDS)-polyacrylamide gel electrophoresis. Western blot analysis was performed as described previously [33,35].

2.5. Animal Models

All animal studies were carried out under protocols approved by the Institutional Animal Care and Use Committee at City of Hope (IACUC 11013). MDA-MB-468 cells (5×10^6 in Matrigel) and 4T1 (1×10^5) were inoculated into the mammary fat pad of 6- to 8-week-old female NOD/SCID/IL2Rgamma null (NSG) mice (MDA-MB-468) or Balb/C mice (4T1). Once the tumors were palpable, animals were randomized into four groups with 7–10 mice for each group to achieve an equal distribution of tumor sizes in all treatment groups. Mice were then treated with a vehicle, eribulin (retro-orbital), everolimus (oral gavage), or a combination of both agents. Tumor sizes were assessed using calipers once to twice a week. Tumor volumes were calculated using the formula $\text{Width}^2 \times \text{Length} \times 0.52$. Body weight was monitored weekly as an indicator of the overall health of mice.

2.6. Immunohistochemistry (IHC)

Tumor tissues were fixed in 10% buffered formalin and embedded in paraffin. IHC was performed by the Pathology Core at City of Hope using VENTANA Ultra IHC automated stainer (VENTANA Medical Systems, Roche Diagnostics, Indianapolis, IN, USA). Briefly, tissue samples were sectioned at a thickness of 5 µm and put on positively charged glass slides. The slides were loaded on the machine and followed by deparaffinization, rehydration, endogenous peroxydase activity inhibition, and antigen retrieval. The slides were first incubated with primary antibody against Ki67 (Clone 30-9, Roche Diagnostics, Indianapolis, IN, USA) for proliferation and cleaved-caspase 3 (clone ASP175, Cell Signaling, Danvers, MA, USA) for apoptosis, and then incubated with DISCOVERY anti-Rabbit HQ and anti-HQ-HRP, visualized with the DISCOVERY ChromoMap DAB Kit, and counterstained with haematoxylin (Roche Diagnostics, Indianapolis, IN, USA). The immunoreactivity is evident as a dark brown color. Slides were scanned with VENTANA iScan HT using VENTANA Image Viewer (VENTANA Medical Systems, Roche Diagnostics, Indianapolis, IN, USA). The images were taken at 40x magnification.

2.7. Statistical Analysis

Data are presented as the mean + S.D. A comparison of the means of two groups was determined by a Student's t-test. Each experiment was carried out in triplicate or more. p values less than 0.05 were considered statistically significant.

3. Results

3.1. Eribulin Inhibits the Phosphorylation of AKT in Triple Negative Breast Cancer Cells

We first studied the anti-tumor activity of eribulin in several TNBC lines. Cells were incubated with serial dilutions of eribulin. Cell viability was determined 72 h later. As shown in Figure 1A, eribulin inhibited cell viability, with an IC_{50} ranging from 0.07 to 71 nM in TNBC.

Activation of the PI3K/AKT pathway by some anti-cancer drugs has been previously shown to cause drug resistance [36]. To study the effect of eribulin on the PI3K/AKT pathway, MDA-MB-468 and 4T1 breast cancer cells were incubated with increasing concentrations of eribulin for 24 h, followed by Western blot analysis. We found that eribulin significantly decreased p-AKT expression in a dose-dependent manner (Figure 1B). The reduced expression of p-AKT by eribulin was seen as early as 4 h in both MDA-MB-468 and BT549 cells (Figure 1C,D).

We next compared the effect of eribulin on the PI3K/AKT pathway with two other microtubule targeting agents, vinblastine and paclitaxel, as well as a conventional DNA damage chemotherapeutic agent, cisplatin. Treatment with vinblastine, a microtubule depolymerizing agent similar to eribulin, resulted in a dose-dependent decrease in p-AKT expression in MDA-MB-468 cells. Treatment with paclitaxel, a microtubule stabilizing agent, resulted in a dose-dependent increase in p-AKT expression. Incubation of cisplatin with MDA-MB 468 also resulted in a dose-dependent increase in p-AKT expression in MDA-MB-468 cells (Figure 2).

Taken together, these results showed that p-AKT expression was suppressed in the presence of microtubule targeting agents that block tubulin polymerization, such as eribulin and vinblastine, in TNBC.

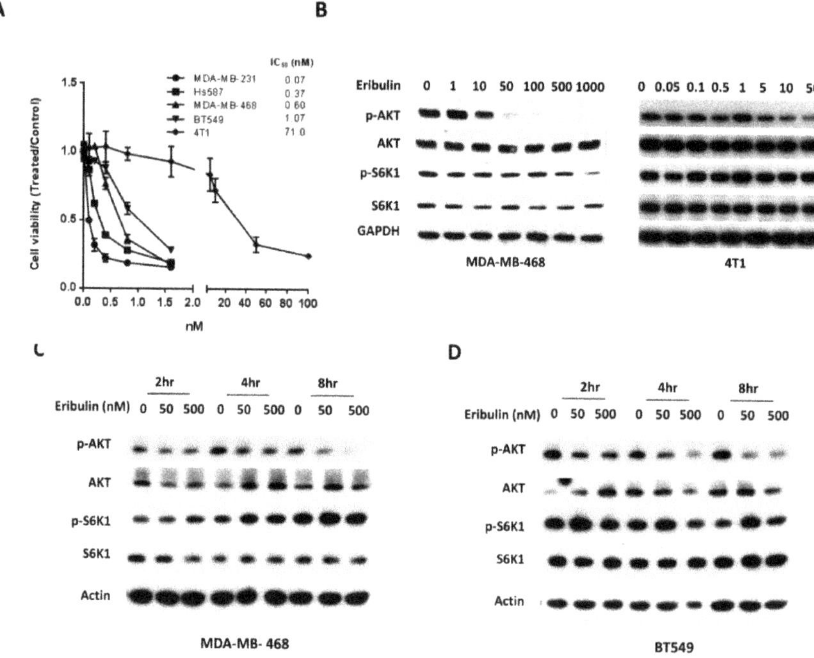

Figure 1. Eribulin inhibits cell viability and AKT phosphorylation in triple negative breast cancer (TNBC) cells. (**A**) TNBC cells were treated with various concentrations of eribulin. Cell viability was determined 72 h later. The IC_{50} was determined by the Chou-Talalay method. (**B**) Cells were treated with eribulin at concentrations of 1–1000 nM for MDA-MB-468 and 0.05–50 μM for 4T1 cells. Cells were harvested at 24 h and measured for the expression of p-AKT, AKT, p-S6K1, and S6K1 by Western blot analysis. (**C–D**) MDA-MB-468 and BT549 cells were treated with eribulin for the indicated times and concentrations. Cells were collected and measured for the expression of p-AKT and p-S6K1 by Western blot analysis.

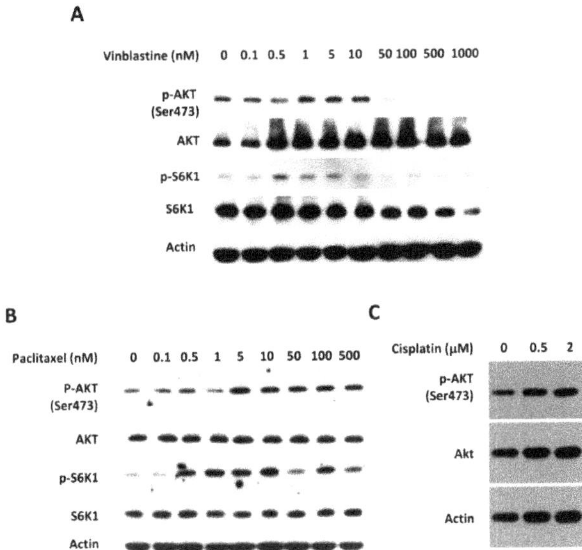

Figure 2. The effect of commonly used cytotoxic agents on AKT phosphorylation. MDA-MB-468 cells were treated with vinblastine (**A**), paclitaxel (**B**), and cisplatin (**C**) at indicated concentrations. Cells were harvested at 24 h, and the expression of p-AKT and p-S6K1 was measured by Western blot analysis.

3.2. Combined Treatment of Eribulin and Everolimus Enhances the Reduction of p-S6K1 and p-S6

Given the capability of eribulin to inhibit p-AKT and tumor growth, we next studied the benefit of combining eribulin with everolimus in TNBC. Everolimus, an inhibitor of mTOR, has emerged as a potential combination therapy drug for cancer treatment, although everolimus alone only exerts modest anti-cancer effects. Everolimus often increases the expression of p-AKT in human cancer cells when used alone. To investigate the effect of combined treatment of everolimus and eribulin on the PI3K/AKT/mTOR pathway, we incubated MDA-MB-468 cells with eribulin and everolimus at various concentrations, either alone or in combination. As shown in Figure 3, Western blot analysis for MDA-MB-468 cells treated with the combination of eribulin and everolimus showed a dose-related suppression of p-AKT expression, along with a greater inhibition of p-S6K1 and p-S6 expression. Combination treatment also caused a greater inhibition of p-S6K1 and p-S6 in 4T1, a highly metastatic mouse TNBC cell line.

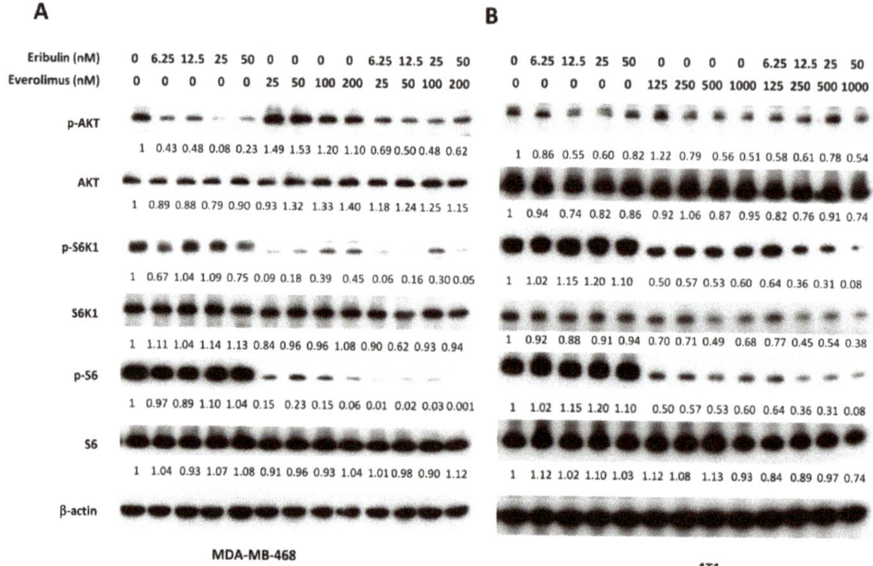

Figure 3. Combined treatment of eribulin and everolimus enhances the reduction of p-S6. MDA-MB-468 (**A**) and 4T1 (**B**) cells were treated with eribulin and everolimus at indicated concentrations, either alone or in combination. Cells were collected 24 h later and analyzed by Western blot for the expression of p-AKT, p-S6K, and p-S6. Numbers below the corresponding blot represent densitometric analysis normalized to β-actin.

3.3. *Combined Treatment of Eribulin and Everolimus Synergistically Inhibits Cell Viability*

We next evaluated whether the combination of eribulin and everolimus resulted in more effective ant-tumor activity. To address this, MDA-MB-468, 4T1, and BT549 cells were treated with eribulin and everolimus, either alone or in combination. Cell viability was determined 72 h later. The combination treatment decreased cell viability more robustly than either agent alone (Figure 4). To determine whether the increased activity was additive or synergistic, the combination index (CI) was calculated according to the Chou-Talalay method. As shown in Figure 4, the combined treatment of erubulin and everolimus caused very strong synergism in all three cell lines.

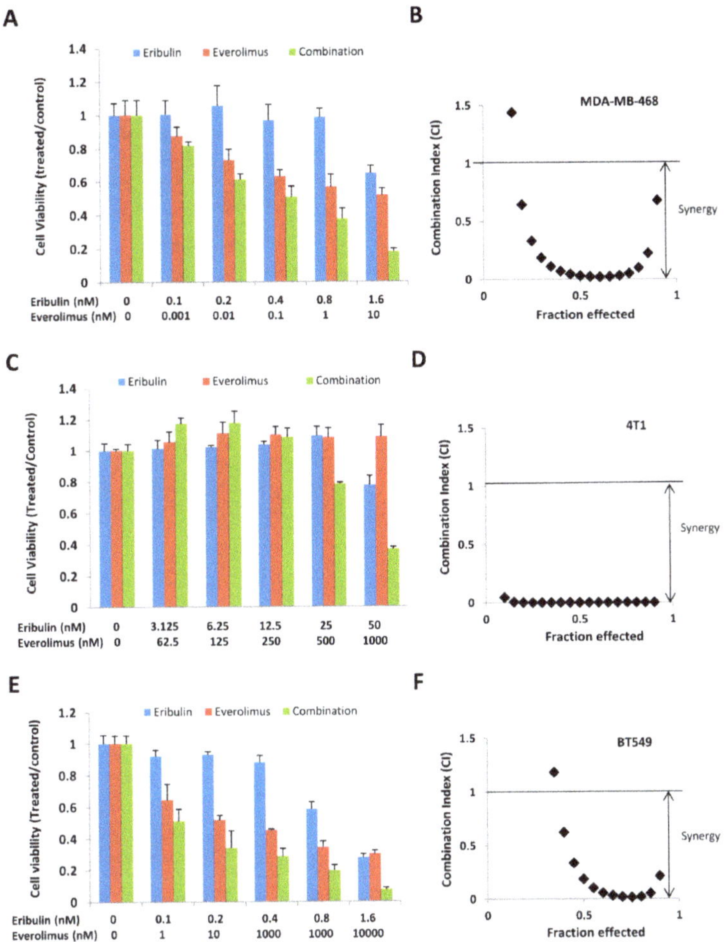

Figure 4. Combined treatment of eribulin and everolimus resulted in synergistic cell growth inhibition. MDA-MB-468 (**A**,**B**), 4T1 (**C**,**D**), and BT549 (**E**,**F**) cells were treated with eribulin or everolimus, either alone or in combination, at indicated concentrations. Cell viability was determined 72 h later. The combination index (CI) was calculated by the Chou-Talalay method.

3.4. Combination of Eribulin with PI3K Inhibitors, BEZ 235 and BKM 120, has a Similar Effect on p-S6K/p-S6 and Cell Viability

Because the combination of eribulin with everolimus enhances anti-tumor activity, we next asked whether a combination of eribulin and other PI3K/AKT/mTOR inhibitors could achieve a similar result. To address this, we combined eribulin with BEZ235 or BKM120, two pan-class PI3K/AKT/mTOR inhibitors that induce pathway inhibition and show anti-tumor activity in PI3K-pathway-dysregulated cancers. As shown in Figure 5A, the combination of eribulin with BEZ235 or BKM120 was more effective in inhibiting the expression of p-AKT, p-S6K1, or p-S6 than either agent alone. We also evaluated whether combination treatment increases anti-tumor activity. MDA-MB-468 cells were incubated with eribulin at various concentrations, either alone or in combination with BEZ235 or BKM120. Cell viability was determined 72 h later. As shown in Figure 5B–5E, the combination of eribulin with either BEZ235 or BKM120 synergistically inhibits growth. Taken together, our study shows significant synergistic growth inhibition when eribulin is combined with PI3K/AKT/mTOR inhibitors.

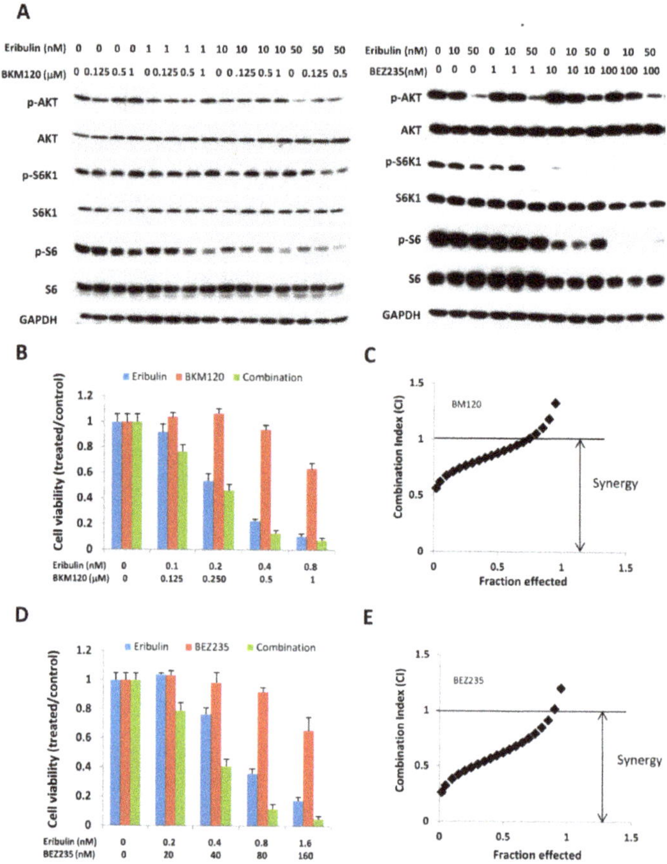

Figure 5. Combined treatment of eribulin with BKM120 and BEZ 235 enhances the inhibition of cell growth. (**A**) MDA-MB-468 cells were treated with eribulin, BKM120, or BEZ 235, either alone or in combination, at indicated concentrations. Cells were collected 24 h later and measured for p-AKT, p-S6K, and p-S6 expression. (**B–E**) MDA-MB-468 cells were treated with eribulin, BKM120 (**B**,**C**), or BEZ235 (**D**,**E**), either alone or in combination, at indicated concentrations. Cell viability was determined 72 h later (**B**,**D**). The combination index (CI) was determined according to the Chou-Talalay method (**C**,**E**).

3.5. *Combined Treatment of Eribulin and Everolimus Enhances Anti-tumor Activity in Mice*

Next, we investigated whether the combination treatment could suppress tumor growth in vivo more effectively than either treatment alone. Mammary fat pads of NSG mice and BALB/c mice were inoculated with MDA-MB-468 human breast cancer cells and 4T1 mouse breast cancer cells, respectively. MDA-MB-468 is a human TNBC cell line. 4T1 is a highly metastatic TNBC cell line derived from a spontaneously arising BALB/c mammary tumor. When the tumors were palpable, we randomized mice into four treatment groups (eribulin, everolimus, eribulin plus everolimus, or vehicle control). No severe toxicity was observed in mice treated with the combination (Figure 6C,D). Treatment with eribulin and everolimus alone had a modest anti-tumor effect. However, the combination of eribulin and everolimus was more effective than any single treatment in both the MDA-MB-468 tumor and 4T1 tumor (Figure 6).

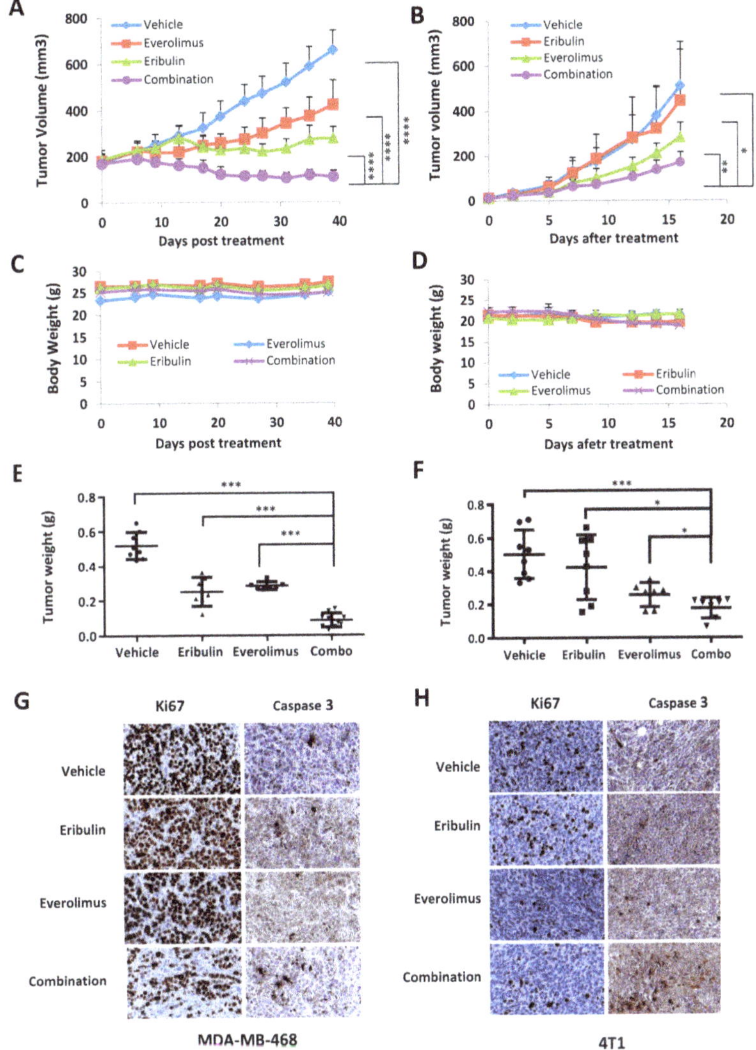

Figure 6. Combined treatment of eribulin and everolimus enhances anti-tumor activity in mice. MDA-MB-468 (**A**,**C**,**E**,**G**) and 4T1 (**B**,**D**,**F**,**H**) were implanted into the mammary fat pad of NOD/SCID/IL2Rgamma null (NSG) mice and BALB/c mice, respectively. NSG mice with MDA-MB-468 tumors were treated with a vehicle, eribulin (0.2 mg/kg for the first week, 0.1 mg/kg for the remainder of the treatment period, via retro-orbital), everolimus (5 mg/kg via oral gavage), or their combination two times a week. BALB/c mice with 4T1 tumors were treated with a vehicle, eribulin (1 mg/kg via retro-orbital), everolimus (5 mg/kg via oral gavage), or their combination three times a week. Tumor growth (**A**,**B**) and body weight (**C**,**D**) were measured once to twice a week. (**E**,**F**) Tumor weight was measured at end of the treatment (**G**,**H**) Shown are representative images of Ki67 and cleaved-caspase 3 immunohistochemstry in MDA-MB-468 and 4T1 tumor tissues. Data represents means ± SD (n = 7–10). *, $p < 0.05$; **, $p < 0.005$. ***, $p < 0.0005$, ****, $p < 0.0001$.

Taken together, our study demonstrates that the combination of eribulin plus everolimus markedly enhances the suppression of tumors compared to treatment with eribulin or everolimus alone in two mouse models of TNBC: a syngeneic model with a well-known highly metastatic TNBC (4T1) and a xenogeneic model with human TNBC (MDA-MB-468). These findings indicate a potential role for eribulin/everolimus combination therapy in the treatment of refractory TNBC.

4. Discussion

Despite advances in breast cancer treatment, patients with TNBC have worse outcomes after chemotherapy than patients with other subtypes of breast cancer [1]. In the era of personalized cancer therapy, molecular characteristics have been sought to identify new therapeutic targets [5,6]. Efforts to profile TNBC tumors have revealed the PI3K/AKT/mTOR pathway as a potential therapeutic target. Activation of this pathway also contributes to the resistance to anti-cancer agents, including microtubule-targeting agents. In this study, we show, for the first time, that the phosphorylation of AKT is suppressed by microtubule depolymerizing agents, eribulin and vinblastine, but activated by microtubule stabilizing agents, such as paclitaxel, or by a conventional DNA damaging chemotherapeutic agent, cisplatin. Dual treatment of eribulin and everolimus results in an increased reduction of p-S6K1 and p-S6, a synergistic suppression of cell survival in a number of breast cancer cell lines in vitro, and an enhanced suppression of tumor growth in two breast cancer mouse models.

Eribulin mesylate is a microtubule-targeting agent used to treat taxane and anthracycline refractory breast cancer [18–24,29]. Phase I clinical trials of eribulin in patients with previously treated solid malignancies demonstrated a dose escalation response that was limited by neutropenia and fatigue. In these trials, eribulin had linear pharmacokinetics with a rapid distribution phase, extensive volume distribution with slow to moderate clearance, and slow elimination [37–39]. Phase II clinical trials were conducted in patients with heavily pretreated metastatic breast cancer. In these trials, eribulin exhibited antitumor activity with a manageable tolerability profile, with side effects consisting of neutropenia, fatigue, alopecia, nausea, and anemia. In addition, there was a low incidence of peripheral neuropathy [40–42]. In the Eisai Metastatic Breast Cancer Study Assessing Physician's Choice Verses E7389 clinical trial, a phase III trial of patients with heavily pretreated metastatic breast cancer, participants received eribulin monotherapy or treatment of physician's choice (TPC). Enrolled patients had received a median of four prior therapies. Improvement was seen in overall survival (OS) with hazard ratio (HR) 0.81 (95% CI: 0.66–0.99, $p = 0.041$). Median OS was 13.1 months in patients receiving eribulin versus 10.6 months in TPC [29]. This study led to the FDA approval of eribulin mesylate for the treatment of breast cancer in patients who had failed taxane- or anthracycline-based therapies.

The PI3K/AKT/mTOR signaling pathway has been implicated in the regulation of microtubule stability by growth factors and drug resistance [43]. In this study, we compared the ability of eribulin to inhibit the PI3K/AKT/mTOR pathway and cell growth with that of two other microtubule-targeting agents, paclitaxel and vinblastine. We found that both eribulin and vinblastine decreased the expression of p-AKT in TNBC cells. Growth inhibition was also seen with treatment with eribulin or vinblastine. Interestingly, treatment of TNBC with paclitaxel or the conventional chemotherapeutic cisplatin resulted in an increased expression of PI3K downstream proteins. Although the mechanism by which eribulin inhibits the phosphorylation of AKT remains to be elucidated, it is possible that depolymerization of the microtubule may interfere with the localization of AKT in the cells. It has been previously shown that the localization of AKT to microtubules is important for sustaining AKT phosphorylation [44,45].

Eribulin was found to inhibit cell growth in MDA-MB-435 triple negative cancer cells at a lower concentration than paclitaxel or vinblastine in vitro and suppressed 95% of growth in breast cancer xenografts in vivo [20]. In a study of paclitaxel-resistant human cancer cells in vitro, eribulin and vinblastine maintained their full potency to inhibit cell proliferation [46]. Enhancement of AKT activity is likely a survival response by cancer cells to chemotherapy, yet AKT activity appears to be suppressed when microtubule polymerization is blocked. These findings suggest a mechanism through which eribulin inhibits the cell growth of refractory triple negative and HER2 expressing breast cancer.

Resistance to microtubule-targeting agents may be due to a multidrug resistant phenotype or the activation of growth signaling pathways [32]. A better understanding of the mechanisms behind drug resistance and the call for more personalized medicine have sparked interest in combination therapy regimens aimed at multiple targets. Drug combination aims to decrease the drug dose and toxicities, achieve a synergistic effect, or overcome drug resistance [34]. Loss of the tumor suppressor PTEN (phosphatase and tensin homolog) and activation of the PI3K pathway have been implicated in resistance to endocrine therapy and trastuzumab [11,47–50]. PTEN works to antagonize the PI3K pathway activation of downstream targets AKT and mTOR. When PTEN is lost, the pathway goes unregulated, resulting in enhanced tumorigenesis [49,51]. The downstream PI3K inhibitors can be used to overcome the drug resistance.

Everolimus, an inhibitor of mTOR, has emerged as a potential combination therapy drug for the treatment of cancer that does not respond to conventional therapy [52]. When used alone, everolimus can induce increased levels of p-AKT via a negative feedback loop, leading to the resistance of cells to mTOR inhibitors [13,14]. A dual blockade of mTOR and other PI3K pathway inhibitors results in synergistic decreases in cancer cell growth [13,14,53,54]. Our findings that eribulin treatment decreases activation of the PI3K/AKT/mTOR pathway led us to investigate the possible synergy between eribulin and everolimus. Our results demonstrate that dual treatment of eribulin and everolimus increases the reduction of p-S6K1 and p-S6 expression, a synergistic suppression of cell survival in vitro, and an enhanced suppression of tumor growth in mouse models. Therefore, targeting both the microtubule cytoskeleton and the PI3K/AKT/mTOR pathway can lead to a synergistic anti-tumor effect.

The combination of everolimus with endocrine therapy was effective in the treatment of hormone receptor positive breast cancers. A phase III clinical trial in patients with hormone receptor positive metastatic breast cancer previously treated with aromatase inhibitors demonstrated that combination therapy with exemestane plus everolimus showed improvement in progression-free survival (HR for progression or death=0.43; 95% CI: 0.35 to 0.54; $p < 0.001$) versus exemestane alone [55,56]. This led to the FDA approval of everolimus for advanced or metastatic aromatase inhibitor-resistant ER + breast cancer [57]. The combination of tamoxifen and everolimus was active in hormone receptor positive breast cancer [58]. In addition, the efficacy of everolimus in Her2-neu overexpressed breast cancer was also confirmed in a phase I/II clinical trial of trastuzumab plus everolimus, with a clinical benefit rate of 34% [59]. These findings from the clinical trials suggest a possible mechanism of drug resistance through continued PI3K/AKT/mTOR pathway activation and that combination treatment with everolimus may re-sensitize cancer to the targeted drug. The efficacy of the combination of everolimus and carboplatin was tested in a phase I trial with a tolerable safety profile and modest clinical activity [60]. An ongoing clinical trial is comparing the combination of carboplatin and everolimus versus carboplatin alone (NCT02531932).

5. Conclusions

We have demonstrated preclinical results to support the use of the microtubule-targeting agent, eribulin, in combination with the mTOR inhibitor, everolimus, against TNBC. This combination therapy is currently being tested in a phase I clinical trial in patients with TNBC who have progressed on anthracyclines and/or taxanes (NCT02120469).

Author Contributions: Conceptualization, T.L., J.H.Y.; methodology, W.W., E.S.H., J.H.Y.; formal analysis, W.W., E.M., D.L., D.T., J.H.Y.; investigation, W.W., Q.X., J.Y., Y.W., J.W., Y.G.; resources, E.M., D.L., T.L., D.T., E.S.H., S.E.Y., Y.Y.; writing—original draft preparation, W.W., E.M., D.L., S.E.Y., Y.Y., J.H.Y.; writing—review and editing, W.W., Y.Y., J.H.Y.; visualization, W.W., J.H.Y.; supervision, J.H.Y.; project administration, W.W., E.M., D.L., J.H.Y.; funding acquisition, T.L., J.H.Y.

Funding: This work was supported in part by the Eisai IIS grant HAL-IIS-022-11. Research reported in this publication included work performed in the Small Animal Studies Core and Pathology Core supported by the National Cancer Institute of the National Institutes of Health under award number P30CA033572. "The content is solely the responsibility of the authors and does not necessarily represent the official views of the National Institutes of Health."

Acknowledgments: We thank Eisai for providing eribulin; Novartis for providing everolimus, BKM120, and BEZ235; Pathology Core and Animal Facility for their technical assistance; and Dr. Chris Gandhi for critical reading of this manuscript.

Conflicts of Interest: The authors declare no conflicts of interest. The sponsors had no role in the design, execution, interpretation, or writing of the study.

References

1. Foulkes, W.D.; Smith, I.E.; Reis-Filho, J.S. Triple-Negative Breast Cancer. *N. Engl. J. Med.* **2010**, *363*, 1938–1948. [CrossRef]
2. Ganesan, P.; Moulder, S.; Lee, J.J.; Janku, F.; Valero, V.; Zinner, R.G.; Naing, A.; Fu, S.; Tsimberidou, A.M.; Hong, D.; et al. Triple-negative breast cancer patients treated at MD anderson cancer center in phase I trials: Improved outcomes with combination chemotherapy and targeted agents. *Mol. Cancer Ther.* **2014**, *13*, 3175–3184. [CrossRef]
3. Tan, D.S.; Marchio, C.; Jones, R.L.; Savage, K.; Smith, I.E.; Dowsett, M.; Reis-Filho, J.S. Triple negative breast cancer: Molecular profiling and prognostic impact in adjuvant anthracycline-treated patients. *Breast Cancer Res. Treat.* **2008**, *111*, 27–44. [CrossRef]
4. Sohn, J.; Do, K.A.; Liu, S.; Chen, H.; Mills, G.B.; Hortobagyi, G.N.; Meric-Bernstam, F.; Gonzalez-Angulo, A.M. Functional proteomics characterization of residual triple-negative breast cancer after standard neoadjuvant chemotherapy. *Ann. Oncol.* **2013**, *24*, 2522–2526. [CrossRef]
5. Mohamed, A.; Krajewski, K.; Cakar, B.; Ma, C.X. Targeted therapy for breast cancer. *Am. J. Pathol.* **2013**, *183*, 1096–1112. [CrossRef]
6. Rivenbark, A.G.; O'Connor, S.M.; Coleman, W.B. Molecular and cellular heterogeneity in breast cancer: Challenges for personalized medicine. *Am. J. Pathol.* **2013**, *183*, 1113–1124. [CrossRef]
7. Gonzalez-Angulo, A.M.; Blumenschein, G.R., Jr. Defining biomarkers to predict sensitivity to PI3K/Akt/mTOR pathway inhibitors in breast cancer. *Cancer Treat. Rev.* **2013**, *39*, 313–320. [CrossRef]
8. Grunt, T.W.; Mariani, G.L. Novel approaches for molecular targeted therapy of breast cancer: Interfering with PI3K/AKT/mTOR signaling. *Curr. Cancer Drug Targets* **2013**, *13*, 188–204. [CrossRef]
9. Bartlett, J.M. Biomarkers and patient selection for PI3K/Akt/mTOR targeted therapies: Current status and future directions. *Clin. Breast Cancer* **2010**, *10*, S86–S95. [CrossRef]
10. McAuliffe, P.F.; Meric-Bernstam, F.; Mills, G.B.; Gonzalez-Angulo, A.M. Deciphering the role of PI3K/Akt/mTOR pathway in breast cancer biology and pathogenesis. *Clin. Breast Cancer* **2010**, *10*, S59–S65. [CrossRef]
11. Perez-Tenorio, G.; Stal, O. Activation of AKT/PKB in breast cancer predicts a worse outcome among endocrine treated patients. *Br. J. Cancer* **2002**, *86*, 540–545. [CrossRef]
12. Mercatali, L.; Spadazzi, C.; Miserocchi, G.; Liverani, C.; De Vita, A.; Bongiovanni, A.; Recine, F.; Amadori, D.; Ibrahim, T. The Effect of Everolimus in an In Vitro Model of Triple Negative Breast Cancer and Osteoclasts. *Int. J. Mol. Sci.* **2016**, *17*, 1827. [CrossRef]
13. Wan, X.; Harkavy, B.; Shen, N.; Grohar, P.; Helman, L.J. Rapamycin induces feedback activation of Akt signaling through an IGF-1R-dependent mechanism. *Oncogene* **2007**, *26*, 1932–1940. [CrossRef]
14. Wang, X.; Yue, P.; Kim, Y.A.; Fu, H.; Khuri, F.R.; Sun, S.Y. Enhancing mammalian target of rapamycin (mTOR)-targeted cancer therapy by preventing mTOR/raptor inhibition-initiated, mTOR/rictor-independent Akt activation. *Cancer Res.* **2008**, *68*, 7409–7418. [CrossRef]
15. Agarwal, R.; Carey, M.; Hennessy, B.; Mills, G.B. PI3K pathway-directed therapeutic strategies in cancer. *Curr. Opin. Investig. drugs* **2010**, *11*, 615–628.
16. Dienstmann, R.; Rodon, J.; Serra, V.; Tabernero, J. Picking the Point of Inhibition: A Comparative Review of PI3K/AKT/mTOR Pathway Inhibitors. *Mol. Cancer Ther.* **2014**, *13*, 1021–1031. [CrossRef]
17. Engebraaten, O.; Vollan, H.K.; Borresen-Dale, A.L. Triple-negative breast cancer and the need for new therapeutic targets. *Am. J. Pathol.* **2013**, *183*, 1064–1074. [CrossRef]
18. Dybdal-Hargreaves, N.F.; Risinger, A.L.; Mooberry, S.L. Eribulin Mesylate: Mechanism of Action of a Unique Microtubule Targeting Agent. *Clin. Cancer Res.* **2015**, *21*, 2445–2452. [CrossRef]
19. Cortes, J.; Schöffski, P.; Littlefield, B.A. Multiple modes of action of eribulin mesylate: Emerging data and clinical implications. *Cancer Treat. Rev.* **2018**, *70*, 190–198. [CrossRef]
20. Towle, M.J.; Salvato, K.A.; Budrow, J.; Wels, B.F.; Kuznetsov, G.; Aalfs, K.K.; Welsh, S.; Zheng, W.; Seletsky, B.M.; Palme, M.H.; et al. In vitro and in vivo anticancer activities of synthetic macrocyclic ketone analogues of halichondrin B. *Cancer Res.* **2001**, *61*, 1013–1021.

21. Jordan, M.A.; Kamath, K.; Manna, T.; Okouneva, T.; Miller, H.P.; Davis, C.; Littlefield, B.A.; Wilson, L. The primary antimitotic mechanism of action of the synthetic halichondrin E7389 is suppression of microtubule growth. *Mol. Cancer Ther.* **2005**, *4*, 1086–1095. [CrossRef]
22. Kuznetsov, G.; Towle, M.J.; Cheng, H.; Kawamura, T.; TenDyke, K.; Liu, D.; Kishi, Y.; Yu, M.J.; Littlefield, B.A. Induction of morphological and biochemical apoptosis following prolonged mitotic blockage by halichondrin B macrocyclic ketone analog E7389. *Cancer Res.* **2004**, *64*, 5760–5766. [CrossRef]
23. Smith, J.A.; Wilson, L.; Azarenko, O.; Zhu, X.; Lewis, B.M.; Littlefield, B.A.; Jordan, M.A. Eribulin binds at microtubule ends to a single site on tubulin to suppress dynamic instability. *Biochemistry* **2010**, *49*, 1331–1337. [CrossRef]
24. Okouneva, T.; Azarenko, O.; Wilson, L.; Littlefield, B.A.; Jordan, M.A. Inhibition of centromere dynamics by eribulin (E7389) during mitotic metaphase. *Mol. Cancer Ther.* **2008**, *7*, 2003–2011. [CrossRef]
25. Towle, M.J.; Salvato, K.A.; Wels, B.F.; Aalfs, K.K.; Zheng, W.; Seletsky, B.M.; Zhu, X.; Lewis, B.M.; Kishi, Y.; Yu, M.J.; et al. Eribulin Induces Irreversible Mitotic Blockade: Implications of Cell-Based Pharmacodynamics for In vivo Efficacy under Intermittent Dosing Conditions. *Cancer Res.* **2011**, *71*, 496–505. [CrossRef]
26. Stehle, A.; Hugle, M.; Fulda, S. Eribulin synergizes with Polo-like kinase 1 inhibitors to induce apoptosis in rhabdomyosarcoma. *Cancer Lett.* **2015**, *365*, 37–46. [CrossRef]
27. Towle, M.J.; Nomoto, K.; Asano, M.; Kishi, Y.; Yu, M.J.; Littlefield, B.A. Broad Spectrum Preclinical Antitumor Activity of Eribulin (Halaven®): Optimal Effectiveness under Intermittent Dosing Conditions. *Anticancer Res.* **2012**, *32*, 1611–1619.
28. Asano, M.; Matsui, J.; Towle, M.J.; Wu, J.; Mcgonigle, S.; De Boisferon, M.H.; Uenaka, T.; Nomoto, K.; Littlefield, B.A. Broad-spectrum Preclinical Antitumor Activity of Eribulin (Halaven®): Combination with Anticancer Agents of Differing Mechanisms. *Anticancer Res.* **2018**, *38*, 3375–3385. [CrossRef]
29. Cortes, J.; O'Shaughnessy, J.; Loesch, D.; Blum, J.L.; Vahdat, L.T.; Petrakova, K.; Chollet, P.; Manikas, A.; Dieras, V.; Delozier, T.; et al. Eribulin monotherapy versus treatment of physician's choice in patients with metastatic breast cancer (EMBRACE): A phase 3 open-label randomised study. *Lancet* **2011**, *377*, 914–923. [CrossRef]
30. Ghayad, S.E.; Vendrell, J.A.; Ben Larbi, S.; Dumontet, C.; Bieche, I.; Cohen, P.A. Endocrine resistance associated with activated ErbB system in breast cancer cells is reversed by inhibiting MAPK or PI3K/Akt signaling pathways. *Int. J. Cancer.* **2010**, *126*, 545–562. [CrossRef]
31. Gonzalez-Angulo, A.M.; Morales-Vasquez, F.; Hortobagyi, G.N. Overview of resistance to systemic therapy in patients with breast cancer. *Adv. Exp. Med. Biol.* **2007**, *608*, 1–22.
32. Marquette, C.; Nabell, L. Chemotherapy-resistant metastatic breast cancer. *Curr. Treat. Options Oncol.* **2012**, *13*, 263–275. [CrossRef]
33. Wang, Y.; Han, E.; Xing, Q.; Yan, J.; Arrington, A.; Wang, C.; Tully, D.; Kowolik, C.M.; Lu, D.M.; Frankel, P.H.; et al. Baicalein upregulates DDIT4 expression which mediates mTOR inhibition and growth inhibition in cancer cells. *Cancer Lett.* **2015**, *358*, 170–179. [CrossRef]
34. Chou, T.C. Drug combination studies and their synergy quantification using the Chou-Talalay method. *Cancer Res.* **2010**, *70*, 440–446. [CrossRef]
35. Wen, W.; Wu, J.; Liu, L.; Tian, Y.; Buettner, R.; Hsieh, M.Y.; Horne, D.; Dellinger, T.H.; Han, E.S.; Jove, R.; et al. Synergistic anti-tumor effect of combined inhibition of EGFR and JAK/STAT3 pathways in human ovarian cancer. *Mol. Cancer* **2015**, *14*, 100. [CrossRef]
36. Rozengurt, E.; Soares, H.P.; Sinnet-Smith, J. Suppression of Feedback Loops Mediated by PI3K/mTOR Induces Multiple Overactivation of Compensatory Pathways: An Unintended Consequence Leading to Drug Resistance. *Mol. Cancer Ther.* **2014**, *13*, 2477–2488. [CrossRef]
37. Goel, S.; Mita, A.C.; Mita, M.; Rowinsky, E.K.; Chu, Q.S.; Wong, N.; Desjardins, C.; Fang, F.; Jansen, M.; Shuster, D.E.; et al. A phase I study of eribulin mesylate (E7389), a mechanistically novel inhibitor of microtubule dynamics, in patients with advanced solid malignancies. *Clin. Cancer Res.* **2009**, *15*, 4207–4212. [CrossRef]
38. Tan, A.R.; Rubin, E.H.; Walton, D.C.; Shuster, D.E.; Wong, Y.N.; Fang, F.; Ashworth, S.; Rosen, L.S. Phase I Study of Eribulin Mesylate Administered Once Every 21 Days in Patients with Advanced Solid Tumors. *Clin. Cancer Res.* **2009**, *15*, 4213–4219. [CrossRef]
39. Synold, T.W.; Morgan, R.J.; Newman, E.M.; Lenz, H.J.; Gandara, D.R.; Colevas, A.D.; Lewis, M.D.; Doroshow, J.H. A phase I pharmacokinetic and target validation study of the novel anti-tubulin agent E7389: A California Cancer Consortium trial. *J. Clin. Oncol.* **2005**, *23*, 3036. [CrossRef]

40. Cortes, J.; Vahdat, L.; Blum, J.L.; Twelves, C.; Campone, M.; Roche, H.; Bachelot, T.; Awada, A.; Paridaens, R.; Goncalves, A.; et al. Phase II study of the halichondrin B analog eribulin mesylate in patients with locally advanced or metastatic breast cancer previously treated with an anthracycline, a taxane, and capecitabine. *J. Clin. Oncol.* **2010**, *28*, 3922–3928. [CrossRef]
41. Vahdat, L.T.; Pruitt, B.; Fabian, C.J.; Rivera, R.R.; Smith, D.A.; Tan-Chiu, E.; Wright, J.; Tan, A.R.; Dacosta, N.A.; Chuang, E.; et al. Phase II study of eribulin mesylate, a halichondrin B analog, in patients with metastatic breast cancer previously treated with an anthracycline and a taxane. *J. Clin. Oncol.* **2009**, *27*, 2954–2961. [CrossRef] [PubMed]
42. Aogi, K.; Iwata, H.; Masuda, N.; Mukai, H.; Yoshida, M.; Rai, Y.; Taguchi, K.; Sasaki, Y.; Takashima, S. A phase II study of eribulin in Japanese patients with heavily pretreated metastatic breast cancer. *Ann. Oncol.* **2012**, *23*, 1441–1448. [CrossRef] [PubMed]
43. Onishi, K.; Higuchi, M.; Asakura, T.; Masuyama, N.; Gotoh, Y. The PI3K-Akt pathway promotes microtubule stabilization in migrating fibroblasts. *Genes Cells* **2007**, *12*, 535–546. [CrossRef] [PubMed]
44. Giustiniani, J.; Daire, V.; Cantaloube, I.; Durand, G.; Pous, C.; Perdiz, D.; Baillet, A. Tubulin acetylation favors Hsp90 recruitment to microtubules and stimulates the signaling function of the Hsp90 clients Akt/PKB and p53. *Cell Signal.* **2009**, *21*, 529–539. [CrossRef] [PubMed]
45. Jo, H.; Loison, F.; Luo, H.R. Microtubule dynamics regulates Akt signaling via dynactin p150. *Cell Signal.* **2014**, *26*, 1707–1716. [CrossRef] [PubMed]
46. Kuznetsov, G.; TenDyke, K.; Yu, M.; Littlefield, B. Antiproliferative effects of halichondrin B analog eribulin mesylate (E7389) against paclitaxel-resistant human cancer cells in vitro. *Mol. Cancer Ther.* **2007**, *6*, C58.
47. Miller, T.W.; Perez-Torres, M.; Narasanna, A.; Guix, M.; Stal, O.; Perez-Tenorio, G.; Gonzalez-Angulo, A.M.; Hennessy, B.T.; Mills, G.B.; Kennedy, J.P.; et al. Loss of Phosphatase and Tensin homologue deleted on chromosome 10 engages ErbB3 and insulin-like growth factor-I receptor signaling to promote antiestrogen resistance in breast cancer. *Cancer Res.* **2009**, *69*, 4192–4201. [CrossRef] [PubMed]
48. Miller, T.W.; Balko, J.M.; Arteaga, C.L. Phosphatidylinositol 3-kinase and antiestrogen resistance in breast cancer. *J. Clin. Oncol.* **2011**, *29*, 4452–4461. [CrossRef] [PubMed]
49. Nagata, Y.; Lan, K.H.; Zhou, X.; Tan, M.; Esteva, F.J.; Sahin, A.A.; Klos, K.S.; Li, P.; Monia, B.P.; Nguyen, N.T.; et al. PTEN activation contributes to tumor inhibition by trastuzumab, and loss of PTEN predicts trastuzumab resistance in patients. *Cancer Cell* **2004**, *6*, 117–127. [CrossRef]
50. Berns, K.; Horlings, H.M.; Hennessy, B.T.; Madiredjo, M.; Hijmans, E.M.; Beelen, K.; Linn, S.C.; Gonzalez-Angulo, A.M.; Stemke-Hale, K.; Hauptmann, M.; et al. A functional genetic approach identifies the PI3K pathway as a major determinant of trastuzumab resistance in breast cancer. *Cancer Cell* **2007**, *12*, 395–402. [CrossRef] [PubMed]
51. Di Cristofano, A.; Pandolfi, P.P. The Multiple Roles of PTEN in Tumor Suppression. *Cell* **2000**, *100*, 387–390. [CrossRef]
52. Seto, B. Rapamycin and mTOR: A serendipitous discovery and implications for breast cancer. *Clin. Transl. Med.* **2012**, *1*, 29. [CrossRef] [PubMed]
53. Sun, S.Y.; Rosenberg, L.M.; Wang, X.; Zhou, Z.; Yue, P.; Fu, H.; Khuri, F.R. Activation of Akt and eIF4E survival pathways by rapamycin-mediated mammalian target of rapamycin inhibition. *Cancer Res.* **2005**, *65*, 7052–7058. [CrossRef] [PubMed]
54. O'Reilly, K.E.; Rojo, F.; She, Q.B.; Solit, D.; Mills, G.B.; Smith, D.; Lane, H.; Hofmann, F.; Hicklin, D.J.; Ludwig, D.L.; et al. mTOR inhibition induces upstream receptor tyrosine kinase signaling and activates Akt. *Cancer Res.* **2006**, *66*, 1500–1508. [CrossRef] [PubMed]
55. Beaver, J.A.; Park, B.H. The BOLERO-2 trial: The addition of everolimus to exemestane in the treatment of postmenopausal hormone receptor-positive advanced breast cancer. *Future Oncol.* **2012**, *8*, 651–657. [CrossRef] [PubMed]
56. Hortobagyi, G.N. Everolimus plus exemestane for the treatment of advanced breast cancer: A review of subanalyses from BOLERO-2. *Neoplasia* **2015**, *17*, 279–288. [CrossRef] [PubMed]
57. Baselga, J.; Campone, M.; Piccart, M.; Burris, H.A., 3rd; Rugo, H.S.; Sahmoud, T.; Noguchi, S.; Gnant, M.; Pritchard, K.I.; Lebrun, F.; et al. Everolimus in postmenopausal hormone-receptor-positive advanced breast cancer. *N. Engl. J. Med.* **2012**, *366*, 520–529. [CrossRef] [PubMed]

58. Bachelot, T.; Bourgier, C.; Cropet, C.; Ray-Coquard, I.; Ferrero, J.M.; Freyer, G.; Abadie-Lacourtoisie, S.; Eymard, J.C.; Debled, M.; Spaeth, D.; et al. Randomized phase II trial of everolimus in combination with tamoxifen in patients with hormone receptor-positive, human epidermal growth factor receptor 2-negative metastatic breast cancer with prior exposure to aromatase inhibitors: A GINECO study. *J. Clin. Oncol.* **2012**, *30*, 2718–2724. [CrossRef] [PubMed]
59. Morrow, P.K.; Wulf, G.M.; Ensor, J.; Booser, D.J.; Moore, J.A.; Flores, P.R.; Xiong, Y.; Zhang, S.; Krop, I.E.; Winer, E.P.; et al. Phase I/II study of trastuzumab in combination with everolimus (RAD001) in patients with HER2-overexpressing metastatic breast cancer who progressed on trastuzumab-based therapy. *J. Clin. Oncol.* **2011**, *29*, 3126–3132. [CrossRef] [PubMed]
60. Singh, J.C.; Novik, Y.; Stein, S.; Volm, M.; Meyers, M.; Smith, J.; Omene, C.; Speyer, J.; Schneider, R.; Jhaveri, K.; et al. Phase 2 trial of everolimus and carboplatin combination in patients with triple negative metastatic breast cancer. *Breast Cancer Res.* **2014**, *16*, R32. [CrossRef] [PubMed]

MDPI
St. Alban-Anlage 66
4052 Basel
Switzerland
Tel. +41 61 683 77 34
Fax +41 61 302 89 18
www.mdpi.com

Cells Editorial Office
E-mail: cells@mdpi.com
www.mdpi.com/journal/cells

www.ingramcontent.com/pod-product-compliance
Lightning Source LLC
LaVergne TN
LVHW071007170726
843515LV00006B/1099

9783039435531